Selected Titles in This Series

(Continued in the back of this publication)

External Memory
Algorithms

DIMACS
Series in Discrete Mathematics and Theoretical Computer Science

Volume 50

External Memory Algorithms

DIMACS Workshop
External Memory Algorithms and Visualization
May 20–22, 1998

James M. Abello
Jeffrey Scott Vitter
Editors

NSF Science and Technology Center
in Discrete Mathematics and Theoretical Computer Science
A consortium of Rutgers University, Princeton University,
AT&T Labs–Research, Bell Labs (Lucent Technologies),
Telcordia Technologies (formerly Bellcore),
and NEC Research Institute

American Mathematical Society

This DIMACS volume contains papers from a DIMACS workshop on External Memory Algorithms and Visualization, held May 20–22, 1998.

2000 *Mathematics Subject Classification*. Primary 68–00, 68–06, 68W01, 68W40, 68Q05, 68Q25, 68P05, 68P10, 68P20, 68R01, 68R10, 65–06, 65F10, 65F15, 65F50, 65Y20, 68U05, 68M20, 68N01.

Library of Congress Cataloging-in-Publication Data

DIMACS Workshop External Memory Algorithms and Visualization (1998 : Rutgers University)
    External memory algorithms : DIMACS Workshop External Memory Algorithms and Visualization, May 20–22, 1998 / James M. Abello, Jeffrey Scott Vitter, editors.
        p. cm. — (DIMACS series in discrete mathematics and theoretical computer science ; v. 50)
    Includes bibliographical references and index. (p.).
    ISBN 0-8218-1184-3 (alk. paper)
    1. Computer algorithms—Congresses. 2. Memory management (Computer science)—Congresses. I. Title. II. Series. III. Abello, James M. IV. Vitter, Jeffrey Scott, [date].
QA76.9A43D56 1998
005.4′221—dc21   99-043684
                                                          CIP

Contents

Foreword

The DIMACS Workshop on External Memory Algorithms and Visualization was held on May 20–22, 1998 at Rutgers University. We would like to express our appreciation to James Abello and Jeffrey Vitter for their efforts to organize and plan this successful workshop.

The workshop was part of the Special Year on Massive Data Sets. We extend our thanks to Jon Bentley, Martin Farach-Colton, Joan Feigenbaum, and Helene Kulsrud for their work over many months as special year organizers.

When processing large amounts of information, input/output between main and external memory is becoming an increasingly significant bottleneck. The central questions are: How, where, and in what circumstances is it possible to incorporate locality directly into the algorithms? These were the themes of this workshop. The corresponding workshop program can be accessed online by visiting http://dimacs.rutgers.edu/Workshops/Visualization.

DIMACS gratefully acknowledges the generous support that makes these programs possible. The National Science Foundation, through its Science and Technology Centers program, the New Jersey Commission on Science and Technology, DIMACS's partners at Rutgers, Princeton, AT&T Labs-Research, Bell Labs, NEC Research Institute, and Telcordia Technologies (formerly Bellcore) generously supported the special year.

Fred S. Roberts
Director

Robert Sedgewick
Co-Director for Princeton

Preface

Loosely speaking, *extended memory algorithms* involve the use of techniques from computer science and mathematics in the solution of combinatorial problems whose associated data require the use of a hierarchy of storage devices. The *input/output* communication (or simply, I/O) between the different levels of the hierarchy is often a significant bottleneck in applications that process massive amounts of data. Substantial gains in performance may be possible by incorporating locality directly into the algorithms and by explicitly managing the contents of each storage level.

Because the relative difference in access speeds is most apparent between random access memory and magnetic disks, a major portion of research has been dedicated to algorithms that focus on this particular I/O bottleneck. These algorithms are usually referred to as *external memory, out-of-core,* or *I/O algorithms*. This volume is devoted to a sampling of the major existing techniques and new research results for the design and analysis of external memory algorithms. It grew out of the Workshop on External Memory Algorithms and Visualization, organized by the Editors and held at DIMACS during May 20–22, 1998. The workshop, made possible by funding provided by DIMACS and AT&T Labs–Research, attracted an audience of approximately 85 attendees from 17 countries. The workshop program consisted of invited lectures and contributed presentations. The topics covered included problems in computational geometry, graph theory, data compression, disk scheduling, linear algebra, statistics, software libraries, text and string processing, visualization, wavelets, and industrial applications.

The vitality of research in this area was clearly demonstrated during the meeting. Its interdisciplinary nature made it a fruitful ground where a variety of ideas and methods came together. We trust that the refereed papers included in this volume will give a good indication of this rich interaction.

Acknowledgments

We are grateful to the DIMACS staff for their assistance with the workshop logistics, the DIMACS directorate and AT&T Labs–Research for the funding, the Info-Lab members of the AT&T Network Services Laboratory for providing a stimulating environment, and D. Belanger, F. Roberts, R. Sedgewick, S. Sudarsky, and K. P. Vo for their continued support.

All the invited speakers (Y. Matias, T. Munzner, W. Sweldens, S. Toledo, K. P. Vo, and A. Wilks), contributing authors, referees, and workshop participants deserve special credit for helping in one way or another make this volume possible.

J. Abello, Shannon Laboratories, AT&T Labs–Research
J. S. Vitter, Duke University and INRIA Sophia Antipolis

DIMACS Series in Discrete Mathematics
and Theoretical Computer Science
Volume **50**, 1999

External Memory Algorithms and Data Structures

Jeffrey Scott Vitter

ABSTRACT. Data sets in large applications are often too massive to fit completely inside the computer's internal memory. The resulting input/output communication (or I/O) between fast internal memory and slower external memory (such as disks) can be a major performance bottleneck. In this paper, we survey the state of the art in the design and analysis of *external memory algorithms and data structures* (which are sometimes referred to as "EM" or "I/O" or "out-of-core" algorithms and data structures). EM algorithms and data structures are often designed and analyzed using the parallel disk model (PDM). The three machine-independent measures of performance in PDM are the number of I/O operations, the CPU time, and the amount of disk space. PDM allows for multiple disks (or disk arrays) and parallel CPUs, and it can be generalized to handle tertiary storage and hierarchical memory.

We discuss several important paradigms for how to solve batched and online problems efficiently in external memory. Programming tools and environments are available for simplifying the programming task. The TPIE system (Transparent Parallel I/O programming Environment) is both easy to use and efficient in terms of execution speed. We report on some experiments using TPIE in the domain of spatial databases. The newly developed EM algorithms and data structures that incorporate the paradigms we discuss are significantly faster than methods currently used in practice.

1. Introduction

The *Input/Output* communication (or simply *I/O*) between the fast internal memory and the slow external memory (such as disk) can be a bottleneck in applications that process massive amounts of data [**68**]. One promising approach is to design algorithms that bypass the virtual memory system and explicitly manage their own I/O. We refer to such algorithms as *external memory algorithms*, or more

1991 *Mathematics Subject Classification.* 68–02, 68Q10, 68Q20, 68Q25, 65Y20, 65Y25.

Key words and phrases. external memory, secondary storage, disk, block, input/output, I/O, out-of-core, hierarchical memory, multilevel memory, batched, online, external sorting.

Supported in part by Army Research Office MURI grant DAAH04–96–1–0013 and by National Science Foundation research grants CCR–9522047 and EIA–9870734. Part of this work was done at BRICS, University of Aarhus, Århus, Denmark and at INRIA, Sophia Antipolis, France.

Earlier versions of this paper, entitled "External Memory Algorithms", appeared as an invited tutorial in *Proceedings of the 17th Annual ACM Symposium on Principles of Database Systems*, Seattle, WA, June 1998, and as an invited paper in *Proceedings of the 6th Annual European Symposium on Algorithms*, Venice, August 1998. An updated version of this paper is available electronically on the author's web page at http://www.cs.duke.edu/~jsv/.

simply *EM algorithms*. (The terms *out-of-core algorithms* and *I/O algorithms* are also sometimes used.)

In this paper we survey several paradigms for solving problems efficiently in external memory. The problems we consider fall into two general categories:

1. *Batched problems*, in which no preprocessing is done and the entire file of data items must be processed, often in stream mode with one or more passes over the data.

2. *Online problems*, in which computation is done in response to a continuous series of query operations. A common technique for online problems is to organize the data items via a hierarchical index, so that only a very small portion of the data needs to be examined in response to each query. The data being queried can be either *static*, which can be preprocessed for efficient query processing, or *dynamic*, where the queries are intermixed with updates such as insertions and deletions.

We base our approach on the *parallel disk model* (PDM) described in the next section. PDM provides an elegant and reasonably accurate model for analyzing the relative performance of EM algorithms and data structures. The three main performance measures of PDM are number of I/O operations, disk space usage, and CPU time. For reasons of brevity, we focus on the first two measures. Most of the algorithms we discuss are also efficient in terms of CPU time.

In Section 3, we look at the canonical batched EM problem of external sorting and the related problems of permuting and Fast Fourier Transform. The two important paradigms of distribution and merging account for all well-known external sorting algorithms. We provide fundamental lower bounds on the number of I/Os needed to perform sorting and several other batched problems in external memory.

We briefly discuss grid and linear algebra batched computations in Section 4. In Section 5 we mention several effective paradigms for batched EM problems in computational geometry. The paradigms include distribution sweep (for spatial join and finding all nearest neighbors), persistent B-trees (batched point location and graph drawing), batched filtering (for 3-D convex hulls and batched point location), external fractional cascading (for red-blue line segment intersection), online filtering (for cooperative search in fractionally cascaded data structures), external marriage-before-conquest (for output-sensitive convex hulls), and randomized incremental construction with gradations (for line segment intersections and other geometric problems). In Section 6 we look at EM algorithms for combinatorial problems on graphs. In many cases, I/O-efficient algorithms can be obtained by using sorting to simulate some well-known parallel algorithms.

In Sections 7 and 8 we consider spatial data structures in the online setting. Section 7 begins with a discussion of B-trees, the most important dynamic online EM data structure. B-trees are the method of choice for dictionary operations and one-dimensional range queries. Weight-balanced B-trees provide a uniform mechanism for dynamically rebuilding substructures, and level-balanced B-trees permit maintenance of parent pointers. They are useful for building interval trees and doing dynamic point location in external memory. The buffer tree is a so-called "batched dynamic" version of the B-tree for efficient implementation of search trees and priority queues in EM sweep line applications, We also consider multidimensional extensions of the B-tree. R-trees and variants work well in practice for several multidimensional spatial applications such as range searching and spatial

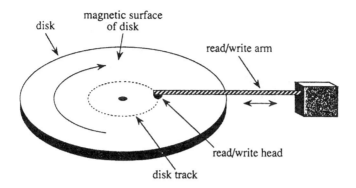

FIGURE 1. Platter of a magnetic disk drive.

joins. In Section 8, we discuss specialized spatial structures for online multidimensional range search, some of which yield optimal bounds for different cases of 2-D range searching. Nonlinear disk space is required in order to achieve optimal query performance for general 2-D range searching. In contrast, R-trees use linear space, but have bad worst-case performance.

Data structures for sorting, searching, and finding matches in strings are the focus of Section 9. In Section 10 we discuss programming environments and tools that facilitate high-level development of efficient EM algorithms. In Section 11, we demonstrate for two problems arising in spatial databases that significant speedups can be obtained in practice by use of efficient EM techniques. We use the TPIE system (Transparent Parallel I/O programming Environment) covered in Section 10 for the implementations. In Section 12 we discuss EM algorithms that adapt optimally to dynamically changing memory allocations. We conclude with some final remarks and observations in Section 13.

2. Parallel Disk Model (PDM)

External memory algorithms explicitly control data placement and movement, and thus it is important for algorithm designers to have a simple but reasonably accurate model of the memory system's characteristics. Magnetic disks consist of one or more rotating platters and one read/write head per platter surface. The data are stored in concentric circles on the platters called *tracks*, as shown in Figure 1. To read or write a data item at a certain address on disk, the read/write head must mechanically *seek* to the correct track and then wait for the desired address to pass by. The seek time to move from one random track to another is often on the order of 5–10 milliseconds, and the average rotational latency, which is the time for half a revolution, has the same order of magnitude. In order to amortize this delay, it pays to transfer a large collection of contiguous data items, called a *block*. Similar considerations apply to all levels of the memory hierarchy.

Even if an application can structure its pattern of memory accesses to exploit locality and take full advantage of disk block transfer, there is still a substantial *access gap* between internal memory performance and external memory performance. In fact the access gap is growing, since the speed of memory chips is increasing more quickly than disk bandwidth. Use of parallel processors further widens the

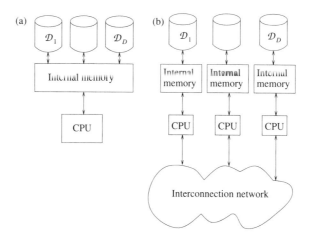

FIGURE 2. Parallel disk model: (a) $P = 1$, (b) $P = D$.

gap. Storage systems such as RAID are being developed that deploy multiple disks to get additional bandwidth [**41, 76**].

In the next section, we describe the high-level parallel disk model (PDM), which we use in this paper for the design and analysis of algorithms and data structures. In Section 2.2, we list four fundamental I/O bounds that pertain to most of the problems considered in this paper. Practical considerations of PDM and alternative memory models are discussed in Sections 2.3 and 2.4.

2.1. PDM and Problem Parameters. We can capture the main properties of magnetic disks and multiple disk systems by the commonly used *parallel disk model* (PDM) introduced by Vitter and Shriver [**138**]:

$$
\begin{aligned}
N &= \text{problem size (in units of data items);} \\
M &= \text{internal memory size (in units of data items);} \\
B &= \text{block transfer size (in units of data items);} \\
D &= \text{\# independent disk drives;} \\
P &= \text{\# CPUs,}
\end{aligned}
$$

where $M < N$, and $1 \leq DB \leq M/2$. In a single I/O, each of the D disks can simultaneously transfer a block of B contiguous data items. If $P \leq D$, each of the P processors can drive about D/P disks; if $D < P$, each disk is shared by about P/D processors. The internal memory size is M/P per processor, and the P processors are connected by an interconnection network. One desired property for the network is the capability to sort the M data items in the collective main memories of the processors in parallel in optimal $O\big((M/P)\log M\big)$ time.[1] The special cases of PDM for $P = 1$ and $P = D$ are pictured in Figure 2.

Queries are naturally associated with online computations, but they can also be done in batched mode. For example, in the batched orthogonal 2-D range searching problem discussed in Section 5, we are given a set of N points in the plane and a set of Q queries in the form of rectangles, and the problem is to report the points

---

[1]We use the notation $\log n$ to denote the binary (base 2) logarithm $\log_2 n$. For bases other than 2, the base will be specified explicitly.

	$\mathcal{D}_0$	$\mathcal{D}_1$	$\mathcal{D}_2$	$\mathcal{D}_3$	$\mathcal{D}_4$
stripe 0	0 1	2 3	4 5	6 7	8 9
stripe 1	10 11	12 13	14 15	16 17	18 19
stripe 2	20 21	22 23	24 25	26 27	28 29
stripe 3	30 31	32 33	34 35	36 37	38 39

FIGURE 3. Initial data layout on the disks, for $D = 5$ disks and block size $B = 2$. The input data items are initially striped block-by-block across the disks. For example, data items 16 and 17 are stored in the second block (i.e., in stripe 1) of disk $\mathcal{D}_3$.

lying in each of the Q query rectangles. In both the batched and online settings, the number of items reported in response to each query may vary. We thus need to define two more performance parameters:

$$Q \quad = \quad \text{\# queries (for a batched problem);}$$
$$Z \quad = \quad \text{query output size (in units of data items).}$$

It is convenient to refer to some of the above PDM parameters in units of disk blocks rather than in units of data items. We define the lower-case notation

$$(2.1) \qquad n = \frac{N}{B}, \qquad m = \frac{M}{B}, \qquad q = \frac{Q}{B}, \qquad z = \frac{Z}{B}$$

to be the problem input size, internal memory size, query specification size, and query output size, respectively, in units of disk blocks. We assume that the input data are initially "striped" across the D disks, in units of blocks, as illustrated in Figure 3, and we require the output data to be similarly striped. Striped format allows a file of N data items to be read or written in $O(N/DB) = O(n/D)$ I/Os, which is optimal.

2.2. Fundamental Bounds and Objectives. The primary measures of performance in PDM are

1. the number of I/O operations performed,
2. the amount of disk space used, and
3. the internal (parallel) computation time.

For reasons of brevity we will focus in this paper on only the first two measures. Most of the algorithms we mention run in optimal CPU time, at least for the single-processor case. Ideally algorithms and data structures should use linear space, which means $O(N/B) = O(n)$ disk blocks of storage.

The I/O performance of many algorithms and data structures can be expressed in terms of the bounds for the following four fundamental operations:

1. *Scanning* (or *streaming* or *touching*) a file of N data items, which takes $\Theta(N/DB) = \Theta(n/D)$ I/Os.
2. *Sorting* N items, which can be done using $\Theta\big((N/DB) \log_{M/B}(N/B)\big) = \Theta\big((n/D) \log_m n\big)$ I/Os.
3. *Online search* among N items, which takes $\Theta(\log_{DB} N)$ I/Os.
4. *Reporting the answers to a query* in blocked fashion onto external memory, which takes $\Theta\big(\lceil Z/DB \rceil\big) = \Theta\big(\lceil z/D \rceil\big)$ I/Os.

The first two of these I/O bounds—scanning and sorting—apply to batched problems. As mentioned earlier, some batched problems also involve queries, in which case the I/O bound for query reporting is relevant. The last two I/O bounds—online search and query reporting—apply to online problems. We typically assume for online problems that there is only one disk, namely, $D = 1$, in which case the bounds for online search and query reporting become simply $\Theta(\log_B N)$ and $\Theta(z)$; multiple disks can generally be used in an optimal way for online problems via the disk striping technique explained in Section 3.

For many of the batched problems we consider, such as sorting, FFTs, triangulation, and computing convex hulls, there are algorithms to solve the corresponding internal memory versions of the problems in $O(N \log N)$ CPU time. But if we deploy such an algorithm naively in an external memory setting (using virtual memory to handle page management), it may require $\Theta(N \log n)$ I/Os, which is excessive. Similarly, in the online setting, many problems can be solved in $O(\log N + Z)$ query time when they fit in internal memory, but the same data structure in an external memory setting may require $\Theta(\log N + Z)$ I/Os per query.

We would like instead to achieve the I/O bounds $O\big((n/D)\log_m n\big)$ in the batched example and $O(\log_{DB} N + z/D)$ for the online case. At the risk of oversimplifying, we can paraphrase the goal of EM algorithm design in the following syntactic way: to derive efficient algorithms so that the N and Z terms in the I/O bounds of the naive algorithms are replaced by n/D and z/D, and so that the base of the logarithm terms is not 2 but instead m (in the case of batched problems) or DB (in the case of online problems). The relative speedup in I/O performance can be very significant, both theoretically and in practice. For example, for batched problems, the I/O performance improvement can be a factor of $(N \log n)/(n/D)\log_m n = DB \log m$, which is very large, even for $D = 1$. For online problems, the performance improvement can be a factor of $(\log N + Z)/(\log_{DB} N + z/D)$, which is at least $(\log N)/\log_{DB} N = \log DB$, which is significant in practice, and it can be as much as $Z/(z/D) = DB$ for large Z.

The I/O bound $\Theta(N/DB) = \Theta(n/D)$ for the trivial batched problem of scanning is considered to be a *linear number of I/Os* in the PDM model. An interesting feature of the PDM model is that almost all nontrivial batched problems require a nonlinear number of I/Os, even those that can be solved easily in linear CPU time in the (internal memory) RAM model. Examples we will discuss later include permuting, transposing a matrix, and several combinatorial graph problems. Sorting is equivalent in I/O complexity to several of these problems.

Often in practice, the nonlinear $\log_m n$ term in the sorting bound and the $\log_{DB} N$ term in the searching bound are small constants. For example, in units of items, we could have $N = 10^{10}$, $M = 10^7$, $B = 10^4$, and $D = 1$, in which case we get $n = 10^6$, $m = 10^3$, and $\log_m n = 2$. If memory is shared with other processes, the $\log_m n$ term will be somewhat larger. In online applications, a smaller B value, such as $B = 10^2$, is more appropriate, as explained in the next section; the corresponding value of $\log_{DB} N$ for the example would be 5.

It still makes sense to identify terms like $\log_m n$ and $\log_{DB} N$ and not hide them within the big-oh factors, since the terms can make a significant difference in practice. (Of course, it is equally important to consider any other constants hidden in big-oh notations!) A nonlinear bound $O\big((n/D)\log_m n\big)$ usually indicates that multiple or extra passes over the data are required. In truly massive problems, the data will reside on tertiary storage. As we mention briefly in Section 2.4, PDM algorithms can often be generalized in a recursive framework to handle multiple levels of memory. A multilevel algorithm developed from a PDM algorithm that does n/D I/Os is likely to run at least an order of magnitude faster in hierarchical memory than a multilevel algorithm generated from a PDM algorithm that does $(n/D)\log_m n$ I/Os [139].

2.3. Practical Modeling Considerations. Track size is a parameter of the disk hardware and cannot be altered; for most disks it is in the range 50–100 kilobytes. For batched applications, the block transfer size B in PDM should be chosen to be a significant fraction of the track size or a small multiple of the track size, so as to better amortize seek time. For online applications, a smaller B value is appropriate; the minimum block transfer size imposed by many systems is 8 kilobytes.

PDM is a good generic programming model that facilitates elegant design of I/O-efficient algorithms, especially when used in conjunction with the programming tools discussed in Section 10. More complex and precise disk models have been developed, such as the ones by Ruemmler and Wilkes [112], Shriver et al. [119], Barve et al. [28], and Farach et al. [60]. They distinguish between sequential reads and random reads and consider the effects on throughput of features such as disk buffer caches and shared buses, which can reduce the time per I/O by eliminating or hiding the seek time. In practice, the effects of more complex models can be realized or approximated by PDM with an appropriate choice of parameters. The bottom line is that programs that perform well in terms of PDM will generally perform well when implemented on real systems.

2.4. Other Memory Models. The study of problem complexity and algorithm analysis when using external memory devices began more than 40 years ago with Demuth's Ph.D. thesis on sorting [55, 88]. In the early 1970s, Knuth [88] did an extensive study of sorting using magnetic tapes and (to a lesser extent) magnetic disks. At about the same time, Floyd [64, 88] considered a disk model akin to PDM for $D = 1$, $P = 1$, $B = M/2 = \Theta(N^c)$, for constant $c > 0$, and developed optimal upper and lower I/O bounds for sorting and matrix transposition. Hong and Kung [78] developed a pebbling model of I/O for straightline computations, and Savage and Vitter [117] extended the model to deal with block transfer. Aggarwal and Vitter [11] generalized Floyd's I/O model to allow simultaneous block transfers, but the model was unrealistic in that the simultaneous transfers were allowed to take place on a single disk. They developed matching upper and lower I/O bounds for all parameter values for a host of problems. Since the PDM model can be thought of as a more restrictive (and more realistic) version of Aggarwal and Vitter's model, their lower bounds apply as well to PDM. Modified versions of PDM that integrate various aspects of parallel computation are developed in [54, 96, 122]. Surveys of I/O models and algorithms appear in [16, 120]. Models of "active disks" augmented with processing capabilities to reduce data traffic to the host, especially during scanning applications, are given in [2, 110].

The same type of bottleneck that occurs between internal memory and external disk storage can also occur at other levels of the memory hierarchy, such as between registers and data cache, between data cache and level 2 cache, between level 2 cache and DRAM, and between disk storage and tertiary devices. The PDM model can be generalized to model the hierarchy of memories ranging from registers at the small end to tertiary storage at the large end. Optimal algorithms for PDM often generalize in a recursive fashion to yield optimal algorithms in the hierarchical memory models. However, the match between theory and practice is harder to establish with hierarchical models; the simpler models are less practical, and the more practical models can be cumbersome to use.

For reasons of brevity and emphasis, we do not consider such hierarchical models in this paper. We refer the reader to the following references: Aggarwal et al. [8] define an elegant hierarchical memory model, and Aggarwal et al. [9] augment it with block transfer capability. Alpern et al. [13] model levels of memory in which the memory size, block size, and bandwidth grow at uniform rates. Vitter and Shriver [139] and Vitter and Nodine [137] discuss parallel versions and variants of the hierarchical models. The parallel model of Li et al. [96] also applies to hierarchical memory. Savage [116] gives a hierarchical pebbling version of [117]. Carter and Gatlin [37] define pebbling models of nonassociative direct-mapped caches.

3. External Sorting and Related Problems

The problem of *external sorting* (or sorting in external memory) is a central problem in the field of EM algorithms, partly because sorting and sorting-like operations account for a significant percentage of computer use [88], and also because sorting is an important paradigm in the design of efficient EM algorithms. With some technical qualifications, many problems that can be solved easily in linear time in internal memory, such as permuting, list ranking, expression tree evaluation, and finding connected components in a sparse graph, require the same number of I/Os in PDM as does sorting.

THEOREM 3.1 ([11, 106]). *The average-case and worst-case number of I/Os required for sorting N data items using D disks is*

$$(3.1) \qquad \Theta\left(\frac{n}{D}\log_m n\right) = \Theta\left(\frac{n}{D}\frac{\log n}{\log m}\right).$$

It is conceptually much simpler to program for the single-disk case ($D = 1$) than for the multiple-disk case. *Disk striping* is a paradigm that can ease the programming task with multiple disks. I/Os are permitted only on entire stripes, one at a time. For example, in the data layout in Figure 3, data items 20–29 can be accessed in a single I/O step because their blocks are grouped into the same stripe. The net effect of striping is that the D disks behave as a single logical disk, but with a larger logical block size DB.

Let us consider what happens if we use the technique of disk striping in conjunction with an optimal sorting algorithm for one disk. The optimal number of I/Os using one disk is

$$(3.2) \qquad \Theta\left(n\frac{\log n}{\log m}\right) = \Theta\left(\frac{N}{B}\frac{\log(N/B)}{\log(M/B)}\right).$$

The effect of disk striping with D disks is to replace B by DB in (3.2), which yields the I/O bound

$$(3.3) \qquad \Theta\left(\frac{N}{DB}\frac{\log(N/DB)}{\log(M/DB)}\right) = \Theta\left(\frac{n}{D}\frac{\log(n/D)}{\log(m/D)}\right).$$

The striping I/O bound (3.3) is larger than the optimal bound (3.1) by a multiplicative factor of about $(\log m)/\log(m/D)$, which is significant when D is on the order of m, causing the $\log(m/D)$ term in the denominator to be very small. In order to attain the optimal sorting bound (3.1) theoretically, we must be able to control the disks independently, so that each disk can access a different stripe in the same I/O step. Sorting via disk striping is often more efficient in practice than more complicated techniques that utilize independent disks, since the $(\log m)/\log(m/D)$ factor may be dwarfed by the additional overhead of using the disks independently [**134**].

In Sections 3.1 and 3.2, we consider some recently developed external sorting algorithms based upon the *distribution* and *merge* paradigms. The SRM method, which uses a randomized merge technique, outperforms disk striping in practice for reasonable values of D (see Section 3.2). In Sections 3.3 and 3.4, we consider the related problems of permuting and Fast Fourier Transform. All the methods we cover, with the exception of Greed Sort in Section 3.2, access the disks independently during parallel read operations, but parallel writes are done in a striped manner, which facilitates the writing of parity error correction information. (We refer the reader to [**41, 76**] for a discussion of error correction issues.) In Section 3.5, we discuss some fundamental lower bounds on the number of I/Os needed to perform sorting and other batched problems in external memory.

3.1. Sorting by Distribution: Simultaneous Online Load Balancings.
Distribution sort is a recursive process in which the data items to be sorted are partitioned by a set of $S - 1$ partitioning elements into S buckets. All the items in one bucket precede all the items in the next bucket. The individual buckets are then sorted recursively and concatenated together to form a single totally sorted list.

The $S - 1$ partitioning elements should be chosen so that the buckets are of roughly equal size. When that is the case, the bucket sizes decrease by a $\Theta(S)$ factor from one level of recursion to the next, and there are $O(\log_S n)$ levels of recursion. During each level of recursion, the data are streamed through internal memory, and the S buckets are written to the disks in an online manner as the streaming proceeds. A double buffer of size $2B$ is allocated to each of the S buckets. When one half of the double buffer fills, its block is written to disk in the next I/O, and the other half is used to store the incoming items. Therefore, the maximum number of buckets (and partitioning elements) is $S = \Theta(M/B) = \Theta(m)$, and the resulting number of levels of recursion is $\Theta(\log_m n)$.

It seems difficult to find $S = \Theta(m)$ partitioning elements using $\Theta(n/D)$ I/Os and guarantee that the bucket sizes are within a constant factor of one another. Efficient deterministic methods exist for choosing $S = \sqrt{m}$ partitioning elements [**105, 138**], which has the effect of doubling the number of levels of recursion. Probabilistic methods based upon random sampling can be found in [**61**].

In order to meet the sorting bound (3.1), the formation of the buckets at each level of recursion must be done in $O(n/D)$ I/Os, which is easy to do for the single-disk case. In the more general multiple-disk case, each read step and each write step during the bucket formation must involve on the average $\Theta(D)$ blocks. The file of items being partitioned was itself one of the buckets formed in the previous level of recursion. In order to read that file efficiently, its blocks must be spread uniformly among the disks, so that no one disk is a bottleneck. The challenge in distribution sort is to write the blocks of the buckets to the disks in an online manner and achieve a global load balance by the end of the partitioning, so that the bucket can be read efficiently during the next level of the recursion.

Partial striping is an effective technique for reducing the amount of information that must be stored in internal memory in order to manage the disks. The disks are grouped into clusters of size C and data are written in "logical blocks" of size CB, one per cluster. Choosing $C = \sqrt{D}$ won't change the optimal sorting time by more than a constant factor, but as pointed out earlier, full striping (in which $C = D$) can be nonoptimal.

Vitter and Shriver [138] use two randomized online techniques during the partitioning so that with high probability each bucket is well balanced across the D disks. (Partial striping is used so that the pointers needed to keep track of the layout of the buckets on the disks can fit in internal memory.) The first technique is used when the size N of the file to partition is sufficiently large or when $M/DB = \Omega(\log D)$, so that the number $\Theta(n/S)$ of blocks in each bucket is $\Omega(D \log D)$. Each parallel write operation writes its D blocks in random order to a disk stripe, with all $D!$ orders equally likely. At the end of the partitioning, with high probability each block is evenly distributed among the disks. This situation is analogous to a hashing scenario in which the number of inserted items is larger by at least a logarithmic factor than the number of bins in the hash table, thereby causing items to be spread fairly evenly, so that the expected maximum bin size is within a constant factor of the expected bin size [136].

If the number of blocks per bucket is not $\Omega(D \log D)$, however, the technique breaks down and the distribution of each bucket among the disks tends to be uneven. For these smaller values of N, Vitter and Shriver use a different technique: In one pass, the file is read, one memoryload at a time. Each memoryload is randomly permuted and written back to the disks in the new order. In a second pass, the file is accessed one memoryload at a time in a "diagonally striped" manner. They show that with very high probability each individual "diagonal stripe" contributes about the same number of items to each bucket, so the blocks of the buckets in each memoryload can be assigned to the disks in a balanced round robin manner using an optimal number of I/Os.

An even better way to do distribution sort, and deterministically at that, is the BalanceSort method developed by Nodine and Vitter [105]. During the partitioning process, the algorithm keeps track of how evenly each bucket has been distributed so far among the disks. For each $1 \le b \le S$ and $1 \le d \le D$, let num_b be the total number of items in bucket b processed so far during the partitioning and let $num_b(d)$ be the number of those items written to disk d; that is, $num_b = \sum_{1 \le d \le D} num_b(d)$. The algorithm is able to write at least half of any given memoryload to the disks and still maintain the invariant for each bucket b that the $\lfloor D/2 \rfloor$ largest values of $num_b(1)$, $num_b(2)$, ..., $num_b(D)$ differ by at most 1, and hence each $num_b(d)$ is at most about twice the ideal value num_b/D.

An alternative sorting technique, with higher overhead, is to use the buffer tree data structure [14] described in Section 7.2, which was developed for batched dynamic applications.

DeWitt et al. [56] present a randomized distribution sort algorithm in a similar model to handle the case when sorting can be done in two passes. They use a sampling technique to find the partitioning elements and route the items in each bucket to a particular processor. The buckets are sorted individually in the second pass.

Matias et al. [99] develop optimal in-place distribution sort algorithms for one disk as a function of the number K of distinct key values. The corresponding I/O bound is $O(n \log_m \min\{K, n\})$. Their technique can be extended within the same I/O bounds to merge sort.

Distribution sort algorithms may have an advantage over the merge approaches presented in the next section in that they typically make better use of lower levels of cache in the memory hierarchy of real systems. Such an intuition comes from analysis of distribution sort and merge sort algorithms on models of hierarchical memory, such as the RUMH model of Vitter and Nodine [**137**].

3.2. Sorting by Merging. The merge paradigm is somewhat orthogonal to the distribution paradigm discussed in Section 3.1. A typical merge sort algorithm works as follows: In the "run formation" phase, the n blocks of data are streamed into memory, one memoryload at a time; each memoryload is sorted into a single "run", which is then output to stripes on disk. At the end of the run formation phase, there are $N/M = n/m$ (sorted) runs, each striped across the disks. (In actual implementations, the "replacement-selection" technique can be used to get runs of $2M$ data items, on the average, when $M \gg B$ [**88**].)

After the initial runs are formed, the merging phase begins. In each pass of the merging phase, groups of R runs are merged together. During each merge, one block from each run resides in internal memory. When the data items of a block expire, the next block for that run is input. Double buffering is used to keep the disks busy. Hence, at most $R = \Theta(m)$ runs can be merged at a time; the resulting number of passes is $O(\log_m n)$.

To achieve the optimal sorting bound (3.1), each merging pass must be done in $O(n/D)$ I/Os, which is easy to do for the single-disk case. In the more general multiple-disk case, each parallel read operation during the merging must on the average bring in the next $\Theta(D)$ blocks needed for the merging. The challenge is to ensure that those blocks reside on different disks so that they can be read in a single I/O (or a small constant number of I/Os). The difficulty lies in the fact that the runs being merged were themselves formed during the previous merge pass. Their blocks were written to the disks in the previous pass without knowledge of how they would interact with other runs in later merges.

A perfect solution, in which the next D blocks needed for the merge are guaranteed to be on distinct disks, can be devised for the binary merging case $R = 2$ based upon the Gilbreath principle [**67, 88**]: The first run is striped in ascending order by disk number, and the other run is striped in descending order. Regardless of how the items in the two runs interleave during the merge, it is always the case that the next D blocks needed for the output can be accessed via a single I/O operation, and thus the amount of internal memory buffer space needed for binary merging can be kept to a minimum. Unfortunately there is no analog to the Gilbreath principle for $R > 2$, and as we have seen above, we need the value of R to be large in order to get an optimal sorting algorithm.

The Greed Sort method of Nodine and Vitter [**106**] was the first optimal deterministic EM algorithm for sorting with multiple disks. It handles the case $R > 2$ by relaxing the condition on the merging process. In each step, the following two blocks from each disk are brought into internal memory: the block b_1 with the smallest data item value and the block b_2 whose largest item value is smallest. If $b_1 = b_2$, only one block is read into memory, and it is added to the next output stripe. Otherwise, the two blocks b_1 and b_2 are merged in memory; the smaller B items are written to the output stripe, and the remaining items are written back to the disk. The resulting run that is produced is only an "approximately" merged

	$D = 5$	$D = 10$	$D = 50$
$k = 5$	0.56	0.47	0.37
$k = 10$	0.61	0.52	0.40
$k = 50$	0.71	0.63	0.51

TABLE 1. The ratio of the number of I/Os used by simple randomized merge sort (SRM) to the number of I/Os used by merge sort with disk striping, during a merge of kD runs. The figures were obtained by simulation; they back up the (more pessimistic) analytic upper bound in [27].

run, but its saving grace is that no two inverted items are too far apart. A final application of Columnsort [94] in conjunction with partial striping suffices to restore total order.

An optimal deterministic merge sort, with somewhat higher constant factors than those of the distribution sort algorithms, was developed by Aggarwal and Plaxton [10], based upon the Sharesort hypercube sorting algorithm [53]. To guarantee even distribution during the merging, it employs two high-level merging schemes in which the scheduling is almost oblivious.

The most practical method for sorting is the simple randomized merge sort (SRM) algorithm of Barve et al. [27] (referred to as "randomized striping" by Knuth [88]). Each run is striped across the disks, but with a random starting point (the only place in the algorithm where randomness is utilized). During the merging process, the next block needed from each disk is read into memory, and if there is not enough room, the least needed blocks are "flushed" (without any I/Os required) to free up space. The expected performance of SRM is not optimal for some parameter values, but it significantly outperforms the use of disk striping for reasonable values of the parameters, as shown in Table 1. Barve et al. [27] derive an upper bound on the I/O performance; the precise analysis is an interesting open problem [88]. Work is beginning on applying the SRM buffer management techniques to distribution sort. The hope is to get better overall sorting performance by means of improved cache utilization, based upon the intuition mentioned at the end of the previous section.

3.3. Permuting and Transposition. Permuting is the special case of sorting in which the key values of the N data items form a permutation of $\{1, 2, \ldots, N\}$.

THEOREM 3.2 ([11]). *The average-case and worst-case number of I/Os required for permuting N data items using D disks is*

$$(3.4) \qquad \Theta\left(\min\left\{\frac{N}{D}, \frac{n}{D}\log_m n\right\}\right).$$

The I/O bound (3.4) for permuting can be realized by using one of the sorting algorithms from Section 3 except in the extreme case $B \log m = o(\log n)$, in which case it is faster to move the data items one by one in a non-blocked way. The one-by-one method is trivial if $D = 1$, but with multiple disks there may be bottlenecks on individual disks; one solution for doing the permuting in $O(N/D)$ I/Os is to apply the randomized balancing strategies of [138].

Matrix transposition is the special case of permuting in which the permutation can be represented as a transposition of a matrix from row-major order into column-major order.

THEOREM 3.3 ([11]). *The number of I/Os required using D disks to transpose a $p \times q$ matrix from row-major order to column-major order is*

$$(3.5) \qquad \Theta\left(\frac{n}{D}\log_m \min\{M, p, q, n\}\right),$$

where $N = pq$ and $n = N/B$.

When B is large compared with M, matrix transposition can be as hard as general sorting, but for smaller B, the special structure of the transposition permutation makes transposition easier. In particular, the matrix can be broken up into square submatrices of B^2 elements such that each submatrix contains B blocks of the matrix in row-major order and also B blocks of the matrix in column-major order. Thus, if $B^2 < M$, the transpositions can be done in a simple one-pass operation by transposing the submatrices one-at-a-time in internal memory.

Matrix transposition is a special case of a more general class of permutations called *bit-permute/complement* (BPC) permutations, which in turn is a subset of the class of *bit-matrix-multiply/complement* (BMMC) permutations. BMMC permutations are defined by a $\log N \times \log N$ nonsingular 0-1 matrix A and a $(\log N)$-length 0-1 vector c. An item with binary address x is mapped by the permutation to the binary address given by $Ax \oplus c$. BPC permutations are the special case of BMMC permutations in which A is a permutation matrix, that is, each row and each column of A contain a single 1. BPC permutations include matrix transposition, bit-reversal permutations (which arise in the FFT), vector-reversal permutations, hypercube permutations, and matrix reblocking. Cormen et al. [50] characterize the optimal number of I/Os needed to perform any given BMMC permutation solely as a function of the associated matrix A, and they give an optimal algorithm for implementing it.

THEOREM 3.4 ([50]). *The number of I/Os required using D disks to perform the BMMC permutation defined by matrix A and vector c is*

$$(3.6) \qquad \Theta\left(\frac{n}{D}\left(1 + \frac{\mathrm{rank}(\gamma)}{\log m}\right)\right),$$

where γ is the lower-left $\log n \times \log B$ submatrix of A.

An interesting theoretical question is whether there is a simple characterization (as a function of the input) of the I/O cost for a general permutation.

3.4. Fast Fourier Transform. Computing the Fast Fourier Transform (FFT) in external memory consists of a series of I/Os that permit each computation implied by the FFT directed graph (or butterfly) to be done while its arguments are in internal memory. A permutation network computation consists of a fixed pattern of I/Os such that any of the $N!$ possible permutations can be realized; data items can only be reordered when they are in internal memory. A permutation network can be realized by a series of three FFTs [145].

THEOREM 3.5. *With D disks, the number of I/Os required for computing the N-input FFT digraph or an N-input permutation network is given by the same bound (3.1) as for sorting.*

Cormen and Nicol [**49**] give some practical implementations for one-dimensional FFTs based upon the optimal PDM algorithm of [**138**]. The algorithms for FFT are faster and simpler than for sorting because the computation is nonadaptive in nature, and thus the communication pattern is oblivious.

3.5. Lower Bounds on I/O. In this section we prove the lower bounds from Theorems 3.1–3.5 and mention some related I/O lower bounds for batched problems in computational geometry and graphs.

The most trivial batched problem is that of *scanning* (or *streaming* or *touching*) a file of N data items, which can be done in a linear number $O(N/DB) = O(n/D)$ of I/Os. Permuting is one of several simple problems that can be done in linear CPU time in the (internal memory) RAM model, but require a nonlinear number of I/Os in PDM because of the locality constraints imposed by the block parameter B.

The following proof of the permutation lower bound (3.4) of Theorem 3.2 is due to Aggarwal and Vitter [**11**]. The idea of the proof is to measure, for each $t \geq 0$, the number of distinct orderings that are realizable by at least one sequence of t I/Os. The value of t for which the number of distinct orderings first exceeds $N!/2$ is a lower bound on the average number of I/Os (and hence the worst-case number of I/Os) needed for permuting.

We assume for the moment that there is only one disk, $D = 1$. Let us consider how the number of realizable orderings can change when we read a given disk block into internal memory. There are at most B data items in the block, and they can intersperse among the M items in internal memory in at most $\binom{M}{B}$ ways, so the number of realizable orderings increases by a factor of $\binom{M}{B}$. If the block has never before resided in internal memory, the number of realizable orderings increases by an extra $B!$ factor, since the items in the block can be permuted among themselves. (This extra contribution of $B!$ can only happen once for each of the N/B original blocks.) The effect of writing the disk block is considerably less than that of reading it. There are at most $n + t \leq N \log N$ ways to choose which disk block is involved in the I/O. (We allow the algorithm to use an arbitrary amount of disk space.) Hence, the number of distinct orderings that can be realized by some sequence of t I/Os is at most

$$(3.7) \qquad (B!)^{N/B} \left(N (\log N) \binom{M}{B} \right)^t .$$

Setting the expression in (3.7) to be at least $N!/2$, and simplifying by taking the logarithm, we get

$$(3.8) \qquad N \log B + t \left(\log N + B \log \frac{M}{B} \right) = \Omega(N \log N).$$

We get the lower bound for the case $D = 1$ by solving for t. The general lower bound (3.4) follows by dividing by D.

Permuting is a special case of sorting, and hence, the permuting lower bound applies also to sorting. In the unlikely case that $B \log m = o(\log n)$, the permuting bound is only $\Omega(N/D)$, and we must resort to the comparison model to get the full lower bound (3.1) of Theorem 3.1 [**11**]. Arge et al. [**19**] show for the comparison model that any problem with an $\Omega(N \log N)$ lower bound in the RAM model requires $\Omega(n \log_m n)$ I/Os in PDM. However, in the typical case in which $B \log m = \Omega(\log n)$, the comparison model is not needed to prove the sorting lower bound; the difficulty of sorting in that case arises not from determining the order of the data but from permuting (or routing) the data.

The proof used above for permuting also works for permutation networks, in which the communication pattern is oblivious. Since the choice of disk block is fixed for each t, there is no $N \log N$ term as there is in (3.7), and correspondingly there is no additive $\log N$ term in the inner expression as there is in (3.8). Hence, when we solve for t, we get the lower bound (3.1) rather than (3.4). The lower bound follows directly from the counting argument; unlike the sorting derivation, it does not require the comparison model for the case $B \log m = o(\log n)$. The lower bound also applies directly to FFTs, since permutation networks can be formed from three FFTs in sequence. The transposition lower bound involves a potential argument based upon a togetherness relation [11]. A related argument demonstrates the optimality of the algorithm in [99] for sorting N items with K distinct key values.

Chiang et al. [43], Arge [15], Arge and Miltersen [20], and Kameshwar and Ranade [83] give models and lower bound reductions for several computational geometry and graph problems. Problems like list ranking and expression tree evaluation have the same nonlinear I/O lower bound as permuting. Other problems like connected components, biconnected components, and minimum spanning trees of sparse graphs with E edges and V vertices require as many I/Os as E/V instances of sorting V items. This situation is in contrast with the RAM model, in which the same problems can all be done in linear CPU time. (The known linear-time RAM algorithm for minimum spanning tree is randomized.) In some cases, there is a gap between the best known upper and lower bounds, which we discuss further in Section 6. The geometry problems discussed in Section 5 are equivalent to sorting in both the internal memory and PDM models.

The lower bounds mentioned above assume that the data items are in some sense "indivisible", in that they are not split up and reassembled in some magic way to get the desired output. It is conjectured that the sorting lower bound (3.1) remains valid even if the indivisibility assumption is lifted. However, for an artificial problem related to transposition, Adler [3] showed that removing the indivisibility assumption can lead to faster algorithms. A similar result is shown by Arge and Miltersen [20] for the decision problem of determining if N data item values are distinct. Whether or not the conjecture is true is a challenging theoretical problem.

4. Matrix and Grid Computations

Dense matrices are generally represented in memory in row-major or column-major order. Matrix transposition, which is the special case of sorting that involves conversion of a matrix from one representation to the other, was discussed in Section 3.3. For certain operations such as matrix addition, both representations work well. However, for standard matrix multiplication (using only semiring operations) and LU decomposition, a better representation is to block the matrix into square $\sqrt{B} \times \sqrt{B}$ submatrices, which gives the upper bound of the following theorem:

THEOREM 4.1 ([78, 117, 138, 144]). *The number of I/Os required for standard matrix multiplication of two $k \times k$ matrices or to compute the LU factorization of a $k \times k$ matrix is $\Theta\left(k^3 / \min\{k, \sqrt{M}\} DB\right)$.*

Hong and Kung [78] and Nodine et al. [104] give optimal EM algorithms for iterative grid computations, and Leiserson et al. [95] reduce the number of I/Os of naive multigrid implementations by a $\Theta(M^{1/5})$ factor. Gupta et al. [73] show

how to derive efficient EM algorithms automatically for computations expressed in tensor form.

If a $k \times k$ matrix A is sparse, that is, if the number N_z of nonzero elements in A is much smaller than k^2, then it may be more efficient to store only the nonzero elements. Each nonzero element $A_{i,j}$ is represented by the triple $(i, j, A_{i,j})$. Unlike the dense case, in which transposition can be easier than sorting (e.g., see Theorem 3.3 when $B^2 \le M$), transposition of sparse matrices is as hard as sorting:

THEOREM 4.2. *For a matrix stored in sparse format and containing* $N_z = n_z B$ *nonzero elements, the number of I/Os required to convert the matrix from row-major order to column-major order, and vice-versa, is*

$$(4.1) \qquad \Theta\left(\frac{n_z}{D} \log_m n_z\right).$$

The lower bound follows by reduction from sorting. If the ith item in the input of the sorting instance has key value $x \ne 0$, there is a nonzero element in matrix position (i, x).

We defer further discussion of numerical EM algorithms and refer the reader to Toledo's survey in this volume [127]. Some issues regarding programming environments are discussed in [48] and Section 10.

5. Batched Problems in Computational Geometry[2]

Problems involving massive amounts of geometric data are ubiquitous in spatial databases [93, 113, 114], geographic information systems (GIS) [93, 113, 130], constraint logic programming [84, 85], object-oriented databases [147], statistics, virtual reality systems, and computer graphics [65]. NASA's Earth Observing System project, the core part of the Earth Science Enterprise (formerly Mission to Planet Earth), produces petabytes (10^{15} bytes) of raster data per year [58]! Microsoft's TerraServer online database of satellite images is over one terabyte in size [125]. A major challenge is to develop mechanisms for processing the data, or else much of it will be useless.

For systems of this size to be efficient, we need fast EM algorithms and data structures for basic problems in computational geometry. Luckily, many problems on geometric objects can be reduced to a small core of problems, such as computing intersections, convex hulls, or nearest neighbors. Useful paradigms have been developed for solving these problems in external memory.

THEOREM 5.1. *The following batched problems and several related problems involving* N *input items,* Q *queries, and* Z *output items can be solved using*

$$(5.1) \qquad O\big((n + q)\log_m n + z\big)$$

I/Os (where Q and Z are set to 0 if they are not relevant for the particular problem):

1. *Computing the pairwise intersections of N orthogonal segments in the plane,*
2. *Answering Q orthogonal 2-D range queries on N points in the plane (i.e., finding all the points within the Q query rectangles),*
3. *Computing the pairwise intersections of N segments in the plane,*

---

[2]For brevity, in the remainder of this paper we deal only with the single-disk case $D = 1$. The single-disk I/O bounds for the batched problems can often be cut by a factor of $\Theta(D)$ for the case $D > 1$ by using the load balancing techniques of Section 3. In practice, disk striping may be sufficient. For online problems, disk striping will convert optimal bounds for the case $D = 1$ into optimal bounds for $D > 1$.

4. *Finding all intersections between N nonintersecting red line segments and N nonintersecting blue line segments in the plane.*
5. *Constructing the 2-D and 3-D convex hull of N points,*
6. *Voronoi diagram and Triangulation of N points in the plane,*
7. *Performing Q point location queries in a planar subdivision of size N,*
8. *Finding all nearest neighbors for a set of N points in the plane,*
9. *Finding the pairwise intersections of N orthogonal rectangles in the plane,*
10. *Computing the measure of the union of N orthogonal rectangles in the plane,*
11. *Computing the visibility of N segments in the plane from a point,*
12. *Performing Q ray-shooting queries in 2-D Constructive Solid Geometry (CSG) models of size N,*

Goodrich et al. [**69**], Zhu [**149**], Arge et al. [**24**], Arge et al. [**22**], and Crauser et al. [**51, 52**] develop EM algorithms for those problems using the following EM paradigms for batched problems:

Distribution sweeping: a generalization of the distribution paradigm of Section 3 for externalizing plane sweep algorithms;

Persistent B-trees: an offline method for constructing an optimal-space persistent version of the B-tree data structure (see Section 7.1), yielding a factor of B improvement over the generic persistence techniques of Driscoll et al. [**57**].

Batched filtering: a general method for performing simultaneous external memory searches in data structures that can be modeled as planar layered directed acyclic graphs and in external fractionally cascaded data structures; it is useful for 3-D convex hulls and batched point location.

External fractional cascading: an EM analog to fractional cascading on a segment tree.

Online filtering: a technique based upon the work of Tamassia and Vitter [**124**] for online queries in data structures with fractional cascading.

External marriage-before-conquest: an EM analog to the well-known technique of Kirkpatrick and Seidel [**87**] for performing output-sensitive convex hull constructions.

Randomized incremental construction with gradations: a localized version of the incremental construction paradigm of Clarkson and Shor [**46**].

The distribution sweep paradigm is fundamental to sweep line processes. For example, we can compute the pairwise intersections of N orthogonal segments in the plane by the following recursive distribution sweep: At each level of recursion, the plane is partitioned into $\Theta(m)$ vertical strips, each containing $\Theta(N/m)$ of the segments' endpoints. We sweep a horizontal line from top to bottom to process the N segments. When a vertical segment is encountered by the sweep line, the segment is inserted into the appropriate strip. When a horizontal segment h is encountered by the sweep line, we report h's intersections with all the "active" vertical segments in the strips that are spanned *completely* by h. (A vertical segment is "active" if it is intersected by the current sweep line; vertical segments that are found to be no longer active are deleted from the strips.) The remaining end portions of h (which partially span a strip) are passed recursively to the next level, along with the vertical segments. After the initial sorting preprocessing, each of the $O(\log_m n)$ levels of recursion requires $O(n)$ I/Os, yielding the desired bound (5.1). Arge et al. [**22**] develop a unified approach to distribution sweep in higher dimensions.

A central operation in spatial databases is spatial join. A common preprocess-ing step is to find the pairwise intersections of the bounding boxes of the objects involved in the spatial join. The problem of intersecting orthogonal rectangles can be solved by combining the previous algorithm for orthogonal segments with one for range searching. A unified approach, extendible to higher dimensions, is taken by Arge et al. [22] using distribution sweep. The objects that are stored in the data structure in this case are rectangles, not vertical segments. The branching factor is chosen to be $\Theta(\sqrt{m})$ rather than $\Theta(m)$. Each rectangle is associated with the largest contiguous range of vertical strips that it spans. Each of the possible $\Theta\left(\binom{\sqrt{m}}{2}\right) = \Theta(m)$ contiguous ranges is called a *multislab*. (The branching factor was chosen to be $\Theta(\sqrt{m})$ rather than $\Theta(m)$ so as to accommodate a buffer in in-ternal memory for each multislab; the height of the tree remains $O(\log_m n)$.) The resulting algorithm outperforms other techniques; empirical timings are given in Section 11.

Arge et al. [22] give an algorithm for finding all intersections among N line segments, but the output component of the I/O bound is slightly nonoptimal: $z \log_m n$ rather than z. Crauser et al. [51, 52] use an incremental randomized construction to attain the optimal I/O bound (5.1) for line segment intersection and other problems. They also show how to compute the trapezoidal decomposition for intersecting segments.

6. Batched Problems on Graphs

The first work on EM graph algorithms was by Ullman and Yannakakis [128] for the problem of transitive closure. Chiang et al. [43] consider a variety of graph problems, several of which have upper and lower I/O bounds related to permuting. One key idea Chiang et al. exploit is that efficient EM algorithms can often be developed by a sequential simulation of a parallel algorithm for the same problem. Sorting is done periodically to reblock the data. In list ranking, which is used as a subroutine in the solution of several other graph problems, the number of working processors in the parallel algorithm decreases geometrically with time, so the number of I/Os for the entire simulation is proportional to the number of I/Os used in the first phase, which is given by the sorting bound $\Theta(n \log_m n)$. Dehne et al. [54] and Sibeyn and Kaufmann [122] show how to get efficient I/O bounds by exploiting coarse-grained parallel algorithms, under certain assumptions on the parameters of the PDM model (such as assuming that $\log_m n \leq 2$ and that the total disk space usage is $O(n)$) so that the periodic sortings can be done in a linear number of I/Os.

For list ranking, the optimality of the EM algorithm in [43] assumes that $\sqrt{m} \log m = \Omega(\log n)$, which is usually true. That assumption can be removed by use of the buffer tree data structure [14] (see Section 7.2). A practical, randomized implementation of list ranking appears in [121]. Recent work on other EM graph algorithms appears in [1, 15, 71, 83, 91]. The problem of how to store graphs on disks for efficient traversal is discussed in [6, 103]. EM problems that arise in data mining and On-Line Analytical Processing include constructing classification trees [142] and computing wavelet decompositions and histograms [140, 141].

The I/O complexity of several of the basic graph problems considered in [43, 83, 128] remain open, including connected components, topological sort-ing, shortest paths, breadth-first search, and depth-first search. For example, for a graph with $V = vB$ vertices and $E = eB$ edges, the best-known EM algorithms for breadth-first search, depth-first search, and transitive closure require $\Theta(e \log_m v + V)$, $\Theta(ve/m + V)$, and $\Theta(Vv\sqrt{e/m})$ I/Os, respectively. Connected components can be determined in $O\big(e(\log_m v)\log\max\{2, \log(vB/e)\}\big)$ I/Os deterministically and in only $O(e \log_m v)$ I/Os probabilistically.

In order for the parallel simulation technique to yield an efficient EM algorithm, the parallel algorithm must not use too many processors, preferably at most N. Unfortunately, the polylog-time algorithms for problems like depth-first search and shortest paths use a polynomial number of processors. The interesting connection between the parallel domain and the EM domain suggests that there may be relationships between computational complexity classes related to parallel computing (such as P-complete problems) and those related to I/O efficiency.

7. Spatial Data Structures

We now turn our attention to some online spatial data structures for massive data applications. For purposes of exposition, we consider dictionary lookup and orthogonal range search as the canonical query operations. That is, we want data structures that can support insert, delete, lookup, and orthogonal range query. Given a value x, the lookup operation returns the item(s), if any, in the structure with key value x. A range query, for a given d-dimensional rectangle, returns all the points in the interior of the rectangle.

Spatial data structures tend to be of two types: space-driven or data-driven. Quad trees, grid files, and hashing are space-driven since they are based upon a partitioning of the embedding space, whereas methods like R-trees and kd-trees are organized by partitioning the data items themselves. We discuss primarily the latter type in this section.

7.1. B-trees and Variants.
Tree-based data structures arise naturally in the dynamic online setting, in which the data can be updated and queries must be processed immediately. Binary trees have a host of applications in the RAM model. In order to exploit block transfer, trees in external memory generally use a block for each node, which can store $\Theta(B)$ pointers and data values. A tree of degree B^c with n leaf nodes has $\lceil \frac{1}{c}\log_B N \rceil$ levels. The well-known *B-tree* due to Bayer and McCreight [**30, 47, 88**] is a balanced multiway tree with height roughly $\log_B N$ and with node degree $\Theta(B)$. (The root node is allowed to have smaller degree.) B-trees support dynamic dictionary operations and one-dimensional range search optimally in linear space, $O(\log_B N + z)$ I/Os per query, and $O(\log_B N)$ I/Os per insert or delete. When a node overflows during an insertion, it splits into two half-full nodes, and if the splitting causes the parent node to overflow, the parent node splits, and so on. Splittings can thus propagate up to the root. Deletions are handled in a symmetric way by merging nodes.

In the B^+-*tree* variant, pictured in Figure 4, all the items are stored in the leaves, and the leaves are linked together in symmetric order to facilitate range queries and sequential access. The internal nodes store only key values and pointers and thus can have a higher branching factor. In the popular variant of B$^+$-trees called B^*-*trees*, splitting can usually be postponed when a node overflows, by instead "sharing" the node's data with one of its adjacent siblings. The node needs to be split only if the sibling is also full; when that happens, the node splits into two, and its data and those of its full sibling are evenly redistributed, making each

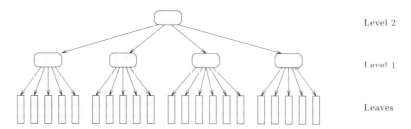

FIGURE 4. B$^+$-tree multiway search tree. Each internal and leaf node corresponds to a disk block. All the items are stored in the leaves. The internal nodes store only key values and pointers, $\Theta(B)$ of them per node. Although not indicated here, the leaf blocks are linked together sequentially.

of the three nodes about 2/3 full. This local optimization reduces how often new nodes must be created, and it increases the relative storage utilization. When no sharing is done (as in B$^+$-trees), Yao [146] shows that nodes are roughly $\ln 2 \approx 69\%$ full on the average, assuming random insertions. With sharing (as in B*-trees), the average storage utilization increases to about $2\ln(3/2) \approx 81\%$ [26, 92]. Storage utilization can be increased further by sharing among several siblings, but insertions and deletions get more complicated.

Persistent versions of B-trees have been developed by Becker et al. [31] and Varman and Verma [131]. Lomet and Salzberg [98] explore mechanisms to add concurrency and recovery to B-trees.

Arge and Vitter [25] give a useful variant of B-trees called *weight-balanced B-trees* with the property that the number of data items in any subtree of height h is $\Theta(a^h)$, for some fixed parameter a of order B. (By contrast, the sizes of subtrees at level h in a regular B-tree can differ by a multiplicative factor that is exponential in h.) When a node on level h gets rebalanced, no further rebalancing is needed until its subtree is updated $\Omega(a^h)$ times. This feature can support applications in which the cost to rebalance a node is $O(w)$, allowing the rebalancing to be done in an amortized (and often worst-case) way with $O(1)$ I/Os. Weight-balanced B-trees were originally conceived as part of optimal dynamic data structures for stabbing queries and segment trees in external memory, which we discuss in Section 8, but they also have applications to the internal memory RAM model [25, 71]. For example, by setting a to a constant, we get a simple, worst-case implementation of interval trees in internal memory. They also serve as a simpler and worst-case alternative to the data structure in [143] for augmenting one-dimensional data structures with range restriction capabilities.

Agarwal et al. [4] develop an interesting variant of B-trees, called *level-balanced B-trees*, that maintain parent pointers. A straightforward modification of conventional B-trees would require $\Theta(B \log_B N)$ I/Os per split to maintain parent parents. Instead, level-balanced B-trees support insert, delete, merge, and split operations in $O\big((1 + (b/B)(\log_m n) \log_b N\big)$ I/Os amortized, for any $2 \leq b \leq B/2$, which is bounded by $O\big((\log_B N)^2\big)$. Agarwal et al. [4] use level-balanced B-trees in a data structure for point location in monotone subdivisions, which supports queries and (amortized) updates in $O\big((\log_B N)^2\big)$ I/Os. They also use it to dynamically maintain planar st-graphs using $O\big((1 + (b/B)(\log_m n) \log_b N\big)$ I/Os (amortized) per update, so that reachability queries can be answered in $O(\log_B N)$ I/Os (worst-case).

It is open as to whether these results can be improved. One question is how to deal with non-monotone subdivisions. Another question is whether level-balanced B-trees can be implemented in $O(\log_B N)$ I/Os per update. Such an improvement would immediately give an optimal dynamic structure for reachability queries in planar st-graphs.

7.2. Buffer Trees. Many batched problems in computational geometry can be solved by plane sweep techniques. For example, in Section 5 we showed how to compute orthogonal segment intersections by keeping track of the vertical segments hit by a horizontal sweep line. If we use a B-tree to store the hit vertical segments, each insertion and query uses $O(\log_B N)$ I/Os, resulting in a huge I/O bound of $O(N \log_B N)$, which can be more than B times larger than the desired bound of $O(n \log_m n)$. One solution suggested in [**135**] is to use a binary tree in which items are pushed lazily down the tree in blocks of B items at a time. The binary nature of the tree results in a data structure of height $\sim \log n$, yielding a total I/O bound of $O(n \log n)$, which is still nonoptimal by a significant $\log m$ factor.

Arge [**14**] developed the elegant *buffer tree* data structure to support *batched dynamic* operations such as in the sweep line example, where the queries do not have to be answered right away or in any particular order. The buffer tree is a balanced multiway tree, but with degree $\Theta(m)$, except possibly for the root. Its key distinguishing feature is that each node has a buffer that can store M items (i.e., m blocks of items). Items in a node are not pushed down to the children until the buffer fills. Emptying the buffer requires $O(m)$ I/Os, which amortizes the cost of distributing the items to the $\Theta(m)$ children. Each item incurs an amortized cost of $O(m/M) = O(1/B)$ I/Os per level. Queries and updates thus take $O\big((1/B) \log_m n\big)$ I/Os amortized. Buffer trees can be used as a subroutine in the standard sweep line algorithm in order to get an optimal EM algorithm for orthogonal segment intersection. Arge [**33**] showed how to extend buffer trees to implement segment trees in external memory in a batched dynamic setting by reducing the node degrees to $\Theta(\sqrt{m})$ and by introducing *multislabs* in each node.

Buffer trees have an ever-expanding list of applications. They provide, for example, a natural amortized implementation of priority queues for use in applications like discrete event simulation, sweeping, and list ranking. Brodal and Katajainen [**35**] provide a worst-case optimal priority queue, in the sense that every sequence of B *insert* and *delete_min* operations requires only $O(\log_m n)$ I/Os.

7.3. Multidimensional Spatial Structures. Grossi and Italiano [**72**] construct a multidimensional version of B-trees, called *cross trees*, that combine the data-driven partitioning of weight-balanced B-trees at the upper levels of the tree with the space-driven partitioning of methods like quad trees at the lower levels of the tree. For $d > 1$, d-dimensional orthogonal range queries can be done in $O(n^{1-1/d} + z)$ I/Os, and inserts and deletes take $O(\log_B N)$ I/Os. The data structure uses linear space and also supports the dynamic operations of split and concatenate in $O(n^{1-1/d})$ I/Os.

One way to get multidimensional EM data structures is to augment known internal memory structures, such as quad trees and kd-trees, with block access capabilities. Examples include *grid files* [**77**, **90**, **101**], kd-B-trees [**111**], *buddy trees* [**118**], and *hB-trees* [**59**, **97**]. Another technique is to "linearize" the multidimensional space by imposing a total ordering on it (a so-called space-filling curve), and then the total order is used to organize the points into a B-tree. All the methods described in this paragraph use linear space, and they work well in certain situations; however, their worst-case range query performance is no better than that of cross trees, and for some methods, like grid files, queries can require $\Theta(n)$ I/Os, even if there are no points satisfying the query. We refer the reader to [**7, 66, 102**] for a broad survey of these methods. Space-filling curves arise again in connection with R-trees, which we describe in the next section.

7.4. R-trees. The *R-tree* of Guttman [**74**] and its many variants are an elegant multidimensional generalization of the B-tree for storing a variety of geometric objects, such as points, segments, polygons, and polyhedra. Internal nodes have degree $\Theta(B)$ (except possibly the root), and leaves store $\Theta(B)$ items. Each node in the tree has associated with it a bounding box (or bounding polygon) of all the elements in its subtree. A big difference between R-trees and B-trees is that in R-trees the bounding boxes of sibling nodes are allowed overlap. If an R-tree is being used for point location, for example, a point may lie within the bounding box of several children of the current node in the search. In that case, the search must proceed to all such children.

Several heuristics for where to insert new items into an R-tree and how to rebalance it are surveyed in [**7, 66, 70**]. The methods perform well in many practical cases, especially in low dimensions, but they have poor worst-case bounds. An interesting open problem is whether nontrivial bounds can be proven for the "typical-case" behavior of R-trees for problems such as range searching and point location. Similar questions apply to the methods discussed in the previous section.

The *R*-tree* variant of Beckmann et al. [**32**] seems to give best overall query performance. Precomputing an R*-tree by repeated insertions, however, is extremely slow. A faster alternative is to use the Hilbert R-tree of Kamel and Faloutsos [**80, 81**]. Each item is labeled with the position of its center on the Hilbert space-filling curve, and a B-tree is built in a bottom-up manner on the totally ordered labels. Bulk loading a Hilbert R-tree is therefore easy once the center points are presorted, but the quality of the Hilbert R-tree in terms of query performance is not as good as that of an R*-tree, especially for higher-dimensional data [**34, 82**].

Arge et al. [**18**] and van den Bercken et al. [**129**] have independently devised fast bulk loading methods for R*-trees that are based upon buffer trees. The former method is especially efficient and can even support dynamic batched updates and queries. Experiments with this technique are discussed in Section 11.

8. Online Multidimensional Range Searching

Multidimensional range search is a fundamental primitive in several online geometric applications, and it provides indexing support for new constraint data models and object-oriented data models. (See [**85**] for background.) We have already discussed multidimensional range searching in a batched setting in Section 5. In this section we concentrate on the important online case.

For many types of range searching problems, it is very difficult to develop theoretically optimal algorithms. We have seen some linear-space online data structures in Sections 7.3 and 7.4, but their query performance is not optimal. Many open problems remain. The primary theoretical challenges are three-fold:

1. to get a combined search and output cost for queries of $O(\log_B N + z)$ I/Os,
2. to use only a linear amount of disk storage space, and
3. to support dynamic updates in $O(\log_B N)$ I/Os.

To develop optimal data structures for queries, it is helpful to combine together the I/O cost $O(\log_B N)$ of the search component with the I/O cost $O(z)$ for reporting the output, as in criterion 1, rather than to consider the search cost separately from the output cost, because when one cost is much larger than the other, the query algorithm has the extra freedom to follow a *filtering* paradigm [**38**], in which both the search component and the output reporting are allowed to use the larger number of I/Os. Subramanian and Ramaswamy [**123**] prove the lower bound that no EM data structure for 2-D range searching can achieve criterion 1 using less than $O\bigl(n(\log n)/\log(\log_B N + 1)\bigr)$ disk blocks, even if we relax 1 to allow $O\bigl((\log_B N)^c + z\bigr)$ I/Os per query, for any constant c. The result holds for an EM version of the pointer machine model, based upon the approach of Chazelle [**39**] for the internal memory model.

Hellerstein et al. [**75**] consider a generalization of the layout-based lower bound argument of Kanellakis et al. [**85**] for studying the tradeoff between disk space usage and query performance. An "efficient" data structure is expected to contain the Z output points to a query compactly within $O\bigl(\lceil Z/B \rceil\bigr) = O\bigl(\lceil z \rceil\bigr)$ blocks. One shortcoming of the model is that it considers only data layout and ignores the search component of queries, and thus it rules out a filtering approach. For example, it is reasonable for any query algorithm to perform at least $\log_B N$ I/Os, so if the output size Z is at most B, an algorithm may still be able to satisfy criterion 1 even if the output is contained within $O(\log_B N)$ blocks rather than $O(z) = O(1)$ blocks. One fix is to consider only output sizes Z larger than $(\log_B N)B$, but then the problem of how to find the relevant blocks is ignored. Despite this shortcoming, the model is elegant and provides insight into the complexity of blocking data in external memory. Further results in this model appear in [**23, 89, 115**].

When the data structure is restricted to contain only a single copy of each item, Kanth and Singh [**86**] show for a restricted class of index-based trees that d-dimensional range queries in the worst case require $\Omega(n^{1-1/d} + z)$ I/Os, and they provide a data structure with a matching bound. Another approach to achieve the same bound is the cross tree data structure [**72**] mentioned in Section 7.3, which in addition supports the operations of split and concatenate.

The lower bounds mentioned above for 2-D range search apply to general rectangular queries. A natural question to ask is whether there are data structures that can meet criteria 1–3 for interesting special cases of 2-D orthogonal range searching. Fortunately, the answer is yes.

To be precise, we define a $(s_1, s_2, \ldots, s_d)$-*sided range query* in d-dimensional space, where each $s_i \in \{1, 2\}$, to be an orthogonal range query with s_i finite limits in the x_i dimension. For example, the 2-D range query $[3, 5] \times [4, \infty)$ is a $(2, 1)$-sided range query, since there are two finite limits in the x_1 dimension (namely, $3 \leq x_1$ and $x_1 \leq 5$) but only one finite limit in the x_2 dimension (namely, $x_2 \geq 4$). A general 2-D range query is a $(2, 2)$-sided query. (See Figure 5.) In the two-dimensional cases studied in [**23, 85, 109, 123**], the authors use the terms "two-sided", "three-sided", and "four-sided" range query to mean what we call $(1, 1)$-sided, $(2, 1)$-sided, and $(2, 2)$-sided queries, respectively.

Arge and Vitter [**25**] design an EM interval tree data structure based upon the weight-balanced B-tree that meets all three criteria. It uses linear disk space and does queries in $O(\log_B N + z)$ I/Os and updates in $O(\log_B N)$ I/Os. It solves the problems of stabbing queries and dynamic interval management, utilizing the optimal static structure of Kanellakis et al. [**85**]. Stabbing queries are equivalent

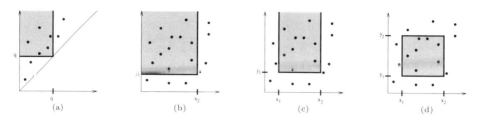

FIGURE 5. Different types of 2-D range queries: (a) Diagonal corner $(1,1)$-sided query, (b) $(1,1)$-sided query, (c) $(2,1)$-sided query,
(d) general $(2,2)$-sided query.

to $(1,1)$-sided range queries where the corner point is on the diagonal. Other
applications arise in graphics and GIS. For example, Chiang and Silva [44] apply
the EM interval tree structure to extract at query time the boundary components
of the isosurface (or contour) of a surface. A data structure for a related problem,
which in addition has optimal output complexity, appears in [6]. The interval tree
approach also yields dynamic EM segment trees with optimal query and update
bound and $O(n \log_B N)$-block space usage.

For nonrestricted $(1,1)$-sided and $(2,1)$-sided 2-D range queries, Ramaswamy
and Subramanian [109] introduce the notion of *path caching* to develop EM data
structures that meet criterion 1 but have higher storage overheads and amortized
and/or nonoptimal update bounds. Subramanian and Ramaswamy [123] present
the *P-range tree* data structure for $(2,1)$-sided queries, which uses optimal linear
disk space and has nearly optimal query and amortized update bounds. They get
a static data structure for general $(2,2)$-sided 2-D range searching with the same
query bound by applying a filtering technique of Chazelle [38]: The outer level of the
structure is a $(\log_B N + 1)$-way one-dimensional search tree; each $(2,2)$-sided query
is reduced to two $(2,1)$-sided queries, a stabbing query, and $\log_B N$ list traversals.
The disk space usage is $O\big(n(\log n)/\log(\log_B N + 1)\big)$, as required by the pointer
machine lower bound. The structure could be modified to perform updates, by
application of a weight-balanced B-tree and the dynamization techniques of [23],
but the resulting update time would be amortized and nonoptimal, as a consequence
of the use of the $(2,1)$-sided data structure.

Arge et al [23] apply notions of persistence to get a simple and optimal static data structure for $(2,1)$-sided range queries; it supports queries in
$O(\log_B N + z)$ I/Os and uses linear disk space. They get a fully dynamic data structure for $(2,1)$-sided queries with the same optimal query and space bounds and with
optimal update bound $O(\log_B N)$, by combining the static structure with an external priority search tree based upon weight-balanced B-trees. The structure can be
generalized using the technique of [38] to handle $(2,2)$-sided queries with optimal
query bound $O(\log_B N)$, optimal disk space usage $O\big(n(\log n)/\log(\log_B N + 1)\big)$,
and update bound $O\big((\log_B N)(\log n)/\log(\log_B N + 1)\big)$.

One intuition from [75] is that less disk space is needed to efficiently answer
2-D queries when the queries have bounded aspect ratio (i.e., when the ratio of the
longest side length to the shortest side length of the query rectangle is bounded). An
interesting question is whether R-trees and the linear-space structures of Sections
7.3 and 7.4 can be shown to perform provably well for such queries.

For other types of range searching, such as in higher dimensions and for nonorthogonal queries, different filtering techniques are needed. So far, relatively little work has been done. Vengroff and Vitter [**133**] develop the first theoretically near-optimal EM data structure for static three-dimensional orthogonal range searching. They create a hierarchical partitioning in which all the items that dominate a query point are densely contained in a set of blocks. With some recent modifications by the author, queries can be done in $O(\log_B N + z)$ I/Os, which is optimal, and the space usage is $O\big(n(\log n)^k/(\log(\log_B N + 1))^k\big)$ disk blocks to support 3-D range queries in which k of the dimensions ($0 \le k \le 3$) have finite ranges. The space bounds are optimal for $(1, 1, 1)$-sided queries (i.e., $k = 0$) and $(2, 1, 1)$-sided queries (i.e., $k = 1$). The result also provides optimal $O(\log N + Z)$-time query performance in the RAM model using linear space for answering $(1, 1, 1)$-sided queries, improving upon the result in [**40**]. Agarwal et al. [**5**] give optimal bounds for static halfspace range searching in two dimensions and some variants in higher dimensions. The number of I/Os needed to build the 3-D and halfspace data structures is rather large (more than order N). Still, the structures shed useful light on the complexity of range searching. An open problem is to design efficient construction and update algorithms and to improve upon the constant factors. Some other types of range searching, such as simplex range searching, have not yet been investigated in the external memory setting.

Callahan et al. [**36**] develop dynamic EM data structures for several online problems such as finding an approximately nearest neighbor and maintaining the closest pair of vertices. Numerous other data structures have been developed for range queries and related problems on spatial data. We refer to [**7, 66, 102**] for a broad survey.

9. String Processing

Digital trie-based structures, in which branching decisions at each node are made based upon the values of particular bits in strings, are effective for string processing in internal memory. In EM applications, what is needed is a multiway digital structure. Unfortunately, if the strings are long, there is no space to store them completely in each node, and if pointers to strings are stored in each node, the number of I/Os per node access will be large.

Ferragina and Grossi [**62, 63**] develop an elegant generalization of the B-tree for storing strings, called the *String B-tree* or simply *SB-tree*. An SB-tree differs from a conventional B-tree in the way that each $\Theta(B)$-way branching node is represented. In a conventional B-tree, $\Theta(B)$ unit-sized keys are stored in each internal node to guide the searching, and thus the entire node fits in one or two blocks. However, strings can be arbitrarily long, so there may not be enough space to store $\Theta(B)$ strings per node. Pointers to $\Theta(B)$ strings could be stored instead in each node, but access to the strings during search would require more than a constant number of I/Os per node.

Ferragina and Grossi's solution for how to represent each node of the SB-tree is based upon a variant of the *Patricia trie* character-based data structure [**88, 100**] along the lines of Ajtai et al. [**12**]. The Patricia trie achieves B-way branching with a total storage of $O(B)$ characters. Each of its internal nodes stores an index (a number from 0 to N) and a one-character label for each of its outgoing edges. For example, in the example in Figure 6, the right child of the root has index 4 and

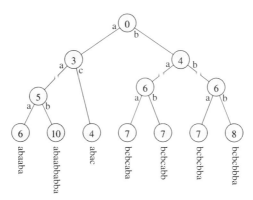

FIGURE 6. Patricia trie representation of a single node of an SB-tree, with branching factor $B = 8$. The seven strings used for partitioning are pictured at the leaves; in the actual data structure, pointers to the strings, not the strings themselves, are stored at the leaves. The pointers to the B children of the SB-tree node are also stored at the leaves.

characters "a" and "b", which means that the node's left subtrie consists of strings whose fifth character (character 4) is "a", and its right subtrie consists of strings whose fifth character is "b". The preceding (four) characters in all the strings in the node's subtrie are identically "bcbc". To find which of the B branches to take for a search string, a binary search is done in the Patricia trie; each binary branching decision is based upon the character indexed at that node. For search string "bcbabcba", a binary search on the trie in Figure 6 traverses the far-right path of the Patricia trie, examining character positions 0, 4, and 6.

Unfortunately, the leaf node that is eventually reached (in our example, the leaf at the far right, corresponding to "bcbcbbba") is not in general at the correct branching point, since only certain character !la positions in the string were examined during the search. The key idea to fix this situation is to sequentially compare the search string with the string associated with the leaf, and if they differ, the index where they differ can be found. In the example, the search string "bcbabcba" differs from "bcbcbbba" in the fourth character (character 3), and the search string therefore is lexicographically smaller than the entire right subtrie of the root. It fits in between the leaves "abac" and "bcbcaba".

Searching the Patricia trie requires one I/O to load it into memory, plus additional I/Os to do the sequential scan of the string after the leaf of the Patricia trie is reached. Each block of the search string that is examined during a sequential scan does not have to be read in again for lower levels of the SB-tree, so the I/Os for the sequential scan can be charged to the blocks of the search string. The resulting query time to search in an SB-tree for a string of ℓ characters is therefore $O(\log_B N + \ell/B)$, which is optimal.

Ferragina and Grossi apply SB-trees to string matching, prefix search, and substring search. Farach et al. [60] show how to construct SB-trees, suffix trees, and suffix arrays on strings of length N using $O(n \log_m n)$ I/Os, which is optimal. Clark and Munro [45] give an alternate approach to suffix trees.

Arge et al. [17] consider several models for the problem of sorting K strings of total length N in external memory. They develop efficient sorting algorithms in these models, making use of the SB-tree, buffer tree techniques, and a simplified version of the SB-tree for merging called the *lazy trie*. In the RAM model, the problem can be solved in $O(K \log K + N)$ time. By analogy to the problem of sorting integers, it would be natural to expect that the I/O complexity would be $O(k \log_m k + n)$ time, where $k = \lceil K/B \rceil$. Arge et al. show somewhat counterintuitively that for sorting short strings (i.e., strings of length at most B) the I/O complexity depends upon the total *number of characters*, whereas for long strings the complexity depends upon the total *number of strings*.

THEOREM 9.1 ([**17**]). *We can sort K strings of total length N, where N_1 is the total length of the short strings and K_2 is the number of long strings, using the following number of I/Os:*

$$(9.1) \qquad O\left(\frac{N_1}{B} \log_m \left(\frac{N_1}{B} + 1 \right) + K_2 \log_M (K_2 + 1) + \frac{N}{B} \right),$$

Lower bounds for various models of how strings can be manipulated are given in [**17**]. There are gaps in some cases between the upper and lower bounds.

10. The TPIE External Memory Programming Environment

There are three basic approaches to supporting development of I/O-efficient code, which we call array-oriented systems (such as PASSION and ViC*), access-oriented systems (such as the UNIX file system, Panda, and MPI-IO), and framework-oriented systems (such as TPIE). We refer the reader to [**68**] and its references for background.

In this section we describe TPIE (Transparent Parallel I/O programming Environment)[3] [**18, 132, 134**], which is used as the implementation platform for the experiments in the next section. TPIE is a comprehensive software package that helps programmers to develop high-level, portable, and efficient implementations of EM algorithms.

TPIE takes a somewhat non-traditional approach to batched computation: Instead of viewing it as an enterprise in which code reads data, operates on it, and writes results, TPIE views computation as a continuous process during which a program is fed streams of data from an outside source and leaves trails of results behind. Programmers do not need to worry about making explicit calls to I/O routines; instead, they merely specify the functional details of the desired computation, and TPIE automatically choreographs a sequence of data movements to keep the computation fed.

TPIE is written in C++ as a set of templated classes and functions. It consists of three main components: a block transfer engine (BTE), a memory manager (MM), and an access method interface (AMI). The BTE is responsible for moving blocks of data to and from the disk. It is also responsible for scheduling asynchronous read-ahead and write-behind when necessary to allow computation and I/O to overlap. The MM is responsible for managing main memory in coordination with the AMI and BTE. The AMI provides the high-level uniform interface for application programs. The AMI is the only component that programmers normally need to interact with directly. Applications that use the AMI are portable across hardware platforms, since they do not have to deal with the underlying details of how I/O

---

[3]The TPIE software distribution is available at no charge on the World Wide Web at http://www.cs.duke.edu/TPIE/.

is performed on a particular machine. We have seen in the previous sections that many batched problems in spatial databases, GIS, scientific computing, graphs, and string processing can be solved optimally using a relatively small number of basic paradigms like scanning, multiway distribution, and merging, which TPIE supports as access mechanisms. TPIE also supports block-oriented operations on trees for online problems.

11. Empirical Comparisons

In this section we examine the empirical performance of algorithms for two problems that arise in spatial databases. The TPIE system described in the previous section is used as the common implementation platform. Other recent experiments involving the paradigms discussed in this paper appear in [**42**, **79**].

11.1. Rectangle Intersection and Spatial Join. In the first experiment, three algorithms are implemented in TPIE for the problem of rectangle intersection, which is typically the first step in a spatial join computation. The first method, called Scalable Sweeping-Based Spatial Join (SSSJ) [**21**], is a robust new algorithm based upon the distribution sweep paradigm of Section 5. The other two methods are Partition-Based Spatial-Merge (QPBSM) used in Paradise [**108**] and a new modification called MPBSM that uses an improved dynamic data structure for intervals [**21**].

The algorithms were tested on several data sets. The timing results for the two data sets in Figures 7(a) and 7(b) are given in Figures 7(c) and 7(d), respectively. The first data set is the worst case for sweep line algorithms; a large fraction of the line segments in the file are active (i.e., they intersect the current sweep line). The second data set is a best case for sweep line algorithms. The two PBSM algorithms have the disadvantage of making extra copies. SSSJ shows considerable improvement over the PBSM-based methods. On more typical data, such as TIGER/line road data sets [**126**], experiments indicate that SSSJ and MPBSM run about 30% faster than QPBSM.

11.2. Batched Operations on R-trees. In the second experiment, three methods for building R-trees are evaluated in terms of their bulk loading time and the resulting query performance. The three methods tested are a newly developed buffer R-tree method [**18**] (labeled "buffer"), the naive sequential method for construction into R*-trees (labeled "naive"), and the best update algorithm for Hilbert R-trees (labeled "Hilbert") [**82**].

The experimental data came from TIGER/line road data sets from four U.S. states [**126**]. One experiment involved building an R-tree on the road data for each state and for each of four possible buffer sizes. The four buffer sizes were capable of storing 0, 600, 1,250, and 5,000 rectangles, respectively. The query performance of each resulting R-tree was measured by posing rectangle intersection queries, using rectangles taken from TIGER hydrographic data. The results, depicted in Figure 8, show that buffer R-trees, even with relatively small buffers, achieve a tremendous speedup in construction time without any worsening in query performance, compared with the naive method (which corresponds to a buffer size of 0).

In another experiment, a single R-tree was built for each of the four U.S. states, containing 50% of the road data objects for that state. Using each of the three algorithms, the remaining 50% of the objects were inserted into the R-tree, and the

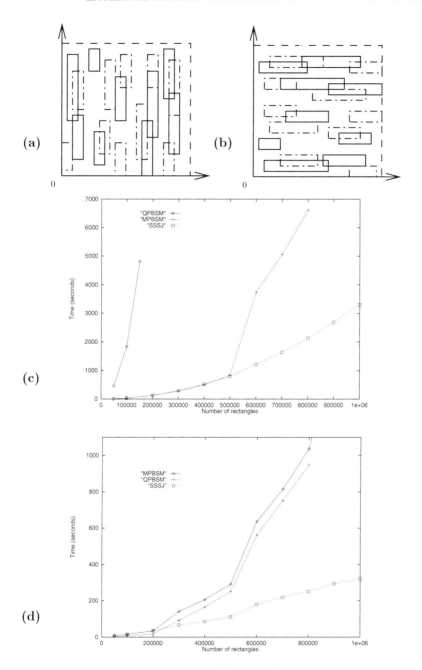

FIGURE 7. Comparison of Scalable Sweeping-Based Spatial Join (SSSJ) with the original PBSM (QPBSM) and a new variant (MPBSM) (a) Data set 1 consists of tall and skinny (vertically aligned) rectangles. (b) Data set 2 consists of short and wide (horizontally aligned) rectangles. (c) Running times on data set 1. (d) Running times on data set 2.

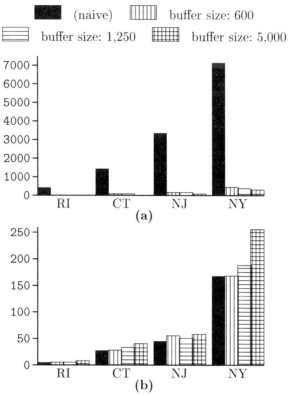

FIGURE 8. Costs for R-tree processing (in units of 1000 I/Os) using the naive repeated insertion method and the buffer R-tree for various buffer sizes: (a) cost for bulk-loading the R-tree, (b) query cost.

Data Set	Update Method	Update with 50% of the data		
		Building	Querying	Packing
RI	naive	259, 263	6, 670	64%
	Hilbert	15, 865	7, 262	92%
	buffer	13, 484	5, 485	90%
CT	naive	805, 749	40, 910	66%
	Hilbert	51, 086	40, 593	92%
	buffer	42, 774	37, 798	90%
NJ	naive	1, 777, 570	70, 830	66%
	Hilbert	120, 034	69, 798	92%
	buffer	101, 017	65, 898	91%
NY	naive	3, 736, 601	224, 039	66%
	Hilbert	246, 466	230, 990	92%
	buffer	206, 921	227, 559	90%

TABLE 2. Summary of the costs (in number of I/Os) for R-tree updates and queries. Packing refers to the percentage storage utilization.

construction time was measured. Query performance was then tested as before. The results in Table 2 show that the buffer R-tree has faster construction time than the Hilbert R-tree (the previous best method for construction time) and similar or better query performance than repeated insertions (the previous best method for query performance).

12. Dynamic Memory Allocation

The amount of memory allocated to a program may fluctuate during the course of execution because of demands placed on the system by other users and processes. EM algorithms must be able to adapt dynamically to whatever resources are available so as to preserve good performance [107]. The algorithms in the previous sections assume a fixed memory allocation; they must resort to virtual memory if the memory allocation is reduced, often causing a severe performance hit.

Barve and Vitter [29] discuss the design and analysis of EM algorithms that adapt gracefully to changing memory allocations. In their model, without loss of generality, a program $\mathcal{P}$ is allocated memory in phases: During the ith phase, $\mathcal{P}$ is allocated m_i blocks of internal memory, and this memory remains allocated to $\mathcal{P}$ until $\mathcal{P}$ completes $2m_i$ I/O operations, at which point the next phase begins. The process continues until $\mathcal{P}$ finishes execution. The model makes the reasonable assumption that the duration for each memory allocation phase is long enough to allow all the memory in that phase to be used by the program.

For sorting, the lower bound approach of (3.7) implies that

$$\sum_i 2m_i \log m_i = \Omega(n \log n).$$

We say that $\mathcal{P}$ is *dynamically optimal* for sorting if

$$\sum_i 2m_i \log m_i = O(n \log n)$$

for all possible sequences $m_1, m_2, \ldots$ of memory allocation. Intuitively, if $\mathcal{P}$ is dynamically optimal, no other program can perform more than a constant number of sorts in the worst-case for the same sequence of memory allocations.

Barve and Vitter [29] define the model in generality and give dynamically optimal strategies for sorting, matrix multiplication, and buffer trees operations. Their work represents the first theoretical model of dynamic allocation for EM algorithms. Pang et al. [107] and Zhang and Larson [148] give memory-adaptive merge sort algorithms, but their algorithms handle only special cases and can be made to perform poorly for certain patterns of memory allocation.

13. Conclusions

In this paper we have described several useful paradigms for the design and implementation of efficient external memory algorithms and data structures. The problem domains we have considered include sorting, permuting, FFT, scientific computing, computational geometry, graphs, databases, geographic information systems, and text and string processing. Interesting challenges remain in virtually all these problem domains. One difficult problem is to prove lower bounds for permuting and sorting without the indivisibility assumption. Another promising area is the design and analysis of algorithms for efficient use of multiple disks. Optimal bounds have not yet been determined for several basic graph problems like topological sorting, shortest paths, breadth-first and depth-first search, and connected components. There is an intriguing connection between problems that have good I/O speedups and problems that have fast and work-efficient parallel algorithms.

A continuing goal is to develop optimal EM algorithms and to translate theoretical gains into observable improvements in practice. For some of the problems that can be solved optimally up to a constant factor, the constant overhead is too large for the algorithm to be of practical use, and simpler approaches are needed. In practice, algorithms cannot assume a static internal memory allocation; they must adapt in a robust way when the memory allocation changes.

New architectures, such as networks of workstations, hierarchical storage devices, and disk drives with processing capabilities present many interesting challenges and opportunities. Work is beginning, for example, on extensions of TPIE to such domains and on applying the buffer management techniques of the SRM method in Section 3.2 to cache-friendly distribution sort algorithms. Active (or intelligent) disks, in which disk drives have some processing capability and can filter information sent to the host, have recently been proposed to further reduce the I/O bottleneck, especially in large database applications [**2, 110**].

Acknowledgements. The author wishes to thank the members of the Center for Geometric Computing at Duke University and the anonymous referees for very helpful comments and suggestions.

References

[1] J. Abello, A. Buchsbaum, and J. Westbrook. A functional approach to external memory graph algorithms. In *Proceedings of the 6th Annual European Symposium on Algorithms*, volume 1461 of *Lecture Notes in Computer Science*, 332–343, Venice, Italy, August 1998. Springer-Verlag.

[2] A. Acharya, M. Uysal, and J. Saltz. Active disks: Programming model, algorithms and evaluation. *ACM SIGPLAN Notices*, 33(11), 81–91, November 1998.

[3] M. Adler. New coding techniques for improved bandwidth utilization. In *37th IEEE Symposium on Foundations of Computer Science*, 173–182, Burlington, VT, October 1996.

[4] P. K. Agarwal, L. Arge, G. S. Brodal, and J. S. Vitter. I/O-efficient dynamic point location in monotone planar subdivisions. In *Proceedings of the ACM-SIAM Symposium on Discrete Algorithms*, 11–20, 1999.

[5] P. K. Agarwal, L. Arge, J. Erickson, P. G. Franciosa, and J. S. Vitter. Efficient searching with linear constraints. In *Proc. 17th ACM Symposium on Principles of Database Systems*, 169–178, 1998.

[6] P. K. Agarwal, L. Arge, T. M. Murali, K. Varadarajan, and J. S. Vitter. I/O-efficient algorithms for contour line extraction and planar graph blocking. In *Proceedings of the ACM-SIAM Symposium on Discrete Algorithms*, 117–126, 1998.

[7] P. K. Agarwal and J. Erickson. Geometric range searching and its relatives. In B. Chazelle, J. E. Goodman, and R. Pollack, editors, *Advances in Discrete and Computational Geometry*, volume 23 of *Contemporary Mathematics*, 1–56. American Mathematical Society Press, Providence, RI, 1999.

[8] A. Aggarwal, B. Alpern, A. K. Chandra, and M. Snir. A model for hierarchical memory. *Proceedings of the 19th ACM Symposium on Theory of Computation*, 305–314, 1987.

[9] A. Aggarwal, A. Chandra, and M. Snir. Hierarchical memory with block transfer. *Proceedings of the 28th Annual IEEE Symposium on Foundations of Computer Science*, 204–216, 1987.

[10] A. Aggarwal and C. G. Plaxton. Optimal parallel sorting in multi-level storage. *Proceedings of the Fifth Annual ACM-SIAM Symposium on Discrete Algorithms*, 659–668, 1994.

[11] A. Aggarwal and J. S. Vitter. The Input/Output complexity of sorting and related problems. *Communications of the ACM*, 31(9), 1116–1127, 1988.

[12] M. Ajtai, M. Fredman, and J. Komlos. Hash functions for priority queues. *Information and Control*, 63(3), 217–225, 1984.

[13] B. Alpern, L. Carter, E. Feig, and T. Selker. The uniform memory hierarchy model of computation. *Algorithmica*, 12(2-3), 72–109, 1994.

[14] L. Arge. The buffer tree: A new technique for optimal I/O-algorithms. In *Proceedings of the Workshop on Algorithms and Data Structures*, volume 955 of *Lecture Notes in Computer Science*, 334–345. Springer-Verlag, 1995. A complete version appears as BRICS technical report RS–96–28, University of Aarhus.

[15] L. Arge. The I/O-complexity of ordered binary-decision diagram manipulation. In *Proceedings of the International Symposium on Algorithms and Computation*, volume 1004 of *Lecture Notes in Computer Science*, 82–91. Springer-Verlag, 1995.

[16] L. Arge. External-memory algorithms with applications in geographic information systems. In M. van Kreveld, J. Nievergelt, T. Roos, and P. Widmayer, editors, *Algorithmic Foundations of GIS*, volume 1340 of *Lecture Notes in Computer Science*. Springer-Verlag, 1997.

[17] L. Arge, P. Ferragina, R. Grossi, and J. Vitter. On sorting strings in external memory. In *Proceedings of the ACM Symposium on Theory of Computation*, 540–548, 1997.

[18] L. Arge, K. H. Hinrichs, J. Vahrenhold, and J. S. Vitter. Efficient bulk operations on dynamic R-trees. In *Proceedings of the 1st Workshop on Algorithm Engineering and Experimentation*, Baltimore, January 1999.

[19] L. Arge, M. Knudsen, and K. Larsen. A general lower bound on the I/O-complexity of comparison-based algorithms. In *Proceedings of the 3rd Workshop on Algorithms and Data Structures*, volume 709 of *Lecture Notes in Computer Science*, 83–94. Springer-Verlag, 1993.

[20] L. Arge and P. Miltersen. On showing lower bounds for external-memory computational geometry problems. In J. Abello and J. S. Vitter, editors, *External Memory Algorithms and Visualization*. American Mathematical Society Press, Providence, RI, this volume.

[21] L. Arge, O. Procopiuc, S. Ramaswamy, T. Suel, and J. S. Vitter. Scalable sweeping-based spatial join. In *Proceedings of the 24th International Conference on Very Large Databases*, 570–581, New York, August 1998.

[22] L. Arge, O. Procopiuc, S. Ramaswamy, T. Suel, and J. S. Vitter. Theory and practice of I/O-efficient algorithms for multidimensional batched searching problems. In *Proceedings of the ACM-SIAM Symposium on Discrete Algorithms*, 685–694, 1998.

[23] L. Arge, V. Samoladas, and J. S. Vitter. Two-dimensional indexability and optimal range search indexing. In *Proceedings of the ACM Symposium Principles of Database Systems*, Philadelphia, PA, May–June 1999.

[24] L. Arge, D. E. Vengroff, and J. S. Vitter. External-memory algorithms for processing line segments in geographic information systems. *Algorithmica*, to appear. Special issue on cartography and geographic information systems. An earlier version appeared in *Proceedings of the Third European Symposium on Algorithms*, volume 979 of *Lecture Notes in Computer Science*, 295–310, Springer-Verlag, September 1995.

[25] L. Arge and J. S. Vitter. Optimal dynamic interval management in external memory. In *Proceedings of the IEEE Symposium on Foundations of Computer Science*, 560–569, Burlington, VT, October 1996.

[26] R. A. Baeza-Yates. Expected behaviour of B^+-trees under random insertions. *Acta Informatica*, 26(5), 439–472, 1989.

[27] R. D. Barve, E. F. Grove, and J. S. Vitter. Simple randomized mergesort on parallel disks. *Parallel Computing*, 23(4), 601–631, 1997.

[28] R. D. Barve, E. A. M. Shriver, P. B. Gibbons, B. K. Hillyer, Y. Matias, and J. S. Vitter. Modeling and optimizing I/O throughput of multiple disks on a bus. In *Joint International Conference on Measurement and Modeling of Computer Systems*, Atlanta, GA, May 1999.

[29] R. D. Barve and J. S. Vitter. External memory algorithms with dynamically changing memory allocations: Long version. Technical Report CS–1998–09, Duke University, 1998.

[30] R. Bayer and E. McCreight. Organization of large ordered indexes. *Acta Inform.*, 1, 173–189, 1972.

[31] B. Becker, S. Gschwind, T. Ohler, B. Seeger, and P. Widmayer. An asymptotically optimal multiversion B-tree. *The VLDB Journal*, 5(4), 264–275, December 1996.

[32] N. Beckmann, H.-P. Kriegel, R. Schneider, and B. Seeger. The R*-tree: An efficient and robust access method for points and rectangles. In *Proceedings of the SIGMOD International Conference on Management of Data*, 322–331, 1990.

[33] J. L. Bentley. Multidimensional divide and conquer. *Communications of the ACM*, 23(6), 214–229, 1980.

[34] S. Berchtold, C. Böhm, and H.-P. Kriegel. Improving the query performance of high-dimensional index structures by bulk load operations. In *Proceedings of the International Conference on Extending Database Technology*, 1998.

[35] G. S. Brodal and J. Katajainen. Worst-case efficient external-memory priority queues. In *Proceedings of the Scandinavian Workshop on Algorithms Theory*, volume 1432 of *Lecture Notes in Computer Science*, 107–118, Stockholm, Sweden, July 1998. Springer-Verlag.

[36] P. Callahan, M. T. Goodrich, and K. Ramaiyer. Topology B-trees and their applications. In *Proceedings of the Workshop on Algorithms and Data Structures*, volume 955 of *Lecture Notes in Computer Science*, 381–392. Springer-Verlag, 1995.

[37] L. Carter and K. S. Gatlin. Towards an optimal bit-reversal permutation program. In *Proceedings of the IEEE Symposium on Foundations of Comp. Sci.*, Palo Alto, CA, November 1998.

[38] B. Chazelle. Filtering search: a new approach to query-answering. *SIAM Journal on Computing*, 15, 703–724, 1986.

[39] B. Chazelle. Lower bounds for orthogonal range searching: I. the reporting case. *Journal of the ACM*, 37(2), 200–212, April 1990.

[40] B. Chazelle and H. Edelsbrunner. Linear space data structures for two types of range search. *Discrete & Computational Geometry*, 2, 113–126, 1987.

[41] P. M. Chen, E. K. Lee, G. A. Gibson, R. H. Katz, and D. A. Patterson. RAID: high-performance, reliable secondary storage. *ACM Computing Surveys*, 26(2), 145–185, June 1994.

[42] Y.-J. Chiang. Experiments on the practical I/O efficiency of geometric algorithms: Distribution sweep vs. plane sweep. *Computational Geometry: Theory and Applications*, 8(4), 211–236, 1998.

[43] Y.-J. Chiang, M. T. Goodrich, E. F. Grove, R. Tamassia, D. E. Vengroff, and J. S. Vitter. External-memory graph algorithms. In *Proceedings of the ACM-SIAM Symposium on Discrete Algorithms*, 139–149, January 1995.

[44] Y.-J. Chiang and C. T. Silva. External memory techniques for isosurface extraction in scientific visualization. In J. Abello and J. S. Vitter, editors, *External Memory Algorithms and Visualization*, Providence, RI, this volume. American Mathematical Society Press.

[45] D. R. Clark and J. I. Munro. Efficient suffix trees on secondary storage. In *Proceedings of the ACM-SIAM Symposium on Discrete Algorithms*, 383–391, Atlanta, GA, June 1996.

[46] K. L. Clarkson and P. W. Shor. Applications of random sampling in computational geometry, II. *Discrete and Computational Geometry*, 4, 387–421, 1989.

[47] D. Comer. The ubiquitous B-tree. *Comput. Surveys*, 11(2), 121–137, 1979.

[48] P. Corbett, D. Feitelson, S. Fineberg, Y. Hsu, B. Nitzberg, J.-P. Prost, M. Snir, B. Traversat, and P. Wong. Overview of the MPI-IO parallel I/O interface. In R. Jain, J. Werth, and J. C. Browne, editors, *Input/Output in Parallel and Distributed Computer Systems*, volume 362 of *The Kluwer International Series in Engineering and Computer Science*, chapter 5, 127–146. Kluwer Academic Publishers, 1996.

[49] T. H. Cormen and D. M. Nicol. Performing out-of-core FFTs on parallel disk systems. *Parallel Computing*, 24(1), 5–20, January 1998.

[50] T. H. Cormen, T. Sundquist, and L. F. Wisniewski. Asymptotically tight bounds for performing BMMC permutations on parallel disk systems. *SIAM Journal on Computing*, 28(1), 105–136, 1999.

[51] A. Crauser, P. Ferragina, K. Mehlhorn, U. Meyer, and E. Ramos. Randomized external-memory algorithms for geometric problems. In *Proceedings of the 14th ACM Symposium on Computational Geometry*, June 1998.

[52] A. Crauser, P. Ferragina, K. Mehlhorn, U. Meyer, and E. Ramos. I/O-optimal computation of segment intersections. In J. Abello and J. S. Vitter, editors, *External Memory Algorithms and Visualization*. American Mathematical Society Press, Providence, RI, this volume.

[53] R. Cypher and G. Plaxton. Deterministic sorting in nearly logarithmic time on the hypercube and related computers. *Journal of Computer and System Sciences*, 47(3), 501–548, 1993.

[54] F. Dehne, D. Hutchinson, and A. Maheshwari. Reducing I/O complexity by simulating coarse grained parallel algorithms. In *Proceedings of the International Parallel Processing Symmposium*, April 1999.

[55] H. B. Demuth. *Electronic Data Sorting*. Ph.d., Stanford University, 1956. A shortened version appears in *IEEE Transactions on Computing*, C-34(4), 296–310, April 1985, special issue on sorting, E. E. Lindstrom, C. K. Wong, and J. S. Vitter, editors.

[56] D. J. DeWitt, J. F. Naughton, and D. A. Schneider. Parallel sorting on a shared-nothing architecture using probabilistic splitting. In *Proceedings of the First International Conference on Parallel and Distributed Information Systems*, 280–291, December 1991.

[57] J. R. Driscoll, N. Sarnak, D. D. Sleator, and R. E. Tarjan. Making data structures persistent. *Journal of Computer and System Sciences*, 38, 86–124, 1989.

[58] NASA's Earth Observing System (EOS) web page, NASA Goddard Space Flight Center, http://eospso.gsfc.nasa.gov/.

[59] G. Evangelidis, D. B. Lomet, and B. Salzberg. The hB$^{\Pi}$-tree: A multi-attribute index supporting concurrency, recovery and node consolidation. *VLDB Journal*, 6, 1–25, 1997.

[60] M. Farach, P. Ferragina, and S. Muthukrishnan. Overcoming the memory bottleneck in suffix tree construction. In *Proceedings of the IEEE Symposium on Foundations of Comp. Sci.*, Palo Alto, CA, November 1998.

[61] W. Feller. *An Introduction to Probability Theory and its Applications*, volume 1. John Wiley & Sons, New York, third edition, 1968.

[62] P. Ferragina and R. Grossi. Fast string searching in secondary storage: Theoretical developments and experimental results. In *Proceedings of the ACM-SIAM Symposium on Discrete Algorithms*, 373–382, Atlanta, June 1996.

[63] P. Ferragina and R. Grossi. The String B-tree: A new data structure for string search in external memory and its applications. *Journal of the ACM*, to appear. An earlier version appeared in *Proceedings of the 27th Annual ACM Symposium on Theory of Computing*, 693–702, Las Vegas, NV, May 1995.

[64] R. W. Floyd. Permuting information in idealized two-level storage. In R. Miller and J. Thatcher, editors, *Complexity of Computer Computations*, 105–109. Plenum, 1972.

[65] T. A. Funkhouser, C. H. Sequin, and S. J. Teller. Management of large amounts of data in interactive building walkthroughs. In *Proceedings of the 1992 ACM SIGGRAPH Symposium on Interactive 3D Graphics*, 11–20, Boston, March 1992.

[66] V. Gaede and O. Günther. Multidimensional access methods. *Computing Surveys*, 30(2), 170–231, June 1998.

[67] M. Gardner. *Magic Show*, chapter 7. Knopf, New York, 1977.

[68] G. A. Gibson, J. S. Vitter, and J. Wilkes. Report of the working group on storage I/O issues in large-scale computing. *ACM Computing Surveys*, 28(4), 779–793, December 1996.

[69] M. T. Goodrich, J.-J. Tsay, D. E. Vengroff, and J. S. Vitter. External-memory computational geometry. In *IEEE Foundations of Computer Science*, 714–723, Palo Alto, CA, November 1993.

[70] D. Greene. An implementation and performance analysis of spatial data access methods. In *Proceedings of the IEEE International Conference on Data Engineering*, 606–615, 1989.

[71] R. Grossi and G. F. Italiano. Efficient splitting and merging algorithms for order decomposable problems. *Information and Computation*, in press. An earlier version appears in *Proceedings of the 24th International Colloquium on Automata, Languages and Programming*, volume 1256 of Lecture Notes in Computer Science, Springer Verlag, 605–615, 1997.

[72] R. Grossi and G. F. Italiano. Efficient cross-trees for external memory. In J. Abello and J. S. Vitter, editors, *External Memory Algorithms and Visualization*. American Mathematical Society Press, Providence, RI, this volume.

[73] S. K. S. Gupta, Z. Li, and J. H. Reif. Generating efficient programs for two-level memories from tensor-products. In *Proceedings of the Seventh IASTED/ISMM International Conference on Parallel and Distributed Computing and Systems*, 510–513, Washington, D.C., October 1995.

[74] A. Guttman. R-trees: A dynamic index structure for spatial searching. In *Proceedings of the ACM SIGMOD Conference on Management of Data*, 47–57, 1985.

[75] J. M. Hellerstein, E. Koutsoupias, and C. H. Papadimitriou. On the analysis of indexing schemes. In *Proceedings of the 16th ACM Symposium on Principles of Database Systems*, 249–256, Tucson, AZ, May 1997.

[76] L. Hellerstein, G. Gibson, R. M. Karp, R. H. Katz, and D. A. Patterson. Coding techniques for handling failures in large disk arrays. *Algorithmica*, 12(2–3), 182–208, 1994.

[77] K. H. Hinrichs. *The grid file system: Implementation and case studies of applications.* PhD thesis, Dept. Information Science, ETH, Zürich, 1985.

[78] J. W. Hong and H. T. Kung. I/O complexity: The red-blue pebble game. *Proceedings of the 13th Annual ACM Symposium on Theory of Computation,* 326–333, May 1981

[79] D. Hutchinson, A. Maheshwari, J.-R. Sack, and R. Velicescu. Early experiences in implementing the buffer tree. Workshop on Algorithm Engineering, 1997.

[80] I. Kamel and C. Faloutsos. On packing R-trees. In *Proceedings of the 2nd International Conference on Information and Knowledge Management,* 490–499, 1993.

[81] I. Kamel and C. Faloutsos. Hilbert R-tree: An improved R-tree using fractals. In *Proceedings of the 20th International Conference on Very Large Databases,* 500–509, 1994.

[82] I. Kamel, M. Khalil, and V. Kouramajian. Bulk insertion in dynamic R-trees. In *Proceedings of the 4th International Symposium on Spatial Data Handling,* 3B, 31–42, 1996.

[83] M. V. Kameshwar and A. Ranade. I/O-complexity of graph algorithms. In *Proceedings of the ACM-SIAM Symposium on Discrete Algorithms,* Baltimore, MD, January 1999.

[84] P. C. Kanellakis, G. M. Kuper, and P. Z. Revesz. Constraint query languages. *Proceedings of the 9th ACM Conference on Principles of Database Systems,* 299–313, 1990.

[85] P. C. Kanellakis, S. Ramaswamy, D. E. Vengroff, and J. S. Vitter. Indexing for data models with constraints and classes. *Journal of Computer and System Science,* 52(3), 589–612, 1996.

[86] K. V. R. Kanth and A. K. Singh. Optimal dynamic range searching in non-replicating index structures. In *Proceedings of the 7th International Conference on Database Theory,* Jerusalem, January 1999.

[87] D. G. Kirkpatrick and R. Seidel. The ultimate planar convex hull algorithm? *SIAM Journal on Computing,* 15, 287–299, 1986.

[88] D. E. Knuth. *Sorting and Searching,* volume 3 of *The Art of Computer Programming.* Addison-Wesley, Reading MA, second edition, 1998.

[89] E. Koutsoupias and D. S. Taylor. Tight bounds for 2-dimensional indexing schemes. In *Proceedings of the 17th ACM Symposium on Principles of Database Systems,* Seattle, WA, June 1998.

[90] R. Krishnamurthy and K.-Y. Wang. Multilevel grid files. Tech. report, IBM T. J. Watson Center, Yorktown Heights, NY, November 1985.

[91] V. Kumar and E. Schwabe. Improved algorithms and data structures for solving graph problems in external memory. In *Proceedings of the 8th IEEE Symposium on Parallel and Distributed Processing,* 169–176, October 1996.

[92] K. Küspert. Storage utilization in B*-trees with a generalized overflow technique. *Acta Informatica,* 19, 35–55, 1983.

[93] R. Laurini and D. Thompson. *Fundamentals of Spatial Information Systems.* Academic Press, 1992.

[94] F. T. Leighton. Tight bounds on the complexity of parallel sorting. *IEEE Transactions on Computers,* C-34(4), 344–354, April 1985. Special issue on sorting, E. E. Lindstrom and C. K. Wong and J. S. Vitter, editors.

[95] C. E. Leiserson, S. Rao, and S. Toledo. Efficient out-of-core algorithms for linear relaxation using blocking covers. In *Proceedings of the IEEE Symposium on Foundations of Comp. Sci.,* 704–713, 1993.

[96] Z. Li, P. H. Mills, and J. H. Reif. Models and resource metrics for parallel and distributed computation. *Parallel Algorithms and Applications,* 8, 35–59, 1996.

[97] D. B. Lomet and B. Salzberg. The hB-tree: a multiattribute indexing method with good guaranteed performance. *ACM Transactions on Database Systems,* 15(4), 625–658, 1990.

[98] D. B. Lomet and B. Salzberg. Concurrency and recovery for index trees. *The VLDB Journal,* 6(3), 224–240, 1997.

[99] Y. Matias, E. Segal, and J. S. Vitter. Efficient bundle sorting, 1999. Manuscript.

[100] D. R. Morrison. Patricia: Practical algorithm to retrieve information coded in alphanumeric. *Journal of the ACM,* 15, 514–534, 1968.

[101] J. Nievergelt, H. Hinterberger, and K. C. Sevcik. The grid file: An adaptable, symmetric multi-key file structure. *ACM Trans. Database Syst.,* 9, 38–71, 1984.

[102] J. Nievergelt and P. Widmayer. Spatial data structures: Concepts and design choices. In M. van Kreveld, J. Nievergelt, T. Roos, and P. Widmayer, editors, *Algorithmic Foundations of GIS,* volume 1340 of *Lecture Notes in Computer Science.* Springer-Verlag, 1997.

[103] M. H. Nodine, M. T. Goodrich, and J. S. Vitter. Blocking for external graph searching. *Algorithmica*, 16(2), 181–214, August 1996.

[104] M. H. Nodine, D. P. Lopresti, and J. S. Vitter. I/O overhead and parallel VLSI architectures for lattice computations. *IEEE Transactions on Computers*, 40(7), 843–852, July 1991.

[105] M. H. Nodine and J. S. Vitter. Deterministic distribution sort in shared and distributed memory multiprocessors. In *Proceedings of the 5th Annual ACM Symposium on Parallel Algorithms and Architectures*, 120–129, Velen, Germany, June–July 1993.

[106] M. H. Nodine and J. S. Vitter. Greed Sort: An optimal sorting algorithm for multiple disks. *Journal of the ACM*, 42(4), 919–933, July 1995.

[107] H. Pang, M. Carey, and M. Livny. Memory-adaptive external sorts. *Proceedings of the 19th Conference on Very Large Data Bases*, 1993.

[108] J. M. Patel and D. J. DeWitt. Partition based spatial-merge join. In *Proceedings of the ACM SIGMOD International Conference on Management of Data*, 259–270, June 1996.

[109] S. Ramaswamy and S. Subramanian. Path caching: a technique for optimal external searching. *Proceedings of the 13th ACM Conference on Principles of Database Systems*, 1994.

[110] E. Riedel, G. A. Gibson, and C. Faloutsos. Active storage for large-scale data mining and multimedia. In *Proceedings of the IEEE International Conference on Very Large Databases*, 62–73, 24–27 August 1998.

[111] J. T. Robinson. The k-d-b-tree: a search structure for large multidimensional dynamic indexes. In *Proc. ACM Conference Principles Database Systems*, 10–18, 1981.

[112] C. Ruemmler and J. Wilkes. An introduction to disk drive modeling. *IEEE Computer*, 17–28, March 1994.

[113] H. Samet. *Applications of Spatial Data Structures: Computer Graphics, Image Processing, and GIS*. Addison-Wesley, 1989.

[114] H. Samet. *The Design and Analysis of Spatial Data Structures*. Addison-Wesley, 1989.

[115] V. Samoladas and D. Miranker. A lower bound theorem for indexing schemes and its application to multidimensional range queries. In *Proc. 17th ACM Conf. on Princ. of Database Systems*, Seattle, WA, June 1998.

[116] J. E. Savage. Extending the Hong-Kung model to memory hierarchies. In *Proceedings of the 1st Annual International Conference on Computing and Combinatorics*, volume 959 of *Lecture Notes in Computer Science*, 270–281. Springer-Verlag, August 1995.

[117] J. E. Savage and J. S. Vitter. Parallelism in space-time tradeoffs. In F. P. Preparata, editor, *Advances in Computing Research, Volume 4*, 117–146. JAI Press, 1987.

[118] B. Seeger and H.-P. Kriegel. The buddy-tree: An efficient and robust access method for spatial data base systems. In *Proc. 16th VLDB Conference*, 590–601, 1990.

[119] E. Shriver, A. Merchant, and J. Wilkes. An analytic behavior model for disk drives with readahead caches and request reordering. In *Joint International Conference on Measurement and Modeling of Computer Systems*, June 1998.

[120] E. A. M. Shriver and M. H. Nodine. An introduction to parallel I/O models and algorithms. In R. Jain, J. Werth, and J. C. Browne, editors, *Input/Output in Parallel and Distributed Computer Systems*, chapter 2, 31–68. Kluwer Academic Publishers, 1996.

[121] J. F. Sibeyn. From parallel to external list ranking. Technical Report MPI–I–97–1–021, Max-Planck-Institut, September 1997.

[122] J. F. Sibeyn and M. Kaufmann. BSP-like external-memory computation. In *Proceedings of the 3rd Italian Conference on Algorithms and Complexity*, 229–240, 1997.

[123] S. Subramanian and S. Ramaswamy. The P-range tree: a new data structure for range searching in secondary memory. *Proceedings of the ACM-SIAM Symposium on Discrete Algorithms*, 1995.

[124] R. Tamassia and J. S. Vitter. Optimal cooperative search in fractional cascaded data structures. *Algorithmica*, 15(2), 154–171, February 1996.

[125] Microsoft's TerraServer online database of satellite images, available on the World-Wide Web at `http://terraserver.microsoft.com/`.

[126] TIGER/Line (tm). 1992 technical documentation. Technical report, U. S. Bureau of the Census, 1992.

[127] S. Toledo. A survey of out-of-core algorithms in numerical linear algebra. In J. Abello and J. S. Vitter, editors, *External Memory Algorithms and Visualization*. American Mathematical Society Press, Providence, RI, this volume.

[128] J. D. Ullman and M. Yannakakis. The input/output complexity of transitive closure. *Annals of Mathematics and Artificial Intellegence*, 3, 331–360, 1991.

[129] J. van den Bercken, B. Seeger, and P. Widmayer. A generic approach to bulk loading multidimensional index structures. In *Proceedings 23rd VLDB Conference*, 406–415, 1997.

[130] M. van Kreveld, J. Nievergelt, T. Roos, and P. W. (Eds.). *Algorithmic Foundations of GIS*, volume 1340 of *Lecture Notes in Computer Science*. Springer-Verlag, 1997.

[131] P. J. Varman and R. M. Verma. An efficient multiversion access structure. *IEEE Transactions on Knowledge and Data Engineering*, 9(3), 391–409, May/June 1997.

[132] D. E. Vengroff. *TPIE User Manual and Reference*. Duke University, 1997. The manual and software distribution are available on the web at `http://www.cs.duke.edu/TPIE/`.

[133] D. E. Vengroff and J. S. Vitter. Efficient 3-d range searching in external memory. In *Proceedings of the ACM Symposium on Theory of Computation*, 192–201, Philadelphia, PA, May 1996.

[134] D. E. Vengroff and J. S. Vitter. I/O-efficient scientific computation using TPIE. In *Proceedings of the Fifth NASA Goddard conference on Mass Storage Systems*, II, 553–570, September 1996.

[135] J. S. Vitter. Efficient memory access in large-scale computation. In *Proceedings of the 1991 Symposium on Theoretical Aspects of Computer Science*, Lecture Notes in Computer Science. Springer-Verlag, 1991. Invited paper.

[136] J. S. Vitter and P. Flajolet. Average-case analysis of algorithms and data structures. In J. van Leeuwen, editor, *Handbook of Theoretical Computer Science, Volume A: Algorithms and Complexity*, chapter 9, 431–524. North-Holland, 1990.

[137] J. S. Vitter and M. H. Nodine. Large-scale sorting in uniform memory hierarchies. *Journal of Parallel and Distributed Computing*, 17, 107–114, 1993.

[138] J. S. Vitter and E. A. M. Shriver. Algorithms for parallel memory I: Two-level memories. *Algorithmica*, 12(2–3), 110–147, 1994.

[139] J. S. Vitter and E. A. M. Shriver. Algorithms for parallel memory II: Hierarchical multilevel memories. *Algorithmica*, 12(2–3), 148–169, 1994.

[140] J. S. Vitter and M. Wang. Approximate computation of multidimensional aggregates of sparse data using wavelets. In *Proceedings of the ACM SIGMOD International Conference on Management of Data*, Philadelphia, PA, June 1999.

[141] J. S. Vitter, M. Wang, and B. Iyer. Data cube approximation and histograms via wavelets. In *Proceedings of the Seventh International Conference on Information and Knowledge Management*, 96–104, Washington, November 1998.

[142] M. Wang, J. S. Vitter, and B. R. Iyer. Scalable mining for classification rules in relational databases. In *Proceedings of the International Database Engineering & Application Symposium*, 58–67, Cardiff, Wales, July 1998.

[143] D. Willard and G. Lueker. Adding range restriction capability to dynamic data structures. *Journal of the ACM*, 32(3), 597–617, 1985.

[144] D. Womble, D. Greenberg, S. Wheat, and R. Riesen. Beyond core: Making parallel computer I/O practical. In *Proceedings of the 1993 DAGS/PC Symposium*, 56–63, Hanover, NH, June 1993. Dartmouth Institute for Advanced Graduate Studies.

[145] C. Wu and T. Feng. The universality of the shuffle-exchange network. *IEEE Transactions on Computers*, C-30, 324–332, May 1981.

[146] A. C. Yao. On random 2-3 trees. *Acta Informatica*, 9, 159–170, 1978.

[147] S. B. Zdonik and D. Maier, editors. *Readings in Object-Oriented Database Systems*. Morgan Kauffman, 1990.

[148] W. Zhang and P.-A. Larson. Dynamic memory adjustment for external mergesort. *Proceedings of the Twenty-third International Conference on Very Large Data Bases*, 1997.

[149] B. Zhu. Further computational geometry in secondary memory. In *Proceedings of the International Symposium on Algorithms and Computation*, 1994.

CENTER FOR GEOMETRIC COMPUTING, DEPARTMENT OF COMPUTER SCIENCE, DUKE UNIVERSITY, DURHAM, NC 27708–0129, USA

E-mail address: `jsv@cs.duke.edu`

URL: `http://www.cs.duke.edu/~jsv/`

DIMACS Series in Discrete Mathematics
and Theoretical Computer Science
Volume **50**, 1999

Synopsis Data Structures for Massive Data Sets

Phillip B. Gibbons and Yossi Matias

ABSTRACT. Massive data sets with terabytes of data are becoming common-place. There is an increasing demand for algorithms and data structures that provide fast response times to queries on such data sets. In this paper, we describe a context for algorithmic work relevant to massive data sets and a framework for evaluating such work. We consider the use of "synopsis" data structures, which use very little space and provide fast (typically approximated) answers to queries. The design and analysis of effective synopsis data structures offer many algorithmic challenges. We discuss a number of concrete examples of synopsis data structures, and describe fast algorithms for keeping them up-to-date in the presence of online updates to the data sets.

1. Introduction

A growing number of applications demand algorithms and data structures that enable the efficient processing of data sets with gigabytes to terabytes to petabytes of data. Such massive data sets necessarily reside on disks or tapes, making even a few accesses of the base data set comparably slow (e.g., a single disk access is often $10,000$ times slower than a single memory access). For fast processing of queries to such data sets, disk accesses should be minimized.

This paper focuses on data structures for supporting queries to massive data sets, while minimizing or avoiding disk accesses. In particular, we advocate and study the use of small space data structures. We denote as *synopsis data structures* any data structures that are substantively smaller than their base data sets. Synopsis data structures have the following advantages over non-synopsis data structures:

- *Fast processing*: A synopsis data structure may reside in main memory, providing for fast processing of queries and of data structure updates, by avoiding disk accesses altogether.
- *Fast swap/transfer*: A synopsis data structure that resides on the disks can be swapped in and out of memory with minimal disk accesses, for the purposes of processing queries or updates. A synopsis data structure can

1991 *Mathematics Subject Classification.* Primary 68P05, 68P20; Secondary 68Q05, 68Q20, 62-04, 68Q25.

Research of the second author was supported in part by an Alon Fellowship, by the Israel Science Foundation founded by The Academy of Sciences and Humanities, and by the Israeli Ministry of Science.

be pushed or pulled remotely (e.g., over the internet) at minimal cost, since the amount of data to be transferred is small.

- *Lower cost*: A synopsis data structure has a minimal impact on the overall space requirements of the data set and its supporting data structures, and hence on the overall cost of the system.
- *Better system performance*: A synopsis data structure leaves space in the memory for other data structures. More importantly, it leaves space for other processing, since most processing that involves the disks uses the memory as a cache for the disks. In a data warehousing environment, for example, the main memory is needed for query-processing working space (e.g., building hash tables for hash joins) and for caching disk blocks. The importance of available main memory for algorithms can be seen from the external memory algorithms literature, where the upper and lower time bounds for many fundamental problems are inversely proportional to the logarithm of the available memory size [**Vit98**]. (See Section 2.3 for examples.) Thus although machines with large main memories are becoming increasingly commonplace, memory available for synopsis data structures remains a precious resource.
- *Small surrogate*: A synopsis data structure can serve as a small surrogate for the data set when the data set is currently expensive or impossible to access.

In contrast, linear space data structures for massive data sets can not reside in memory, have very slow swap and transfer times, can increase the space requirements and hence the overall cost of the system by constant factors, can hog the memory when they are in use, and can not serve as a small surrogate. Hence a traditional viewpoint in the algorithms literature — that a linear space data structure is a good one — is not appropriate for massive data sets, as such data structures often fail to provide satisfactory application performance.

On the other hand, since synopsis data structures are too small to maintain a full characterization of their base data sets, they must summarize the data set, and the responses they provide to queries will typically be approximate ones. The challenges are to determine (1) what synopsis of the full data set to keep in the limited space in order to maximize the accuracy and confidence of its approximate responses, and (2) how to efficiently compute the synopsis and maintain it in the presence of updates to the data set.

Due to their importance in applications, there are a number of synopsis data structures in the literature and in existing systems. Examples include uniform and biased random samples, various types of histograms, statistical summary information such as frequency moments, data structures resulting from lossy compression of the data set, etc. Often, synopsis data structures are used in a heuristic way, with no formal properties proved on their performance or accuracy, especially under the presence of updates to the data set. Our ongoing work since 1995 seeks to provide a systematic study of synopsis data structures, including the design and analysis of synopsis data structures with performance and accuracy guarantees, even in the presence of data updates.

In this paper, we describe a context for algorithmic work relevant to massive data sets and a framework for evaluating such work. In brief, we combine the PDM external memory model [**VS94**] with input/output conventions more typical for the study of (online) data structure problems. Two general scenarios are considered: one where the input resides on the disks of the PDM and one where the input arrives online in the PDM memory. We describe some of our work on synopsis data structures, and highlight results on three problem domains from the database literature: frequency moments, hot list queries, and histograms and quantiles.

Outline. Section 2 describes our framework in detail. Results on frequency moments, hot list queries and histograms are described in Sections 3, 4, and 5, respectively. Related work and further results are discussed in Section 6.

2. Framework

In this section, we first describe a context for data structure problems for massive data sets. We then introduce synopsis data structures and present a cost model for their analysis. Finally, we discuss two example application domains.

2.1. Problem set-up. In the data structure questions we consider, there are a number of data sets, $S_1, S_2, \ldots, S_\ell$, and a set of query classes, $Q_1, \ldots, Q_k$, on these data sets. The query classes are given a priori, and may apply to individual data sets or to multiple data sets. Data structure performance is analyzed on a model of computation that distinguishes between two types of storage, fast and slow, where the fast storage is of limited size. We equate the fast storage with the computer system's main memory and the slow storage with its disks, and use a relevant model of computation (details are in Section 2.3). However, the framework and results in this paper are also relevant to scenarios where (1) the fast storage is the disks and the slow storage is the tapes, or (2) the fast storage is the processor cache memory and the slow storage is the main memory.

In the *static* or *offline* scenario, the data sets are given as input residing on the disks. Given a class of queries Q, the goal is to design a data structure for the class Q that minimizes the response time to answer queries from Q, maximizes the accuracy and confidence of the answers, and minimizes the preprocessing time needed to build the data structure.

In the *dynamic* or *online* scenario, which models the ongoing loading of new data into the data set, the data sets arrive online in the memory, and are stored on the disks. Specifically, the input consists of a sequence of operations that arrive online to be processed by the data structure, where an operation is either an insertion of a new data item, a deletion of an existing data item, or a query. Given a class of queries Q, the goal is to design a data structure for the class Q that minimizes the response time to answer queries from Q, maximizes the accuracy and confidence of the answers, and minimizes the update time needed to maintain the data structure. As we are interested in the *additional* overheads for maintaining the data structure, there is no charge for updating the data sets.

This set-up reflects many environments for processing massive data sets. For example, it reflects most data warehousing environments, such as Walmart's multi-terabyte warehouse of its sales transactions. For most data sets, there are far more insertions than deletions. An important exception is a "sliding-window" data set, comprised of the most recent data from a data source (such as the last 15 months of sales transactions). In such data sets, batches of old data are periodically deleted to make room for new data, making the number of insertions comparable to the number of deletions.

To handle many data sets and many query classes, a large number of synopsis data structures may be needed. Thus we will assume that when considering any one data structure problem in isolation, the amount of memory available to the data structure is a small fraction of the total amount of memory. We evaluate the effectiveness of a data structure as a function of its space usage or *footprint*. For example, it is common practice to evaluate the effectiveness of a histogram in range selectivity queries as a function of its footprint (see, e.g., [**PIHS96**]).

Finally, note that in some online environments, the data set is not stored alongside with the data structure, but instead resides in a remote computer system that may be currently unavailable [**FJS97**]. In such cases, the online view of the data is effectively the only view of the data used to maintain the data structure and answer queries. We denote this scenario the *purely online* scenario.

2.2. Synopsis data structures. The above set-up motivates the need for data structures with small footprints. We denote as *synopsis data structures* any data structures that are substantively smaller than their base data sets. Since such data structures are often too small to maintain a full characterization of their base data sets with regards to a class of queries, the responses they provide to queries will typically be approximate ones. Synopsis data structures seek to characterize the data using succinct representations.

A natural synopsis data structure is a uniform random sample, and indeed, it is well known that a random sample of a data set can be quite useful to support a variety of queries on the set. However, for many classes of queries, uniform sampling is not the best choice. A trivial example is the class of "number of items in the set" queries, for which a single counter is much better. More interesting examples can be found in the rest of this paper.

We define an $f(n)$-synopsis data structure as follows.

DEFINITION 2.1. An $f(n)$-*synopsis data structure* for a class Q of queries is a data structure for providing (exact or approximate) answers to queries from Q that uses $O(f(n))$ space for a data set of size n, where $f(n) = o(n^\epsilon)$ for some constant $\epsilon < 1$.

While any sublinear space data structure may be an important improvement over a linear space data structure, the above definition demands at least a polynomial savings in space, since only with such savings can most of the benefits of synopsis data structures outlined in Section 1 be realized. For example, massive data sets typically exceed the available memory size by a polynomial factor, so a data structure residing in memory must have a $o(n^\epsilon)$ footprint.

As with traditional data structures, a synopsis data structure can be evaluated according to five metrics:

- *Coverage*: the range and importance of the queries in Q.
- *Answer quality*: the accuracy and confidence of its (approximate) answers to queries in Q.
- *Footprint*: its space bound (smaller $f(n)$ is better).
- *Query time*.
- *Computation/Update time*: its preprocessing time in the static scenario, or its update time in the dynamic scenario.

Ideally, $f(n)$ is $\log^2 n$ or better, queries and updates require a constant number of memory operations and no disks operations, and the answers are exact.

2.3. Cost model. Query times and computation/update times can be analyzed on any of a number of models of computation, depending on the target computer system, including parallel or distributed models. For concreteness in this paper, we will use the parallel disk model (PDM) of Vitter and Shriver [**VS94, Vit98**], adapted to the scenarios discussed above.

In the PDM, there are P processors, D disks, and an (internal) memory of size M (i.e., M/P per processor). Each disk is partitioned into blocks of size B, and is of unbounded size. The input of size N is partitioned (striped) evenly among the disks, $D_0, D_1, \ldots, D_{D-1}$, such that for $i = 0, 1, \ldots, N/B - 1$, the ith block of input data is the $\lfloor i/D \rfloor$th block of data on the $(i \bmod D)$th disk. The output is required to be similarly striped. The size parameters N, M, and B are in units of the input data items, M is less than N, and $1 \le DB \le M/2$. Thus the internal memory is too small to hold the input but sufficiently large to hold two blocks from each of the disks.

Algorithms are analyzed based on three metrics: the number of I/O operations, the processing time, and the amount of disk space used. In a single I/O read (I/O write), each of the D disks can simultaneously transfer a block to (from, respectively) the internal memory. The processing time is analyzed assuming that each processor is a unit-cost RAM for its in-memory computation times, and that the processors are connected by an interconnection network whose properties depend on the setting. Most of the algorithmic work on the PDM has focused on reducing the number of I/O operations and proving matching lower bounds. As mentioned in the introduction, the I/O bounds are often inversely proportional to the logarithm of the available memory; specifically, they are inversely proportional to $\log(M/B)$. Examples discussed in [**Vit98**] include sorting, permuting, matrix transpose, computing the Fast Fourier Transform, and various batched problems in computational geometry. For other problems, such as matrix multiplication and LU factorization, the I/O bounds are inversely proportional to $\sqrt{M}$.

Our main deviation from the PDM is in the input and output requirements. Query times and computation/update times are analyzed on a PDM with input and output requirements adapted to the set-up described in Section 2.1. Our first deviation is to supplement the PDM with a write-only "output" memory, of unbounded size.

In our static scenario, the input resides on the disks as in the PDM, but we are allowed to preprocess the data and store the resulting data structures in the internal memory. In response to a query, the output is written to the output memory, in contrast to the PDM. Thus processing the query may incur no I/O operations.

In our dynamic scenario, the input arrives online in the internal memory in the form of insertions to the data set, deletions from the data set, or queries. Data structures are maintained for answering queries. As in the static scenario, data structures may be stored in the internal memory and responses to queries are written to the output memory, and hence queries may be answered without incurring any I/O operations.

Reducing the number of I/O operations is important since, as pointed out in Section 1, an I/O operation can take as much time as $10,000$ in-memory operations on modern computer systems.

Note that any insertions and deletions in the dynamic scenario are applied to the base data set, at no charge, so that the current state of the data set resides on the disks at all times. However, the cost of reading this data depends on the setting, and needs to be specified for algorithms that perform such reads. One can consider a variety of settings, such as cases where the base data is striped across the disks or cases where there are various indices such as B-trees that can be exploited. For the purely online scenario, the base data is unavailable.

With massive data sets, it will often be the case that the input size N is not just larger than the memory size M, as assumed by the PDM, but is in fact polynomially larger: $N = M^c$, for a constant $c > 1$. Also, note that any algorithm for the dynamic scenario in which updates incur (amortized) processing time t per update and no I/O operations yields an algorithm for computing the same synopsis data structure in the static scenario in one pass over the data set, i.e., $\frac{N}{DB}$ I/O operations and Nt processing time.

For simplicity, in the remainder of this paper, we will assume that the PDM has only a single processor (i.e., $P = 1$).

2.4. Applications: Approximate query answering and cost estimation. An important application domain for synopsis data structures is approximate query answering for ad hoc queries of large data warehouses [**GM98**]. In large data recording and warehousing environments, it is often advantageous to provide fast, approximated answers to complex decision-support queries (see the TPC-D benchmark [**TPC**] for examples of such queries). The goal is to provide an estimated response in orders of magnitude less time than the time to compute an exact answer, by avoiding or minimizing the number of accesses to the base data.

In the *Approximate query answering (Aqua)* project [**GMP97a, GPA**$^+$**98**] at Bell Labs, we seek to provide fast, approximate answers to queries using synopsis data structures. Unlike the traditional data warehouse set-up depicted in Figure 1, in which each query is answered exactly using the data warehouse, Aqua considers the set-up depicted in Figure 2. In this set-up, new data being loaded into the data warehouse is also observed by the approximate answer engine. This engine maintains various synopsis data structures, for use in answering queries.

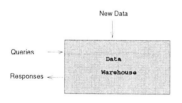

FIGURE 1. Traditional data warehouse

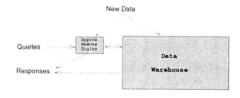

FIGURE 2. Data warehouse set-up for providing approximate query answers

Queries are sent to the approximate answer engine. Whenever possible, the engine promptly returns a response consisting of an approximated answer and a confidence measure (e.g., a 95% confidence interval). The user can then decide whether or not to have an exact answer computed from the base data, based on the user's desire for the exact answer and the estimated time for computing an exact answer as determined by the query optimizer and/or the approximate answer engine. There are a number of scenarios for which a user may prefer an approximate answer in a few seconds over an exact answer that requires tens of minutes or more to compute, e.g., during a drill down query sequence in data mining [**SKS97**].

Moreover, as discussed in Section 2.1, sometimes the base data is remote and currently unavailable, so that an exact answer is not an option, until the data again becomes available.

Another important application domain for synopsis data structures is cost estimation within a query optimizer. In commercial database systems, limited storage is set aside for synopsis data structures such as histograms; these are used by the query optimizer to estimate the cost of the primitive operations comprising a complex SQL query (i.e., estimates of the number of items that satisfy a given predicate, estimates of the size of a join operation [**GGMS96**], etc.). The query optimizer uses these cost estimates to decide between alternative query plans and to make more informed scheduling (allocation, load balancing, etc.) decisions in multi-query and/or multiprocessor database environments, in order to minimize query response times and maximize query throughput.

These two application domains highlight the fact that good synopsis data structures are useful either for providing fast approximate answers to user queries, or for speeding up the time to compute an exact answer, or for both.

The next three sections of this paper highlight in detail our work on synopsis data structures for three problem domains. These sections are not meant to be comprehensive, but instead provide a flavor of the difficulties and the techniques. Much of the details, including most of the proofs and the experimental results, are omitted; the reader is referred to the cited papers for these details. The first problem domain is that of estimating the frequency moments of a data set, such as the number of distinct values or the maximum frequency of any one value. The second problem domain is that of estimating the m most frequently occurring values in a data set. The third problem domain is that of approximating the quantiles and other types of histograms of a data set. Note that the emphasis in these sections will be on what synopses to keep within the limited space, how to maintain these synopses, and what can be proved about the quality of the answers they provide; these are the challenges particular to synopsis data structures. Traditional data structure issues concerning the representation used to store the synopsis and its impact on query time and update time are important, but somewhat secondary to the main emphasis. As can be seen by the techniques presented in these sections, randomization and approximation seem to be essential features in the study of synopsis data structures for many problems, and have been proven to be essential for several problems.

3. Frequency moments

In this section, we highlight our results on synopsis data structures for estimating the frequency moments of a data set.

Let $A = (a_1, a_2, \ldots, a_n)$ be a sequence of elements, where each a_i is a member of $U = \{1, 2, \ldots, u\}$. For simplicity of exposition, we assume that $u \leq n$.[1] Let $m_i = |\{j : a_j = i\}|$ denote the number of occurrences of i in the sequence A, or the *frequency* of i in A. The demographic information of the frequencies in the data set A can be described by maintaining the full histogram over U: $H = (m_1, m_2, \ldots, m_u)$. However, when the desired footprint is substantially smaller than u, then a more succinct representation of the frequencies is required.

---

[1] A more detailed analysis would show that for the $f(n)$-synopsis data structures reported in this section, it suffices that $u \leq 2^{n^\epsilon}$ for some constant $\epsilon < 1$.

Define, for each $k \geq 0$, $F_k = \sum_{i=1}^{u} m_i^k$. In particular, F_0 is the number of distinct elements appearing in the sequence, F_1 ($= n$) is the length of the sequence, and F_2 is the *repeat rate* or *Gini's index of homogeneity* needed in order to compute the *surprise index* of the sequence (see, e.g., [Goo89]). Also define $F_\infty^* = \max_{1 \leq i \leq u} m_i$. (Since the moment F_k is defined as the sum of k-powers of the numbers m_i and not as the k-th root of this sum the last quantity is denoted by F_∞^* and not by F_∞.) The numbers F_k are called the *frequency moments* of A.

The frequency moments of a data set represent useful demographic information about the data, for instance in the context of database applications. They indicate the degree of *skew* in the data, which is of major consideration in many parallel database applications. Thus, for example, the degree of the skew may determine the selection of algorithms for data partitioning, as discussed by DeWitt *et al* [DNSS92] (see also the references therein).

We discuss the estimation of frequency moments when the available memory is smaller than u (i.e., when the full histogram H is not available). We first consider the problem of estimating F_0, which demonstrates the advantages of viewing the input online versus ad hoc sampling from the data set. In particular, we present results showing that F_0 can be effectively estimated using a synopsis data structure with footprint only $O(\log u)$, but it cannot be effectively estimated based solely on a random sample of the data set unless $\Omega(u)$ memory is employed. We then discuss space-efficient algorithms for estimating F_k for all $k \geq 2$, using $(n^{1-1/k} \log n)$-synopsis data structures, and an improved $(\log n)$-synopsis data structure for estimating F_2. Finally, lower bounds on the estimation of F_k and F_∞^* are mentioned, as well as results showing that that both randomization and approximation are essential for evaluating F_k, $k \neq 1$.

3.1. Estimating the number of distinct values. Estimating the number of distinct values in a data set is a problem that frequently occurs in database applications, and in particular as a subproblem in query optimization. Indeed, Haas *et al* [HNSS95] claim that virtually all query optimization methods in relational and object-relational database systems require a means for assessing the number of distinct values of an attribute in a relation, i.e., the function F_0 for the sequence consisting of the attribute values for each item in the relation.

When no synopsis data structure is maintained, then the best methods for estimating F_0 are based on sampling. Haas *et al* [HNSS95] consider sampling-based algorithms for estimating F_0. They propose a hybrid approach in which the algorithm is selected based on the degree of skew of the data, measured essentially by the function F_2. However, they observe that fairly poor performance is obtained when using the standard statistical estimators, and remark that estimating F_0 via sampling is a hard and relatively unsolved problem. This is consistent with Olken's assertion [Olk93] that all known estimators give large errors on at least some data sets. In a recent paper, Chaudhuri *et al* [CMN98] show that "large error is unavoidable even for relatively large samples *regardless of the estimator used*. That is, there does not exist an estimator which can guarantee reasonable error with any reasonable probability unless the sample size is very close to the size of the database itself." Formally, they show the following.

THEOREM 3.1. [CMN98] *Consider any estimator $\hat{d}$ for the number of distinct values d based on a random sample of size r from a relation with n tuples. Let the error of the estimator $\hat{d}$ be defined as $error(\hat{d}) = \max\{\hat{d}/d, d/\hat{d}\}$. Then for any*

$\gamma > e^{-r}$, *there exists a choice of the relation such that with probability at least* γ, *error*$(\hat{d}) \geq \sqrt{\frac{n \ln 1/\gamma}{r}}$.

In contrast, the algorithm given below demonstrates a $(\log n)$-synopsis data structure which enables estimation of F_0 within an arbitrary fixed error bound with high probability, *for any given data set*. Note that the synopsis data structure is maintained while observing the entire data set. In practice, this can be realized while the data set is loaded into the disks, and the synopsis data structure is maintained in main memory with very small overhead.

Flajolet and Martin [**FM83, FM85**] described a randomized algorithm for estimating F_0 using only $O(\log u)$ memory bits, and analyzed its performance assuming one may use in the algorithm an explicit family of hash functions which exhibits some ideal random properties. The $(\log n)$-synopsis data structure consists of a bit vector V initialized to all 0. The main idea of the algorithm is to let each item in the data set select at random a bit in V and set it to 1, with (quasi-)geometric distribution; i.e., $V[i]$ is selected with probability $(\approx)$ $1/2^i$. The selection is made using a random hash function, so that all items of the same value will make the same selection. As a result, the expected number of items selecting $V[i]$ is $\approx F_0/2^i$, and therefore $2^{i'}$, where i' is the largest i such that $V[i] = 1$, is a good estimate for F_0. Alon et al [**AMS96**] adapted the algorithm so that linear hash functions could be used instead, obtaining the following.

THEOREM 3.2. [**FM83, AMS96**] *For every $c > 2$ there exists an algorithm that, given a sequence A of n members of $U = \{1, 2, \ldots, u\}$, computes a number Y using $O(\log u)$ memory bits, such that the probability that the ratio between Y and F_0 is not between $1/c$ and c is at most $2/c$.*

PROOF. Let d be the smallest integer so that $2^d > u$, and consider the members of U as elements of the finite field $F = GF(2^d)$, which are represented by binary vectors of length d. Let a and b be two random members of F, chosen uniformly and independently. When a member a_i of the sequence A appears, compute $z_i = a \cdot a_i + b$, where the product and addition are computed in the field F. Thus z_i is represented by a binary vector of length d. For any binary vector z, let $\rho(z)$ denote the largest r so that the r rightmost bits of z are all 0 and let $r_i = \rho(z_i)$. Let R be the maximum value of r_i, where the maximum is taken over all elements a_i of the sequence A. The output of the algorithm is $Y = 2^R$. Note that in order to implement the algorithm we only have to keep (besides the $d = O(\log u)$ bits representing an irreducible polynomial needed in order to perform operations in F) the $O(\log u)$ bits representing a and b and maintain the $O(\log \log u)$ bits representing the current maximum r_i value.

Suppose, now, that F_0 is the correct number of distinct elements that appear in the sequence A, and let us estimate the probability that Y deviates considerably from F_0. The only two properties of the random mapping $f(x) = ax + b$ that maps each a_i to z_i we need is that for every fixed a_i, z_i is uniformly distributed over F (and hence the probability that $\rho(z_i) \geq r$ is precisely $1/2^r$), and that this mapping is pairwise independent. Thus, for every fixed distinct a_i and a_j, the probability that $\rho(z_i) \geq r$ and $\rho(z_j) \geq r$ is precisely $1/2^{2r}$.

Fix an r. For each element $x \in U$ that appears at least once in the sequence A, let W_x be the indicator random variable whose value is 1 if and only if $\rho(ax + b) \geq r$. Let $Z = Z_r = \sum W_x$, where x ranges over all the F_0 elements x that appear in the sequence A. By linearity of expectation and since the expectation of each W_x is $1/2^r$, the expectation $\mathrm{E}(Z)$ of Z is $F_0/2^r$. By pairwise independence, the variance of Z is $F_0 \frac{1}{2^r}(1 - \frac{1}{2^r}) < F_0/2^r$. Therefore, by Markov's Inequality, if $2^r > cF_0$ then $\mathrm{Prob}(Z_r > 0) < 1/c$, since $\mathrm{E}(Z_r) = F_0/2^r < 1/c$. Similarly, by Chebyshev's Inequality, if $c2^r < F_0$ then $\mathrm{Prob}(Z_r = 0) < 1/c$, since $\mathrm{Var}(Z_r) < F_0/2^r = \mathrm{E}(Z_r)$ and hence $\mathrm{Prob}(Z_r = 0) \le \mathrm{Var}(Z_r)/(\mathrm{E}(Z_r)^2) < 1/\mathrm{E}(Z_r) = 2^r/F_0$. Since our algorithm outputs $Y = 2^R$, where R is the maximum r for which $Z_r > 0$, the two inequalities above show that the probability that the ratio between Y and F_0 is not between $1/c$ and c is smaller than $2/c$, as needed. $\qquad\square$

Thus we have a $(\log n)$-synopsis data structure for the class of F_0 queries, designed for the dynamic scenario of both insertions and queries. Analyzed on the cost model of Section 2, both the query and update times are only $O(1)$ processing time per query/update and no I/O operations.

3.2. Estimating F_k for $k \ge 2$. Alon *et al* [**AMS96**] developed an algorithm which, for every sequence A and a parameter k, can estimate F_k within a small constant factor with high probability, using an $(n^{1-1/k} \log n)$-synopsis data structure. The description below is taken from [**AGMS97**], which considered implementation issues of the algorithm and showed how the algorithm, coined sample-count, could be adapted to support deletions from the data set.

The idea in the sample-count algorithm is rather simple: A random sample of locations is selected in the sequence of data items that are inserted into the data set. This random selection can be easily done as the items are being inserted. Once we reach an item that was chosen to be in the sample, we will count from now on the number of incoming items that have its value. It turns out that the count r for each sample point is a random variable which satisfies $\mathrm{E}(nkr^{k-1}) \approx F_k$, and that the variance is reasonably small, for small k. The desired accuracy and confidence of the final estimate are obtained by applying averaging techniques over the counts of sample items.

More specifically, the number of memory words used by the algorithm is $s = s_1 \cdot s_2$, where s_1 is a parameter that determines the accuracy of the result, and s_2 determines the confidence; e.g., for any input set, the relative error of the estimate Y for F_2 exceeds $4u^{1/4}/\sqrt{s_1}$ with probability at most $2^{-s_2/2}$. The algorithm computes s_2 random variables $Y_1, Y_2, \ldots, Y_{s_2}$ and outputs their median Y. Each Y_i is the average of s_1 random variables $X_{ij} : 1 \le j \le s_1$, where the X_{ij} are independent, identically distributed random variables. Averaging is used to reduce the variance, and hence the error (Chebyshev's inequality), and the median is used to boost the confidence (Chernoff bounds). Each of the variables $X = X_{ij}$ is computed from the sequence in the same way as follows:

- Choose a random member a_p of the sequence A, where the index p is chosen randomly and uniformly among the numbers $1, 2, \ldots, n$; suppose that $a_p = l$ ($\in U = \{1, 2, \ldots, u\}$).
- Let $r = |\{q : q \ge p, a_q = l\}|$ be the number of occurrences of l among the members of the sequence A following a_p (inclusive).
- Define $X = n(r^k - (r-1)^k)$, e.g., for $k = 2$ let $X = n(2r - 1)$.

For $k = 2$, it is shown in [**AMS96**] that the estimate Y computed by the above algorithm satisfies $\mathrm{E}(Y) = F_2$, and $\mathrm{Prob}\left(|Y - F_2| \le 4u^{1/4}/\sqrt{s_1}\right) \ge 1 - 2^{-s_2/2}$. An accurate estimate for F_2 can therefore be guaranteed with high probability by selecting $s_1 = \Theta(\sqrt{u})$ and $s_2 = \Theta(\log n)$. More generally, by selecting $s_1 = \frac{8ku^{1-1/k}}{\lambda^2}$ and $s_2 = 2\log(1/\epsilon)$, one can obtain the following.

THEOREM 3.3. [**AMS96**] *For every $k \geq 1$, every $\lambda > 0$ and every $\epsilon > 0$ there exists a randomized algorithm that computes, given a sequence $A = (a_1, \ldots, a_n)$ of members of $U = \{1, 2, \ldots, u\}$, in one pass and using*

$$O\left(\frac{k\log(1/\epsilon)}{\lambda^2}u^{1-1/k}(\log u + \log n)\right)$$

memory bits, a number Y so that the probability that Y deviates from F_k by more than λF_k is at most ϵ.

Thus for fixed k, λ, and ϵ, we have an $(n^{1-1/k}\log n)$-synopsis data structure for the class of F_k queries, designed for the dynamic scenario. Waiting until query time to compute the averages Y_i would result in $O(s_1) = O(n^{1-1/k})$ query time on our cost model. However, these averages can be maintained as running averages as updates arrive, resulting in $O(1)$ processing time per query, and no I/O operations. Moreover, by representing the samples a_p as a concise sample (defined in Section 4) and using a dynamic dictionary data structure, the update time can likewise be reduced to $O(1)$ processing time per update and no I/O operations.

3.3. Improved estimation for F_2. An improved estimation algorithm for F_2 was also presented in [**AMS96**]. For every sequence A, F_2 can be estimated within a small constant factor with high probability, using a $(\log n)$-synopsis data structure. Again, the description below is taken from [**AGMS97**], which considers implementation issues of the algorithm and shows how the algorithm, coined **tug-of-war**, can be adapted to support deletions from the data set.

The **tug-of-war** algorithm can be illustrated as follows: Suppose that a crowd consists of several groups of varying numbers of people, and that our goal is to estimate the skew in the distribution of people to groups. That is, we would like to estimate F_2 for the set $\{a_i\}_{i=1}^n$, where a_i is the group to which the i'th person belongs. We arrange a tug-of-war, forming two teams by having each group assigned at random to one of the teams. Equating the displacement of the rope from its original location with the difference in the sizes of the two teams, it is shown in [**AMS96**] that the expected square of the rope displacement is exactly F_2, and that the variance is reasonably small. This approach can be implemented in small memory, using the observation that we can have the persons in the crowd come one by one, and contribute their displacement in an incremental fashion. In addition to the updated displacements, the only thing that requires recording in the process is the assignment of groups to teams, which can be done succinctly using an appropriate pseudo-random hash function.

As with **sample-count**, the number of memory words used by **tug-of-war** is $s = s_1 \cdot s_2$, where s_1 is a parameter that determines the accuracy of the result, and s_2 determines the confidence. As before, the output Y is the median of s_2 random variables $Y_1, Y_2, \ldots, Y_{s_2}$, each being the average of s_1 random variables $X_{ij} : 1 \leq j \leq s_1$, where the X_{ij} are independent, identically distributed random variables. Each $X = X_{ij}$ is computed from the sequence in the same way, as follows:

- Select at random a 4-wise independent mapping $i \mapsto \epsilon_i$, where $i \in U = \{1, 2, \ldots, u\}$ and $\epsilon_i \in \{-1, 1\}$.
- Let $Z = \sum_{i=1}^{u} \epsilon_i m_i$.

- Let $X = Z^2$.

For accurate estimates for F_2 of fixed error with guaranteed fixed probability, constant values suffice for s_1 and s_2. Specifically, by selecting $s_1 = \frac{16}{\lambda^2}$ and $s_2 = 2\log(1/\epsilon)$, the following is obtained.

THEOREM 3.4. [**AMS96**] *For every $\lambda > 0$ and $\epsilon > 0$ there exists a randomized algorithm that computes, given a sequence $A = (a_1, \ldots, a_n)$ of members of U, in one pass and using $O\left(\frac{\log(1/\epsilon)}{\lambda^2}(\log u + \log n)\right)$ memory bits, a number Y so that the probability that Y deviates from F_2 by more than λF_2 is at most ϵ. For fixed λ and ϵ, the algorithm can be implemented by performing, for each member of the sequence, a constant number of arithmetic and finite field operations on elements of $O(\log u + \log n)$ bits.*

Thus for fixed λ and ϵ, we have a $(\log n)$-synopsis data structure for the class of F_2 queries, designed for the dynamic scenario. Both the query and update times are only $O(1)$ processing time per query/update and no I/O operations.

3.4. Lower bounds. We mention lower bounds given in [**AMS96**] for the space complexity of randomized algorithms that approximate the frequency moments F_k. The lower bounds are obtained by reducing the problem to an appropriate communication complexity problem [**Yao83, BFS86, KS87, Raz92**], a set disjointness problem, obtaining the following.

THEOREM 3.5. [**AMS96**] *For any fixed $k > 5$ and $\gamma < 1/2$, any randomized algorithm that outputs, given one pass through an input sequence A of at most n elements of $U = \{1, 2, \ldots, n\}$, a number Z_k such that $\mathrm{Prob}(|Z_k - F_k| > 0.1F_k) < \gamma$, requires $\Omega(n^{1-5/k})$ memory bits.*

THEOREM 3.6. [**AMS96**] *Any randomized algorithm that outputs, given one pass through an input sequence A of at most $2n$ elements of $U = \{1, 2, \ldots, n\}$, a number Y such that $\mathrm{Prob}(|Y - F_\infty^*| \geq F_\infty^*/3) < \gamma$, for some fixed $\gamma < 1/2$, requires $\Omega(n)$ memory bits.*

The first theorem above places a lower bound on the footprint of a synopsis data structure that can estimate F_k to within constant factors in the purely online scenario, over all distributions. The second theorem shows that *no synopsis data structure exists* for estimating F_∞^* to within constant factors in the purely online scenario, over all distributions. As will be discussed in the next section, good synopsis data structures exist for skewed distributions, which may be of practical interest.

The number of elements F_1 can be computed deterministically and exactly using a $(\log n)$-synopsis data structure (a simple counter). The following two theorems show that for all $k \neq 1$, both randomization and approximation are essential in evaluating F_k.

THEOREM 3.7. [**AMS96**] *For any nonnegative integer $k \neq 1$, any randomized algorithm that outputs, given one pass through an input sequence A of at most $2n$ elements of $U = \{1, 2, \ldots, n\}$ a number Y such that $Y = F_k$ with probability at least $1 - \epsilon$, for some fixed $\epsilon < 1/2$, requires $\Omega(n)$ memory bits.*

THEOREM 3.8. [**AMS96**] *For any nonnegative integer $k \neq 1$, any deterministic algorithm that outputs, given one pass through an input sequence A of $n/2$ elements* *of* $U = \{1, 2, \ldots, n\}$, *a number* Y *such that* $|Y - F_k| \leq 0.1 F_k$ *requires* $\Omega(n)$ *memory bits.*

PROOF. Let G be a family of $t = 2^{\Omega(n)}$ subsets of U, each of cardinality $n/4$ so that any two distinct members of G have at most $n/8$ elements in common. (The existence of such a G follows from standard results in coding theory, and can be proved by a simple counting argument). Fix a deterministic algorithm that approximates F_k for some fixed nonnegative $k \neq 1$. For every two members G_1 and G_2 of G let $A(G_1, G_2)$ be the sequence of length $n/2$ starting with the $n/4$ members of G_1 (in a sorted order) and ending with the set of $n/4$ members of G_2 (in a sorted order). When the algorithm runs, given a sequence of the form $A(G_1, G_2)$, the memory configuration after it reads the first $n/4$ elements of the sequence depends only on G_1. By the pigeonhole principle, if the memory has less than $\log t$ bits, then there are two distinct sets G_1 and G_2 in G, so that the content of the memory after reading the elements of G_1 is equal to that content after reading the elements of G_2. This means that the algorithm must give the same final output to the two sequences $A(G_1, G_1)$ and $A(G_2, G_1)$. This, however, contradicts the assumption, since for every $k \neq 1$, the values of F_k for the two sequences above differ from each other considerably; for $A(G_1, G_1)$, $F_0 = n/4$ and $F_k = 2^k n/4$ for $k \geq 2$, whereas for $A(G_2, G_1)$, $F_0 \geq 3n/8$ and $F_k \leq n/4 + 2^k n/8$. Therefore, the answer of the algorithm makes a relative error that exceeds 0.1 for at least one of these two sequences. It follows that the space used by the algorithm must be at least $\log t = \Omega(n)$, completing the proof.   □

4. Hot list queries

In this section, we highlight our results on synopsis data structures for answering hot list and related queries.

A *hot list* is an ordered set of m ⟨value, count⟩ pairs for the m most frequently occurring "values" in a data set, for a prespecified m. In various contexts, hot lists are denoted as *high-biased histograms* [IC93] of $m + 1$ buckets, the first m mode statistics, or the m largest itemsets [AS94]. Hot lists are used in a variety of data analysis contexts, including:

- Best sellers lists ("top ten" lists): An example is the top selling items in a database of sales transactions.
- Selectivity estimation in query optimization: Hot lists capture the most skewed (i.e., popular) values in a relation, and hence have been shown to be quite useful for estimating predicate selectivities and join sizes (see [Ioa93, IC93, IP95]).
- Load balancing: In a mapping of values to parallel processors or disks, the most skewed values limit the number of processors or disks for which good load balance can be obtained.
- Market basket analysis: Given a sequence of sets of values, the goal is to determine the most popular k-*itemsets*, i.e., k-tuples of values that occur together in the most sets. Hot lists can be maintained on k-tuples of values for any specified k, and indicate a positive correlation among values in itemsets in the hot list. These can be used to produce association rules, specifying a (seemingly) causal relation among certain values [AS94, BMUT97]. An example is a grocery store, where for a sequence of customers, a set of the

items purchased by each customer is given, and an association rule might
be that customers who buy bread typically also buy butter.

• "Caching" policies based on most-frequently used: The goal is to retain in
the cache the most-frequently-used items and evict the least-frequently used
whenever the cache is full. An example is the most-frequently-called countries list in caller profiles for real-time telephone fraud detection [**Pre97**],
and in fact an early version of the hot list algorithm described below has
been in use in such contexts for several years.

As these examples suggest, the input need not be simply a sequence of individual values, but can be tuples with various fields such that for the purposes of the
hot list, both the value associated with a tuple and the contribution by that tuple to
that value's count are functions on its fields. However, for simplicity of exposition,
we will discuss hot lists in terms of a sequence of values, each contributing one to
its value's count.

Hot lists are trivial to compute and maintain given sufficient space to hold
the full histogram of the data set. However, for many data sets, such histograms
require space linear in the size of the data set. Thus for synopsis data structures
for hot list queries, a more succinct representation is required, and in particular,
counts cannot be maintained for each value. Note that the difficulty in maintaining
hot lists in the dynamic scenario is in detecting when values that were infrequent
become frequent due to shifts in the distribution of arriving data. With only a
small footprint, such detection is difficult since there is insufficient space to keep
track of all the infrequent values, and it is expensive (or impossible, in the purely
online scenario) to access the base data once it is on the disks.

A related, and seemingly simpler problem to hot list queries is that of "popular
items" queries. A *popular items* query returns a set of ⟨value, count⟩ pairs for
all values whose frequency in the data set exceeds a prespecified threshold, such
as 1% of the data set. Whereas hot list queries prespecify the number of pairs
to be output but not a frequency lower bound, popular items queries prespecify a
frequency lower bound but not the number of pairs. An approximate answer for
a popular items query can be readily obtained by sampling, since the sample size
needed to obtain a desired answer quality can be predetermined from the frequency
threshold. For example, if $p < 1$ is the prespecified threshold percentage, then by
Chernoff bounds, any value whose frequency exceeds this threshold will occur at
least $c/2$ times in a sample of size c/p with probability $1 - e^{-c/8}$. A recent paper
by Fang *et al* [**FSGM$^+$98**] presented techniques for improving the accuracy and
confidence for popular items queries. They considered the generalization to tuples
and functions on its fields mentioned above for hot list queries, and denoted this
class of queries as *iceberg queries*. They presented algorithms combining sampling
with the use of multiple hash functions to perform coarse-grained counting, in order
to significantly improve the answer quality over the naive sampling approach given
above.

In the remainder of this section, we describe results in [**GM98**] presenting and
studying two synopsis data structures, *concise samples* and *counting samples*. As
mentioned in Section 3.4, there are no synopsis data structures for estimating the
count of the most frequently occurring value, F_∞^*, to within constant factors in the
purely online scenario, over all distributions. Hence, no synopsis data structure
exists for the more difficult problem of approximating the hot list in the purely

online scenario, over all distributions. On the other hand, concise samples and counting samples are shown in [**GM98**] both analytically and experimentally to produce more accurate approximate hot lists than previous methods, and perform quite well for the skewed distributions that are of interest in practice.

4.1. Concise samples. Consider a hot list query on a data set of size n. One possible synopsis data structure is the set of values in a uniform random sample of the data set, as was proposed above for popular items queries. The m most frequently occurring values in the sample are returned in response to the query, with their counts scaled by n/m. However, note that any value occurring frequently in the sample is a wasteful use of the available space. We can represent k copies of the same value v as the pair $\langle v, k \rangle$, and (assuming that values and counts use the same amount of space), we have freed up space for $k - 2$ additional sample points. This simple observation leads to the following synopsis data structure.

DEFINITION 4.1. In a *concise representation* of a multiset, values appearing more than once in the multiset are represented as a value and a count. A *concise sample* of size m is a uniform random sample of the data set whose concise representation has footprint m.

We can quantify the advantage of concise samples over traditional samples in terms of the number of additional sample points for the same footprint. Let $S = \{\langle v_1, c_1 \rangle, \ldots, \langle v_j, c_j \rangle, v_{j+1}, \ldots, v_\ell\}$ be a concise sample of a data set of n values. We define *sample-size(S)* to be $\ell - j + \sum_{i=1}^{j} c_i$. Note that the footprint of S depends on the number of bits used per value and per count. For example, variable-length encoding could be used for the counts, so that only $\lceil \log x \rceil$ bits are needed to store x as a count; this reduces the footprint but complicates the memory management. Approximate counts [**Mor78**] could be used as well, so that only $\lceil \log \log x \rceil$ bits are needed to store x to within a power of two. For simplicity of exposition, we will consider only fixed-length encoding of $\log n$ bits per count and per value, including any bits needed to distinguish values from counts, so that the footprint of S is $(\ell + j) \log n$. For a traditional sample with m sample points, the sample-size is m and the footprint is $m \log n$.

Concise samples are never worse than traditional samples (given the encoding assumptions above), and can be exponentially or more better depending on the data distribution. For example, if there are at most $m/(2 \log n)$ distinct values in the data set, then a concise sample of size m would have sample-size n (i.e., in this case, the concise sample is the full histogram). Thus, the sample-size of a concise sample may be arbitrarily larger than its footprint:

LEMMA 4.2. [**GM98**] *For any footprint $m \geq 2 \log n$, there exists data sets for which the sample-size of a concise sample is n/m times larger than its footprint, where n is the size of the data set.*

For exponential distributions, the advantage is exponential:

LEMMA 4.3. [**GM98**] *Consider the family of exponential distributions: for $i = 1, 2, \cdots$, $\Pr(v = i) = \alpha^{-i}(\alpha - 1)$, for $\alpha > 1$. For any $m \geq 2$, the expected sample-size of a concise sample with footprint $m \log n$ is at least $\alpha^{m/2}$.*

PROOF. Let $x = m/2$. Note that we can fit at least x values and their counts within the given footprint. The expected sample-size can be lower bounded by the expected number of randomly selected tuples before the first tuple whose attribute value v is greater than x. The probability of selecting a value greater than x is $\sum_{i=x+1}^{\infty} \alpha^{-i}(\alpha - 1) = \alpha^{-x}$, so the expected number of tuples selected before such an event occurs is α^x.   □

The expected gain in using a concise sample over a traditional sample for arbitrary data sets is a function of the frequency moments F_k, for $k \geq 2$, of the data set. Recall from Section 3 that $F_k = \sum_j m_j^k$, where j is taken over the values represented in the set and m_j is the number of set elements of value j.

THEOREM 4.4. [**GM98**] *For any data set, when using a concise sample S with sample-size m, the expected gain is*

$$E[m - \textit{number of distinct values in } S] = \sum_{k=2}^{m}(-1)^k \binom{m}{k} \frac{F_k}{n^k} \ .$$

PROOF. Let $p_j = m_j/n$ be the probability that an item selected at random from the set is of value j. Let X_i be an indicator random variable so that $X_i = 1$ if the ith item selected to be in the traditional sample has a value not represented as yet in the sample, and $X_i = 0$ otherwise. Then, $\Pr(X_i = 1) = \sum_j p_j(1 - p_j)^{i-1}$, where j is taken over the values represented in the set (since $X_i = 1$ if some value j is selected so that it has not been selected in any of the first $i - 1$ steps). Clearly, $X = \sum_{i=1}^{m} X_i$ is the number of distinct values in the traditional sample. We can now evaluate $E[$number of distinct values$]$ as

$$
\begin{aligned}
E[X] &= \sum_{i=1}^{m} E[X_i] = \sum_{i=1}^{m} \sum_j p_j(1 - p_j)^{i-1} = \sum_j \sum_{i=1}^{m} p_j(1 - p_j)^{i-1} \\
&= \sum_j p_j \frac{1 - (1 - p_j)^m}{1 - (1 - p_j)} = \sum_j \left(1 - (1 - p_j)^m\right) \\
&= \sum_j \left(1 - \sum_{k=0}^{m}(-1)^k \binom{m}{k} p_j^k\right) = \sum_j 1 - \sum_{k=0}^{m}(-1)^k \binom{m}{k} \sum_j p_j^k \\
&= \sum_{k=1}^{m}(-1)^{k+1} \binom{m}{k} \frac{F_k}{n^k} \ .
\end{aligned}
$$

□

Note that the footprint for a concise sample is at most $2\log n$ times the number of distinct values, whereas the footprint for a traditional sample of sample-size m is $m \log n$.

Maintaining concise samples. We describe next the algorithm given in [**GM98**] for maintaining a concise sample within a given footprint bound as new data is inserted into the data set. Since the number of sample points provided by a concise sample depends on the data distribution, the problem of maintaining a concise sample as new data arrives is more difficult than with traditional samples. For traditional samples, the reservoir sampling algorithm of Vitter [**Vit85**] can be used to maintain a sample in the presence of insertions of new data (see Section 5.1 for details). However, this algorithm relies heavily on a priori knowledge of the target sample-size (which, for traditional samples, equals the footprint divided by $\log n$). With concise samples, the sample-size depends on the data distribution to date, and any changes in the data distribution must be reflected in the sampling frequency.

Our maintenance algorithm is as follows. Let τ be an entry threshold (initially 1) for new data to be selected for the sample. Let S be the current concise sample and consider an insertion of a data item with value v. With probability $1/\tau$, add v to S, preserving the concise representation. If the footprint for S now exceeds the prespecified footprint bound, raise the threshold to some τ' and then subject each sample point in S to this higher threshold. Specifically, each of the sample-size(S) sample points is evicted with probability τ/τ'. It is expected that sample-size(S) $\cdot (1 - \tau/\tau')$ sample points will be evicted. Note that the footprint is only decreased when a $\langle$value, count$\rangle$ pair reverts to a singleton or when a value is removed altogether. If the footprint has not decreased, repeat with a higher threshold.

There is complete flexibility in this algorithm in selecting the sequence of increasing thresholds, and [GM98] discussed a variety of approaches and their trade-offs, as well as ways to improve the constant factors.

THEOREM 4.5. [GM98] *The above algorithm maintains a concise sample within a prespecified size bound in constant amortized expected update time per insert, and no I/O operations.*

PROOF. The algorithm maintains a uniform random sample since, whenever the threshold is raised, it preserves the invariant that each item in the data set has been treated (probabilistically) as if the threshold were always the new threshold. The look-ups can be done in constant expected time using a dynamic dictionary data structure such as a hash table. Raising a threshold costs $O(m')$ processing time, where m' is the sample-size of the concise sample before the threshold was raised. For the case where the threshold is raised by a constant factor each time, we expect there to be a constant number of coin tosses resulting in sample points being retained for each sample point evicted. Thus we can amortize the retained against the evicted, and we can amortize the evicted against their insertion into the sample (each sample point is evicted only once). □

4.2. Counting samples. Counting samples are a variation on concise samples in which the counts are used to keep track of all occurrences of a value inserted into the data set after the value was selected for the sample. Their definition is motivated by a sampling&counting process of this type from a static data set:

DEFINITION 4.6. A *counting sample* for a data set A with threshold τ is any subset of A stored in a concise representation (as defined in Definition 4.1) that is obtained by a process that is probabilistically equivalent to the following process: For each value v occurring $c > 0$ times in A, we flip a coin with probability $1/\tau$ of heads until the first heads, up to at most c coin tosses in all; if the ith coin toss is heads, then v occurs $c - i + 1$ times in the subset, else v is not in the subset.

A counting sample differs from the approach used in Section 3.2 in not allowing multiple counts for the same value and in its use of a threshold (that will adapt to a data distribution) versus a prespecified sample size. Although counting samples are not uniform random samples of the data set, a concise sample can be obtained from a counting sample by considering each pair $\langle v, c \rangle$ in the counting sample in turn, and flipping a coin with probability $1/\tau$ of heads $c - 1$ times and reducing the count by the number of tails.

Maintaining counting samples. The following algorithm is given in [**GM98**] for maintaining a counting sample within a given footprint bound for the dynamic scenario. Let τ be an entry threshold (initially 1) for new data to be selected for the sample. Let S be the current counting sample and consider an insertion of a data item with value v. If v is represented by a $\langle$value, count$\rangle$ pair in S, increment its count. If v is a singleton in S, create a pair with count set to 2. Otherwise, add v to S with probability $1/\tau$. If the footprint for S now exceeds the prespecified footprint bound, raise the threshold to some τ' and then subject each value in S to this higher threshold. Specifically, for each value in the counting sample, flip a biased coin, decrementing its observed count on each flip of tails until either the count reaches zero or a heads is flipped. The first coin toss has probability of heads τ/τ', and each subsequent coin toss has probability of heads $1/\tau'$. Values with count zero are removed from the counting sample; other values remain in the counting sample with their (typically reduced) counts.

An advantage of counting samples over concise samples is that one can maintain counting samples in the presence of deletions to the data set. Maintaining concise samples in the presence of such deletions is difficult: If we fail to delete a sample point in response to the delete operation, then we risk having the sample fail to be a subset of the data set. On the other hand, if we always delete a sample point, then the sample may no longer be a random sample of the data set.[2] With counting samples, we do not have this difficulty. For a delete of a value v, it suffices to reverse the increment procedure by decrementing a count, converting a pair to a singleton, or removing a singleton, as appropriate.

As with concise samples, there is complete flexibility in this algorithm in selecting the sequence of increasing thresholds, and [**GM98**] discussed a variety of approaches and their tradeoffs, as well as ways to improve the constant factors.

THEOREM 4.7. [**GM98**] *For any sequence of insertions and deletions in the dynamic scenario, the above algorithm maintains a counting sample within a prespecified footprint in constant amortized expected update time and no I/O operations.*

PROOF. We must show that the requirement in the definition of a counting sample is preserved when an insert occurs, a delete occurs, or the threshold is raised. Let A be the data set and S be the counting sample.

An insert of a value v increases by one its count in A. If v is in S, then one of its coin flips to date was heads, and we increment the count in S. Otherwise, none of its coin flips to date were heads, and the algorithm flips a coin with the appropriate probability. All other values are untouched, so the requirement is preserved.

A delete of a value v decreases by one its count in A. If v is in S, then the algorithm decrements the count (which may drop the count to 0). Otherwise, c coin flips occurred to date and were tails, so the first $c-1$ were also tails, and the value remains omitted from S. All other values are untouched, so the requirement is preserved.

Consider raising the threshold from τ to τ', and let v be a value occurring $c > 0$ times in A. If v is not in S, there were the equivalent of c coin flips with heads probability $1/\tau$ that came up tails. Thus the same c probabilistic events would fail to come up heads with the new, stricter coin (with heads probability only $1/\tau'$). If

---
[2]For some applications of random samples, an effective alternative approach is to collect and make use of two uniform samples: one for the inserted data and one for the deleted data.

v is in S with count c', then there were the equivalent of $c - c'$ coin flips with heads probability $1/\tau$ that came up tails, and these same probabilistic events would come up tails with the stricter coin. This was followed by the equivalent of a coin flip with heads probability $1/\tau$ that came up heads, and the algorithm flips a coin with heads probability τ/τ', so that the result is equivalent to a coin flip with probability $(1/\tau) \cdot (\tau/\tau') = (1/\tau')$. If this coin comes up tails, then subsequent coin flips for this value have heads probability $1/\tau'$. In this way, the requirement is preserved for all values.

The update time bounds are argued as in the proof of Theorem 4.5. $\quad\square$

Note that although both concise samples and counting samples have $O(1)$ amortized update times, counting samples are slower to update than concise samples, since, unlike concise sample, they perform a look-up (into the counting sample) at each update to the data set. On the other hand, with counting samples, the guarantees on the counts are stronger, since exact counting is used on values already in the sample.

4.3. Application to hot list queries. Consider a hot list query requesting k pairs. Given a concise sample S of footprint $m \log n$, $m \geq 2k$, an approximate hot list can be reported by computing the k'th largest count c_k (using a linear time selection algorithm), and then reporting all pairs with counts at least $\max(c_k, \delta)$, scaling the counts by n/m', where $\delta \geq 1$ is a confidence threshold and $m' =$ sample-size(S). Note that when $\delta = 1$, k pairs will be reported, but with larger δ, fewer than k may be reported. The response time for reporting is $O(m)$ processing time and no I/O operations. Alternatively, we can trade-off update time versus query time by keeping the concise sample sorted by counts. This allows for reporting in $\Theta(k)$ time.

Given a counting sample S of footprint $m \log n$ with threshold τ, an approximate hot list can be reported by computing the k'th largest count c_k, and then reporting all pairs with counts at least $\max(c_k, \tau - \hat{c})$, where $\hat{c}$ is a compensation added to each reported count that serves to compensate for inserts of a value into the data set prior to the successful coin toss that placed it in the counting sample. An analysis in [**GM98**] argued for $\hat{c} = \tau\left(\frac{e-2}{e-1}\right) - 1 \approx .418 \cdot \tau - 1$. Given the conversion of counting samples into concise samples discussed in Section 4.2, this can be seen to be similar to taking $\delta = 2 - \frac{\hat{c}+1}{\tau} \approx 1.582$.

Analytical bounds and experimental results are presented in [**GM98**] quantifying the accuracy of the approximate hot lists reported using concise samples or counting samples. An example plot from that paper is given in Figure 3, where the data is drawn from a Zipf distribution with parameter 1.5 and the footprint is measured in memory words.

5. Histograms and quantiles

Histograms approximate a data set by grouping values into "buckets" (subsets) and approximating the distribution of values in the data set based on summary statistics maintained in each bucket (see, e.g., [**PIHS96**]). Histograms are commonly used in practice in various databases (e.g., in DB2, Informix, Ingres, Oracle, Microsoft SQL Server, Sybase, and Teradata). They are used for selectivity estimation purposes within a query optimizer and in query execution, and there is work in progress on using them for approximate query answering.

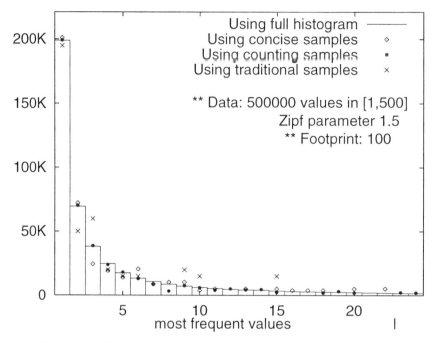

FIGURE 3. Comparison of algorithms for a hot list query, depicting the frequency of the most frequent values as reported using a full histogram, using a concise sample, using a counting sample, and using a traditional sample

Two histogram classes used extensively in database systems are equi-depth histograms and compressed histograms. In an *equi-depth* or *equi-height* histogram, contiguous ranges of values are grouped into buckets such that the number of data items falling into each bucket is the same. The endpoints of the value ranges are denoted the bucket boundaries or *quantiles*. In a *compressed* histogram [**PIHS96**], the highest frequency values are stored separately in single-valued buckets; the rest are partitioned as in an equi-depth histogram. Compressed histograms typically provide more accurate estimates than equi-depth histograms.

A common problem with histograms is their dynamic maintenance. As a data set is updated, its distribution of values might change and the histogram (which is supposed to reflect the distribution) should change as well, since otherwise estimates based on the histogram will be increasingly inaccurate. In this section, we describe our work in [**GMP97b**] on algorithms for maintaining approximate equi-depth and compressed histograms as synopsis data structures in the dynamic scenario. We also discuss recent related work by Manku *et al* [**MRL98**] on computing approximate quantiles.

Another concern for histograms is their construction costs in the static scenario. Sampling can be used to improve the construction times (see, e.g., [**PIHS96**]), and we discuss recent work by Chaudhuri *et al* [**CMN98**] on using sampling to construct approximate equi-depth histograms in the static scenario.

An important feature of our algorithms for maintaining approximate histograms is the use of a "backing sample". Backing samples are interesting for two reasons:

they can be used to convert sampling-based algorithms for the static scenario into algorithms for the dynamic scenario, and their use is an example of a hierarchical approach to synopsis data structures.

5.1. Backing samples. A *backing sample* for a data set, A, is a uniform random sample of A that is kept up-to-date in the presence of updates to A [**GMP97b**]. In most sampling-based estimation techniques, whenever a sample of size m is needed, either the entire relation is scanned to extract the sample, or several random disk blocks are read. In the latter case, the values in a disk block may be highly correlated, and hence to obtain a truly random sample, m disk blocks may need to be read, with only a single value used from each block. In contrast, a backing sample is a synopsis data structure that may reside in main memory, and hence be accessed with no I/O operations. Moreover, if, as is typically the case in databases, each data item is a record ("tuple") comprised of fields ("attributes"), then only the fields desired for the sampling need be retained in the synopsis. In the case of using samples for histograms, for example, only the field(s) needed for the histogram need be retained. If the backing sample is stored on the disks, it can be packed densely into disk blocks, allowing it to be more quickly swapped in and out of memory. Finally, an indexing structure for the sample can be maintained, which would enable fast access of the sample values within a certain range.

Clearly, a backing sample of m sample points can be used to convert a sampling-based algorithm requiring $\frac{m}{D}$ I/O operations for its sampling into an algorithm that potentially requires no I/O operations.

Maintaining backing samples. A uniform random sample of a target size m can be maintained under insertions to the data set using Vitter's *reservoir sampling* technique [**Vit85**]: The algorithm proceeds by inserting the first m items into a "reservoir." Then a random number of new items are skipped, and the next item replaces a randomly selected item in the reservoir. Another random number of items are then skipped, and so forth. The distribution function of the length of each random skip depends explicitly on the number of items so far, and is chosen such that at any point each item in the data set is equally likely to be in the reservoir. Specifically, when the size of the data set is n, the probability for an item to be selected for the backing sample of size m is m/n. Random skipping is employed in order to reduce constant factors in the update times compared with the approach of flipping a coin for each new item. Reservoir sampling maintains a traditional random sample as a backing sample; an alternative is to use a concise sample or a counting sample as a backing sample, and maintain them as discussed in Section 4.

As discussed in Section 4.2, there are difficulties in maintaining uniform random samples under deletions to the data set, with two possible solutions being counting samples and deletion samples. In [**GMP97b**], we assumed that each data item has a unique id (namely, its row id in the database table in which it resides), so that a deletion removes a unique item from the data set. We retained the row id with the sample point (which precludes the use of concise samples or counting samples). With row ids, deletions can be handled by removing the item from the sample, if it is in the sample. However, such deletions decrease the size of the sample from the target size m, and moreover, it is not apparent how to use subsequent insertions to obtain a provably random sample of size m once the sample has dropped below m. Instead, we maintained a sample whose size is initially a prespecified upper bound

U, and allowed for it to decrease as a result of deletions of sample items down to a prespecified lower bound L. If the sample size dropped below L, the data set is read from the disks in order to re-populate the random sample, either by rereading all the data or by reading $U - L + 1$ random disk blocks. Since the sampling is independent of the deletions, the deletion of a fraction of the sample is expected to occur only after the deletion of the same fraction of the data set.

We presented in [GMP97b] several techniques for reducing constant factors in the update times. For example, since the algorithm maintains a random sample independent of the order of the updates to the data set, we postponed the processing of deletes until the next insert selected for the backing sample. This reduced the maintenance to the insert-only case, for which random skipping can be employed (having deletions intermixed with insertions foils random skipping).

Note that since a backing sample is a fixed sample of a prespecified size, it may be desirable to augment the sample and/or refresh the sample, as appropriate for a particular application.

Backing samples in a synopsis hierarchy. In [GMP97b], we used a backing sample in support of dynamically maintaining histograms. In the scenario we considered, the histogram resided in main memory whereas the backing sample, being somewhat larger than the histogram, resided on the disks. The goal was to maintain the histogram under the dynamic scenario, while minimizing the accesses and updates to the backing sample, in order to minimize the number of I/O operations. The backing sample was a traditional random sample maintained using reservoir sampling. When the size of the data set is n, the probability for an item to be selected for the backing sample of size m is m/n, and hence in maintaining the backing sample an I/O operation is expected only once every $\Omega(n/m)$ insertions. Therefore, over the process of maintaining the backing sample, while the data set grows from m to n, an I/O operation is expected (on the average) only once every $\Omega(n/(m\log(n/m)))$ insertions. Thus, since this overhead is small for large n and small m, the goal became to design an algorithm for maintaining histograms that minimized the number of accesses to a given backing sample.

5.2. Equi-depth histograms. An equi-depth histogram partitions the range of possible values into β buckets such that the number of data items whose value falls into a given bucket is the same for all buckets. An *approximate* equi-depth histogram approximates the exact histogram by relaxing the requirement on the number of data items falling in a bucket and/or the accuracy of the counts associated with the buckets. Let N be the number of items in the data set, let B.count be the count associated with a bucket B, and let f_B be the number of items falling in a bucket B.[3] In [GMP97b], we defined two error metrics for evaluating approximate equi-depth histograms. Our first metric, μ_{ed}, was defined to be the standard deviation of the bucket sizes from the mean bucket size, normalized with respect to the mean bucket size:

$$\mu_{ed} = \frac{\beta}{N}\sqrt{\frac{1}{\beta}\sum_{i=1}^{\beta}\left(f_{B_i} - \frac{N}{\beta}\right)^2}.$$

Our second error metric, μ_{count}, was defined to be the standard deviation of the bucket counts from the actual number of items in each bucket, normalized with respect to the mean bucket count:

$$\mu_{\text{count}} = \frac{\beta}{N}\sqrt{\frac{1}{\beta}\sum_{i=1}^{\beta}(f_{B_i} - B_i.\text{count})^2}\ .$$

In [**GMP97b**], we presented the first low overhead algorithms for maintaining highly-accurate approximate equi-depth histograms. Each algorithm relied on using a backing sample, S, of a fixed size dependent on β.

Our simplest algorithm, denoted Equi-depth_Simple, worked as follows. At the start of each phase, compute an approximate equi-depth histogram from S by sorting S and then taking every $(|S|/\beta)$'th item as a bucket boundary. Set the bucket counts to be N'/β, where N' is the number of items in the data set at the beginning of the phase. Let $T = \lceil(2+\gamma)N'/\beta\rceil$, where $\gamma > -1$ is a tunable performance parameter. Larger values for γ allow for greater imbalance among the buckets in order to have fewer phases. As each new item is inserted into the data set, increment the count of the appropriate bucket. When a count exceeds the threshold T, start a new phase.

THEOREM 5.1. [**GMP97b**] *Let $\beta \geq 3$. Let $m = (c\ln^2\beta)\beta$, for some $c \geq 4$. Consider Equi-depth_Simple applied to a sequence of $N \geq m^3$ inserts of items into an initially empty data set. Let S be a random sample of size m of tuples drawn uniformly from the relation, either with or without replacement. Let $\alpha = (c\ln^2\beta)^{-1/6}$. Then Equi-depth_Simple computes an approximate equi-depth histogram such that with probability at least $1 - \beta^{-(\sqrt{c}-1)} - (N/(2+\gamma))^{-1/3}$, $\mu_{\text{ed}} \leq \alpha + (1+\gamma)$ and $\mu_{\text{count}} \leq \alpha$.*

PROOF. Let H be an approximate equi-depth histogram computed by the Equi-depth_Simple algorithm after N items have been inserted into the data set. Let Φ be the current phase of the algorithm, and let $N' \leq N$ be the number of items in the data set at the beginning of phase Φ. Let μ'_{count} and μ'_{ed} be the errors μ_{count} and μ_{ed}, respectively, resulting after extracting an approximate histogram from S at the beginning of phase Φ. Finally, let $\rho' = 1 - \beta^{-(\sqrt{c}-1)} - (N')^{-1/3}$, and let $\rho = 1 - \beta^{-(\sqrt{c}-1)} - (N/(2+\gamma))^{-1/3}$. Since during phase Φ, we have that $N \leq N'(2+\gamma)$, it follows that $\rho \leq \rho'$. We show in [**GMP97b**] that $\mu'_{\text{ed}} = \mu'_{\text{count}} \leq \alpha$ with probability at least ρ', and hence at least ρ.

During phase Φ, a value inserted into bucket B_i increments both f_{B_i} and B_i.count. Therefore, by the definition of μ_{count}, its value does not change during phase Φ, and hence at any time during the phase, $\mu_{\text{count}} = \mu'_{\text{count}} \leq \alpha$ with probability ρ. It remains to bound μ_{ed} for H.

Let f'_{B_i} and B_i.count$'$ be the values of f_{B_i} and B_i.count, respectively, at the beginning of phase Φ. Let $\Delta'_i = f'_{B_i} - N'/\beta$, and let $\Delta_i = f_{B_i} - N/\beta$. We claim that $|\Delta_i - \Delta'_i| \leq (1+\gamma)N'/\beta$. Note that $|\Delta_i - \Delta'_i| \leq \max(f_{B_i} - f'_{B_i}, N/\beta - N'/\beta)$. The claim follows since $f_{B_i} - f'_{B_i} = B_i$.count $- B_i$.count$' \leq T - B_i$.count$' = (2+\gamma)N'/\beta - N'/\beta$, and $N - N' \leq \beta(B_i$.count $- B_i$.count$')$.

By the claim,

$$\Delta_i^2 \leq (\Delta'_i + (1+\gamma)N'/\beta)^2 = \Delta_i'^2 + 2\Delta'_i(1+\gamma)N'/\beta + ((1+\gamma)N'/\beta)^2\ .$$

Note that $\sum_{i=1}^{\beta} \Delta_i' = \sum_{i=1}^{\beta} (f_{B_i} - N'/\beta) = 0$. Hence, substituting for Δ_i^2 in the definition of μ_{ed} we obtain

$$
\begin{aligned}
\mu_{\text{ed}} &= \frac{\beta}{N} \sqrt{\frac{1}{\beta} \left(\sum_{i}^{\beta} \Delta_i'^{\,2} + \sum_{i=1}^{\beta} ((1+\gamma)N'/\beta)^2 \right)} \\
&\leq \mu_{\text{ed}}' + \frac{\beta}{N}(1+\gamma)N'/\beta) \leq \mu_{\text{ed}}' + (1+\gamma) \ .
\end{aligned}
$$

The theorem follows.   □

A second algorithm from [**GMP97b**] reduced the number of recomputations from S by trying to balance the buckets using a local, less expensive procedure. The algorithm, denoted Equi-depth_SplitMerge, worked in phases. As in Equi-depth_Simple, at each phase there is a threshold $T = \lceil (2+\gamma)N'/\beta \rceil$. As each new item is inserted into the data set, increment the count of the appropriate bucket. When a count exceeds the threshold T, split the bucket in half. In order to maintain the number of buckets β fixed, merge two adjacent buckets whose total count is less than T, if such a pair of buckets can be found. When such a merge is not possible, recompute the approximate equi-depth histogram from S.

To merge two buckets, sum the counts of the two buckets and dispose of the boundary between them. To split a bucket B, select an approximate median in B to serve as the bucket boundary between the two new buckets, by selecting the median among the items in S that fall into B. The split and merge operation is illustrated in Figure 4. Note that split and merge can occur only for $\gamma > 0$.

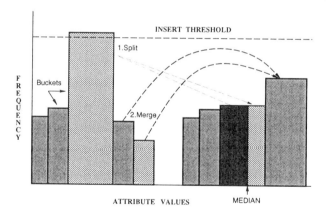

FIGURE 4. Split and merge operation during equi-depth histogram maintenance

The number of splits and the number of phases can be bounded as follows.

THEOREM 5.2. [**GMP97b**] *Consider Equi-depth_SplitMerge with β buckets and performance parameter $-1 < \gamma \leq 2$ applied to a sequence of N inserts. Then the total number of phases is at most $\log_\alpha N$, and the total number of splits is at most $\beta \log_\alpha N$, where $\alpha = 1 + \gamma/2$ if $\gamma > 0$, and otherwise $\alpha = 1 + (1+\gamma)/\beta$.*

To handle deletions to the data set, let $T_\ell = \lfloor N'/(\beta(2+\gamma_\ell)) \rfloor$ be a lower threshold on the bucket counts, where $\gamma_\ell > -1$ is a tunable performance parameter. When an item is deleted from the data set, decrement the count of the appropriate bucket. If a bucket's count drops to the threshold T_ℓ, merge the bucket with one of its adjacent buckets and then split the bucket B' with the largest count, as long as its count is at least $2(T_\ell + 1)$. (Note that B' may be the newly merged bucket.) If no such B' exists, recompute the approximate equi-depth histogram from S. The merge and split operation is illustrated in Figure 5.

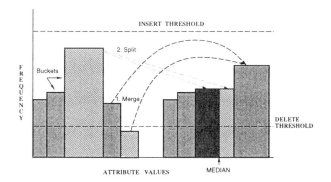

FIGURE 5. Merge and split operation during equi-depth histogram maintenance

Related work. A recent paper by Manku *et al* [**MRL98**] presented new algorithms for computing approximate quantiles of large data sets in a single pass over the data and with limited main memory. Whereas in an equi-depth histogram, the desired ranks for the quantiles are at regular intervals, their paper considered arbitrary prespecified ranks. Compared to an earlier algorithm of Munro and Paterson [**MP80**], their deterministic algorithm restricts attention to a single pass and improves the constants in the memory requirements. Specifically, let an item be an ϵ-approximate ϕ-quantile in a data set of N items if its rank in the sorted data set is between $\lceil (\phi - \epsilon)N \rceil$ and $\lceil (\phi + \epsilon)N \rceil$. Manku *et al* presented a deterministic algorithm that, given $\phi_1, \ldots, \phi_k \in [0, 1]$, computes ϵ-approximate ϕ_i-quantiles for $i = 1, \ldots, k$ in a single pass using only $O(\frac{1}{\epsilon} \log^2(\epsilon N))$ memory. Note that this algorithm performs $\frac{N}{DB}$ I/O operations in the static scenario, for a class of queries where the ranks of the desired quantiles are prespecified.

Manku *et al* also analyzed the approach of first taking a random sample and then running their deterministic algorithm on the sample, in order to reduce the memory requirements for massive data sets. They did not explicitly consider dynamic maintenance of quantiles, and indeed they have not attempted to minimize the query time to output their approximate quantiles, since their output operation occurs only once, after the pass over the data. However, by using a backing sample residing in memory, their algorithm can be used in the dynamic scenario with no I/O operations at update time or query time.

A recent paper by Chaudhuri *et al* [**CMN98**] studied the problem of how much sampling is needed to guarantee an approximate equi-depth histogram of a certain accuracy. The error metric they used to evaluate accuracy is the maximum over all buckets B of $|f_B - \frac{N}{\beta}|$, where N is the number of data items, β is the number of buckets, and f_B is the number of items falling into B. As argued in the paper, this error metric seems more appropriate than the μ_{ed} metric considered above, for providing guarantees on the accuracy of approximate answers to range queries. (See also [**JKM⁺98**] for another approach to providing improved quality guarantees when using histograms to answer range queries.) Chaudhuri *et al* provided a tighter analysis than in [**GMP97b**] for analyzing the accuracy of equi-depth histograms computed from a sample. The paper studied only the static scenario of constructing an equi-depth histogram, including a discussion of techniques for extracting multiple sample points from a sampled disk block. However, by using a backing sample, such issues are no longer a concern, and their analysis can be used to improve the guarantees of the algorithms in [**GMP97b**] for maintaining equi-depth histograms in the dynamic scenario.

5.3. Compressed histograms. In an equi-depth histogram, values with high frequencies can span a number of buckets; this is a waste of buckets since the sequence of spanned buckets for a value can be replaced by a single bucket with a single count, resulting in the same information within a smaller footprint. A *compressed histogram* has a set of such singleton buckets and an equi-depth histogram over values not in singleton buckets [**PIHS96**]. Our target compressed histogram with β buckets has β' equi-depth buckets and $\beta - \beta'$ singleton buckets, where $1 \leq \beta' \leq \beta$, such that the following requirements hold: (i) each equi-depth bucket has N'/β' tuples, where N' is the total number of data items in equi-depth buckets, (ii) no single value "spans" an equi-depth bucket, i.e., the set of bucket boundaries are distinct, and conversely, (iii) the value in each singleton bucket has frequency $\geq N'/\beta'$. An *approximate* compressed histogram approximates the exact histogram by relaxing one or more of the three requirements above and/or the accuracy of the counts associated with the buckets.

In [**GMP97b**], we presented the first low overhead algorithm for maintaining highly-accurate approximate compressed histograms. As in the equi-depth case, the algorithm relied on using a backing sample S. An approximate compressed histogram can be computed from S as follows. Let m, initially $|S|$, be the number of items tentatively in equi-depth buckets. Consider the $\beta - 1$ most frequent values occurring in S, in order of maximum frequency. For each such value, if the frequency f of the value is at least m divided by the number of equi-depth buckets, create a singleton bucket for the value with count $fN/|S|$, and decrease m by f. Otherwise, stop creating singleton buckets and produce an equi-depth histogram on the remaining values, using the approach of the previous subsection, but setting the bucket counts to $(N/|S|) \cdot (m/\beta')$. Our algorithm reduced the number of such recomputations from S by employing a local procedure for adjusting bucket boundaries.

Similar to the equi-depth algorithm, the algorithm worked in phases, where each phase has an upper threshold for triggering equi-depth bucket splits and a lower threshold for triggering bucket merges. The steps for updating the bucket boundaries are similar to those for an equi-depth histogram, but must address several additional concerns:

1. New values added to the data set may be skewed, so that values that did not warrant singleton buckets before may now belong in singleton buckets.
2. The threshold for singleton buckets grows with N', the number of items in equi-depth buckets. Thus values rightfully in singleton buckets for smaller N' may no longer belong in singleton buckets as N' increases.
3. Because of concerns 1 and 2 above, the number of equi-depth buckets, β', grows and shrinks, and hence we must adjust the equi-depth buckets accordingly.

4. Likewise, the number of items in equi-depth buckets grows and shrinks dramatically as sets of items are removed from and added to singleton buckets. The ideal is to maintain N'/β' items per equi-depth bucket, but both N' and β' are growing and shrinking.

Briefly and informally, the algorithm in [GMP97b] addressed each of these four concerns as follows. To address concern 1, it used the fact that a large number of updates to the same value v will suitably increase the count of the equi-depth bucket containing v so as to cause a bucket split. Whenever a bucket is split, if doing so creates adjacent bucket boundaries with the same value v, then a new singleton bucket for v must be created. To address concern 2, the algorithm allowed singleton buckets with relatively small counts to be merged back into the equi-depth buckets. As for concerns 3 and 4, it used our procedures for splitting and merging buckets to grow and shrink the number of buckets, while maintaining approximate equi-depth buckets, until the histogram is recomputed from S. The imbalance between the equi-depth buckets is controlled by the thresholds T and T_ℓ (which depend on the tunable performance parameters γ and γ_ℓ, as in the equi-depth algorithm). When an equi-depth bucket is converted into a singleton bucket or vice-versa, the algorithm ensured that at the time, the bucket is within a constant factor of the average number of items in an equi-depth bucket (sometimes additional splits and merges are required). Thus the average is roughly maintained as such equi-depth buckets are added or subtracted.

The requirements for when a bucket can be split or when two buckets can be merged are more involved than in the equi-depth algorithm: A bucket B is a *candidate split bucket* if it is an equi-depth bucket whose count is at least $2(T_\ell + 1)$ or a singleton bucket whose count is bounded by $2(T_\ell + 1)$ and $T/(2 + \gamma)$. A pair of buckets, B_i and B_j, is a *candidate merge pair* if (1) either they are adjacent equi-depth buckets or they are a singleton bucket and the equi-depth bucket in which its singleton value belongs, and (2) the sum of their counts is less than T. When there are more than one candidate split bucket (candidate merge pair), the algorithm selected the one with the largest (smallest combined, respectively) bucket count.

In [GMP97b], we presented analytical and experimental studies of the algorithms discussed above for maintaining equi-depth histograms and for maintaining compressed histograms in the dynamic scenario.

6. Related work and further results

A concept related to synopsis data structures is that of condensed representations, presented by Mannila and Toivonen [MT96, Man97]: Given a class of structures D, a data collection $d \in D$, and a class of patterns P, a *condensed representation* for d and P is a data structure that makes it possible to answer queries of the form "How many times does $p \in P$ occur in d" approximately correctly and more efficiently than by looking at d itself. Related structures include the data cube [GCB+97], pruned or cached data structures considered in machine learning [Cat92, ML97], and ϵ-nets widely used in computational geometry [Mul94]. Mannila and Toivonen also proposed an approximation metric for their structures, denoted an ϵ-adequate representation.

Approximate data structures that provide fast approximate answers were proposed and studied by Matias *et al* [MVN93, MVY94, MSY96]. For example, a priority queue data structure supports the operations insert, findmin, and deletemin; their approximate priority queue supports these operations with smaller overheads while reporting an approximate min in response to findmin and deletemin operations. The data structures considered have linear space footprints, so are not synopsis data structures. However, they can be adapted to provide a synopsis approximate priority queue, where the footprint is determined by the approximation error.

There have been several papers discussing systems and techniques for providing approximate query answers without the benefit of precomputed/maintained synopsis data structures. Hellerstein *et al* [**HHW97**] proposed a framework for approximate answers of aggregation queries called *online aggregation*, in which the base data is scanned in a certain order at query time and the approximate answer for an aggregation query is updated as the scan proceeds. Bayardo and Miranker [**BM96**] devised techniques for "fast-first" query processing, whose goal is to quickly provide a few tuples of the query answer from the base data. The Oracle Rdb system [**AZ96**] also provides support for fast-first query processing, by running multiple query plans simultaneously. Vrbsky and Liu [**VL93**] (see also the references therein) described a query processor that provides approximate answers to queries in the form of subsets and supersets that converge to the exact answer. The query processor uses various class hierarchies to iteratively fetch blocks of the base data that are relevant to the answer, producing tuples certain to be in the answer while narrowing the possible classes that contain the answer. Since these approaches read from the base data at query time, they incur multiple I/O operations at query time.

A recent survey by Barbará *et al.* [**BDF$^+$97**] describes the state of the art in *data reduction* techniques, for reducing massive data sets down to a "big picture" and for providing quick approximate answers to queries. The data reduction techniques surveyed by the paper are singular value decomposition, wavelets, regression, log-linear models, histograms, clustering techniques, index trees, and sampling. Each technique is described briefly (see the references therein for further details on these techniques and related work) and then evaluated *qualitatively* on to its effectiveness and suitability for various data types and distributions, on how well it can be maintained under insertions and deletions to the data set, and on whether it supports answers that progressively improve the approximation with time.

The list of data structures work that could be considered synopsis data structures is extensive. We have described a few of these works in the paper; here we mention several others. Krishnan *et al* [**KVI96**] proposed and studied the use of a compact suffix tree-based structure for estimating the selectivity of an alphanumeric predicate with wildcards. Manber [**Man94**] considered the use of concise "signatures" to find similarities among files. Broder *et al* [**BCFM98**] studied the use of (approximate) min-wise independent families of permutations for signatures in a related context, namely, detecting and filtering near-duplicate documents. Our work on synopsis data structures also includes the use of multi-fractals and wavelets for synopsis data structures [**FMS96, MVW98**] and *join samples* for queries on the join of multiple sets [**GPA$^+$98**].

7. Conclusions

This paper considers synopsis data structures as an algorithmic framework relevant to massive data sets. For such data sets, the available memory is often substantially smaller than the size of the data. Since synopsis data structures are too small to maintain a full characterization of the base data sets, the responses they provide to queries will typically be approximate ones. The challenges are to determine (1) what synopsis of the data to keep in the limited space in order to maximize the accuracy and confidence of its approximate responses, and (2) how to efficiently compute the synopsis and maintain it in the presence of updates to the data set.

The context of synopsis data structures presents many algorithmic challenges. Problems that may have easy and efficient solutions using linear space data structures may be rather difficult to address when using limited-memory, synopsis data structures. We discussed three such problems: frequency moments, hot list queries, and histograms. Different classes of queries may require different synopsis data structures. While several classes of queries have been recently considered, there is a need to consider many more classes of queries in the context of synopsis data structures, and to analyze their effectiveness in providing accurate or approximate answers to queries. We hope that this paper will motivate others in the algorithms community to study these problems. Due to the increasing prevalence of massive data sets, improvements in this area will likely find immediate applications.

Acknowledgments. We thank Andy Witkowski and Ramesh Bhashyam for discussions on estimation problems in database systems such as the NCR Teradata DBS. We also thank those who collaborated on the results surveyed in this paper. The research described in Section 3 is joint work with Noga Alon and Mario Szegedy. The research described in Section 5 is joint work with Vishy Poosala. The Aqua project, discussed briefly in Section 2.4, is joint work with Swarup Acharya, Vishy Poosala, Sridhar Ramaswamy, and Torsten Suel with additional contributions by Yair Bartal and S. Muthukrishnan. Other collaborators on the synopsis data structures research mentioned briefly in Section 6 are Christos Faloutsos, Avi Silberschatz, Jeff Vitter, and Min Wang. Finally, we thank Torsten Suel for helpful comments on an earlier draft of this paper.

References

[AGMS97] N. Alon, P. B. Gibbons, Y. Matias, and M. Szegedy, *Dynamic probabilistic maintenance of self-join sizes in limited storage*, Manuscript, February 1997.

[AMS96] N. Alon, Y. Matias, and M. Szegedi, *The space complexity of approximating the frequency moments*, Proc. 28th ACM Symp. on the Theory of Computing, May 1996, Full version to appear in JCSS special issue for STOC'96, pp. 20–29.

[AS94] R. Agrawal and R. Srikant, *Fast algorithms for mining association rules in large databases*, Proc. 20th International Conf. on Very Large Data Bases, September 1994, pp. 487–499.

[AZ96] G. Antoshenkov and M. Ziauddin, *Query processing and optimization in Oracle Rdb*, VLDB Journal **5** (1996), no. 4, 229–237.

[BCFM98] A. Z. Broder, M. Charikar, A. M. Frieze, and M. Mitzenmacher, *Min-wise independent permutations*, Proc. 30th ACM Symp. on the Theory of Computing, May 1998, pp. 327–336.

[BDF+97] D. Barbará, W. DuMouchel, C. Faloutsos, P. J. Haas, J. M. Hellerstein, Y. Ioannidis, H. V. Jagadish, T. Johnson, R. Ng, V. Poosala, K. A. Ross, and K. C. Sevcik, *The New Jersey data reduction report*, Bulletin of the Technical Committee on Data Engineering **20** (1997), no. 4, 3–45.

[BFS86] L. Babai, P. Frankl, and J. Simon, *Complexity classes in communication complexity theory*, Proc. 27th IEEE Symp. on Foundations of Computer Science, October 1986, pp. 337–347.

[BM96]   R. J. Bayardo, Jr. and D. P. Miranker, *Processing queries for first-few answers*, Proc. 5th International Conf. on Information and Knowledge Management, November 1996, pp. 45–52.

[BMUT97]   S. Brin, R. Motwani, J. D. Ullman, and S. Tsur, *Dynamic itemset counting and implication rules for market basket data*, Proc. ACM SIGMOD International Conf. on Management of Data, May 1997, pp. 255–264.

[Cat92]   J. Catlett, *Peepholing: Choosing attributes efficiently for megainduction*, Machine Learning: Proc. 9th International Workshop (ML92), July 1992, pp. 49–54.

[CMN98]   S. Chaudhuri, R. Motwani, and V. Narasayya, *Random sampling for histogram construction: How much is enough?*, Proc. ACM SIGMOD International Conf. on Management of Data, June 1998, pp. 436–447.

[DNSS92]   D. J. DeWitt, J. F. Naughton, D. A. Schneider, and S. Seshadri, *Practical skew handling in parallel joins*, Proc. 18th International Conf. on Very Large Data Bases, August 1992, pp. 27–40.

[FJS97]   C. Faloutsos, H. V. Jagadish, and N. D. Sidiropoulos, *Recovering information from summary data*, Proc. 23rd International Conf. on Very Large Data Bases, August 1997, pp. 36–45.

[FM83]   P. Flajolet and G. N. Martin, *Probabilistic counting*, Proc. 24th IEEE Symp. on Foundations of Computer Science, November 1983, pp. 76–82.

[FM85]   P. Flajolet and G. N. Martin, *Probabilistic counting algorithms for data base applications*, J. Computer and System Sciences **31** (1985), 182–209.

[FMS96]   C. Faloutsos, Y. Matias, and A. Silberschatz, *Modeling skewed distribution using multifractals and the '80-20' law*, Proc. 22rd International Conf. on Very Large Data Bases, September 1996, pp. 307–317.

[FSGM⁺98]   M. Fang, N. Shivakumar, H. Garcia-Molina, R. Motwani, and J. D. Ullman, *Computing iceberg queries efficiently*, Proc. 24th International Conf. on Very Large Data Bases, August 1998, pp. 299–310.

[GCB⁺97]   J. Gray, S. Chaudhuri, A. Bosworth, A. Layman, D. Reichart, M. Venkatrao, F. Pellow, and H. Pirahesh, *Data cube: A relational aggregation operator generalizing group-by, cross-tabs, and sub-totals*, Data Mining and Knowledge Discovery **1** (1997), no. 1, 29–53.

[GGMS96]   S. Ganguly, P. B. Gibbons, Y. Matias, and A. Silberschatz, *Bifocal sampling for skew-resistant join size estimation*, Proc. 1996 ACM SIGMOD International Conf. on Management of Data, June 1996, pp. 271–281.

[GM98]   P. B. Gibbons and Y. Matias, *New sampling-based summary statistics for improving approximate query answers*, Proc. ACM SIGMOD International Conf. on Management of Data, June 1998, pp. 331–342.

[GMP97a]   P. B. Gibbons, Y. Matias, and V. Poosala, *Aqua project white paper*, Tech. report, Bell Laboratories, Murray Hill, New Jersey, December 1997.

[GMP97b]   P. B. Gibbons, Y. Matias, and V. Poosala, *Fast incremental maintenance of approximate histograms*, Proc. 23rd International Conf. on Very Large Data Bases, August 1997, pp. 466–475.

[Goo89]   I. J. Good, *Surprise indexes and p-values*, J. Statistical Computation and Simulation **32** (1989), 90–92.

[GPA⁺98]   P. B. Gibbons, V. Poosala, S. Acharya, Y. Bartal, Y. Matias, S. Muthukrishnan, S. Ramaswamy, and T. Suel, *AQUA: System and techniques for approximate query answering*, Tech. report, Bell Laboratories, Murray Hill, New Jersey, February 1998.

[HHW97]   J. M. Hellerstein, P. J. Haas, and H. J. Wang, *Online aggregation*, Proc. ACM SIGMOD International Conf. on Management of Data, May 1997, pp. 171–182.

[HNSS95]   P. J. Haas, J. F. Naughton, S. Seshadri, and L. Stokes, *Sampling-based estimation of the number of distinct values of an attribute*, Proc. 21st International Conf. on Very Large Data Bases, September 1995, pp. 311–322.

[IC93]   Y. E. Ioannidis and S. Christodoulakis, *Optimal histograms for limiting worst-case error propagation in the size of join results*, ACM Transactions on Database Systems **18** (1993), no. 4, 709–748.

[Ioa93]   Y. E. Ioannidis, *Universality of serial histograms*, Proc. 19th International Conf. on Very Large Data Bases, August 1993, pp. 256–267.

[IP95]   Y. E. Ioannidis and V. Poosala, *Balancing histogram optimality and practicality for query result size estimation*, Proc. ACM SIGMOD International Conf. on Management of Data, May 1995, pp. 233–244.

[JKM+98]   H. V. Jagadish, N. Koudas, S. Muthukrishnan, V. Poosala, K. Sevcik, and T. Suel, *Optimal histograms with quality guarantees*, Proc. 24th International Conf. on Very Large Data Bases, August 1998, pp. 275–286.

[KS87]   B. Kalyanasundaram and G. Schnitger, *The probabilistic communication complexity of set intersection*, Proc. 2nd Structure in Complexity Theory Conf., June 1987, pp. 41–49.

[KVI96]   P. Krishnan, J. S. Vitter, and B. Iyer, *Estimating alphanumeric selectivity in the presence of wildcards*, Proc. ACM SIGMOD International Conf. on Management of Data, June 1996, pp. 282–293.

[Man94]   U. Manber, *Finding similar files in a large file system*, Proc. Usenix Winter 1994 Technical Conf., January 1994, pp. 1–10.

[Man97]   H. Mannila, *Inductive databases and condensed representations for data mining*, Proc. International Logic Programming Symposium, 1997, pp. 21–30.

[ML97]   A. Moore and M. S. Lee, *Cached sufficient statistics for efficient machine learning with large datasets*, Tech. Report CMU-RI-TR-97-27, Robotics Institute, Carnegie-Mellon University, 1997, To appear in *J. Artificial Intelligence Research*.

[Mor78]   R. Morris, *Counting large numbers of events in small registers*, Communications of the ACM **21** (1978), 840–842.

[MP80]   J. I. Munro and M. S. Paterson, *Selection and sorting with limited storage*, Theoretical Computer Science **12** (1980), no. 3, 315–323.

[MRL98]   G. S. Manku, S. Rajagopalan, and B. G. Lindsley, *Approximate medians and other quantiles in one pass and with limited memory*, Proc. ACM SIGMOD International Conf. on Management of Data, June 1998, pp. 426–435.

[MSY96]   Y. Matias, S. C. Sahinalp, and N. E. Young, *Performance evaluation of approximate priority queues*, Presented at *DIMACS Fifth Implementation Challenge: Priority Queues, Dictionaries, and Point Sets*, organized by D. S. Johnson and C. McGeoch, October 1996.

[MT96]   H. Mannila and H. Toivonen, *Multiple uses of frequent sets and condensed representations*, Proc. 2nd International Conf. on Knowledge Discovery and Data Mining, August 1996, pp. 189–194.

[Mul94]   K. Mulmuley, *Computational geometry: An introduction through randomized algorithms*, Prentice Hall, Englewood Cliffs, NJ, 1994.

[MVN93]   Y. Matias, J. S. Vitter, and W.-C. Ni, *Dynamic generation of discrete random variates*, Proc. 4th ACM-SIAM Symp. on Discrete Algorithms, January 1993, pp. 361–370.

[MVW98]   Y. Matias, J. S. Vitter, and M. Wang, *Wavelet-based histograms for selectivity estimation*, Proc. ACM SIGMOD International Conf. on Management of Data, June 1998, pp. 448–459.

[MVY94]   Y. Matias, J. S. Vitter, and N. E. Young, *Approximate data structures with applications*, Proc. 5th ACM-SIAM Symp. on Discrete Algorithms, January 1994, pp. 187–194.

[Olk93]   F. Olken, *Random sampling from databases*, Ph.D. thesis, Computer Science, U.C. Berkeley, April 1993.

[PIHS96]   V. Poosala, Y. E. Ioannidis, P. J. Haas, and E. J. Shekita, *Improved histograms for selectivity estimation of range predicates*, Proc. ACM SIGMOD International Conf. on Management of Data, June 1996, pp. 294–305.

[Pre97]   D. Pregibon, *Mega-monitoring: Developing and using telecommunications signatures*, October 1997, Invited talk at the *DIMACS Workshop on Massive Data Sets in Telecommunications*.

[Raz92]   A. A. Razborov, *On the distributional complexity of disjointness*, Theoretical Computer Science **106** (1992), no. 2, 385–390.

[SKS97]   A. Silberschatz, H. F. Korth, and S. Sudarshan, *Database system concepts*, third ed., McGraw-Hill, New York, 1997.

[TPC]   *TPC-Committee. Transaction processing council (TPC)*, http://www.tpc.org.

[Vit85]   J. S. Vitter, *Random sampling with a reservoir*, ACM Transactions on Mathematical Software **11** (1985), no. 1, 37–57.

[Vit98]   J. S. Vitter, *External memory algorithms*, Proc. 17th ACM Symp. on Principles of Database Systems, June 1998, pp. 119–128.

[VL93]   S. V. Vrbsky and J. W. S. Liu, *Approximate—a query processor that produces monotonically improving approximate answers*, IEEE Trans. on Knowledge and Data Engineering **5** (1993), no. 6, 1056–1068.

[VS94]   J. S. Vitter and E. A. M. Shriver, *Algorithms for parallel memory I: Two-level memories*, Algorithmica **12** (1994), no. 2–3, 110–147.

[Yao83]   A. C. Yao, *Lower bounds by probabilistic arguments*, Proc. 24th IEEE Symp. on Foundations of Computer Science, November 1983, pp. 420–428.

INFORMATION SCIENCES RESEARCH CENTER, BELL LABORATORIES, ROOM 2A-341, LUCENT TECHNOLOGIES, 600 MOUNTAIN AVENUE, MURRAY HILL, NEW JERSEY 07974
E-mail address: gibbons@research.bell-labs.com
URL: http://www.bell-labs.com/~pbgibbons/

DEPARTMENT OF COMPUTER SCIENCE, TEL-AVIV UNIVERSITY, TEL-AVIV 69978 ISRAEL, AND INFORMATION SCIENCES RESEARCH CENTER, BELL LABORATORIES
E-mail address: matias@math.tau.ac.il
URL: http://www.math.tau.ac.il/~matias/

DIMACS Series in Discrete Mathematics
and Theoretical Computer Science
Volume **50**, 1999

Calculating Robust Depth Measures for Large Data sets

Ibraheem Al-Furaih, Theodore Johnson, and Sanjay Ranka

ABSTRACT. Data analysis often requires finding the median of the data set. However, multidimensional median finding is difficult, and proposed algorithms for multidimensional median finding are expensive to compute and do not scale well. We present approximate algorithms for computing two measures of a median on a two-dimensional data set, the regression depth, and the half-space depth. Our algorithms require a few passes over the entire data set and provide tight error bounds.

1. Introduction

Statistical measures such as mean and median are widely used for data analysis. The mean of a data set is easy to compute, but it is not a robust measure. An outlier can affect the mean by an arbitrarily large amount. The median is relatively more robust. the mean and the median can be calculated in $O(n)$ time for univariate datasets [**HSM95**].

The mean can be extended trivially for multiple dimensional data. Generally, a mean along all the dimensions is calculated separately. The extension of the median for multiple dimension is not simple, because the rank (depth measure) of an element is hard to define in multiple dimensions. Several definitions of rank based on depth measures have been proposed. The following definition and notation have been adapted from the description in [**RH96**]. Two of the prominent measures used are as follows:

1. Regression Depth: The regression depth [**RH96**] gives the rank of a line with respect to a given data set. The input is a set of observations $Z_n = \{(x_i, y_i) : 1 \le i \le n\}$ in a weak general position ($x_i \ne x_j \ \forall \ i \ne j$). Let $\theta = (\theta_1, \theta_2)$ represent a line $y = \theta_1 \times x + \theta_2$, i.e. θ_1 is the slope and θ_2 is the intercept. θ is called a non-fit if there exists a tilting point v on θ around

---

1991 *Mathematics Subject Classification*. Primary 54C40, 14E20; Secondary 46E25, 20C20.

Ibraheem Al-Furaih is supported by a scholarship from King AbdulAziz City for Science and Technology (KACST), Riyadh, Saudi Arabia. This work was done while visiting University of Florida.

The work of this author was supported in part by AFMC and ARPA under F19628-94-C-0057 and WM-82738-K-19 (subcontract from Syracuse University) and in part by ARO under DAAG 55-97-1-0368 and Q000302 (subcontract from NMSU). The content of the information does not necessarily reflect the position or the policy of the Government and no official endorsement should be inferred.

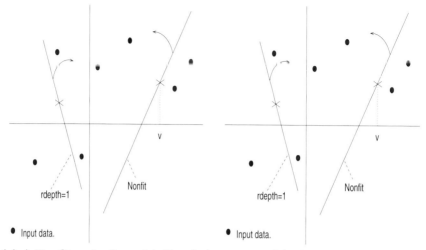

(a) A Nonfit and a line with Depth 1   (b) The r Function

FIGURE 1. Regression Depth

which θ can be rotated until it is vertical without passing any points in Z_n (Figure 1(a)). Formally, θ is a non-fit iff there exists a real number v and $v \neq x_i \ \forall \ 1 \leq i \leq n$ such that:

$$r_i(\theta) < 0 \ \forall \ x_i < v \ \ and \ \ r_i(\theta) > 0 \ \forall \ x_i > v$$

or

$$r_i(\theta) > 0 \ \forall \ x_i < v \ \ and \ \ r_i(\theta) < 0 \ \forall \ x_i > v$$

where $r_i(\theta) = r_i = y_i - \theta_1 \times x_i - \theta_2$; r is positive for points above θ, negative for points under θ, and zero for points on θ (Figure 1(b)).

The regression depth of $\theta = (\theta_1, \theta_2)$ with respect to Z_n, $rdepth(\theta, Z_n)$, is defined as the minimum number of observations that need to be removed to make θ a non-fit.

2. Location Depth: The location depth of a point can be defined in terms of its *half-space depth* or *simplicial depth*. These two error measures are related (cf. Section 4). Given a point $\theta = (x_\theta, y_\theta) \in \mathcal{R}^2$ and a set of points $Z_n = \{(x_i, y_i) : 1 \leq i \leq n, (x_i, y_i) \in \mathcal{R}^2\}$, the half-space depth of θ in Z_n is defined as the smallest number of elements of Z_n that are contained in a closed half space which its boundary passes through θ. The simplicial depth of θ relative to Z_n is defined as the number of triangles $\triangle((x_i, y_i), (x_j, y_j), (x_k, y_k))$; $(x_i, y_i), (x_j, y_j), (x_k, y_k) \in Z_n$ that contain θ, where $\triangle((x_i, y_i), (x_j, y_j), (x_k, y_k))$ is the triangle formed by the three points. The simplicial depth provides a measure of how deep θ is inside the data cloud.

As stated before, multidimensional median finding is a difficult problem. Proposed algorithms for multidimensional median finding based on regression depth [**RH96**] and location depth [**RR96**] are expensive to compute and do not scale well. We present approximate algorithms for computing two measures of a median on a two-dimensional data set, the regression depth, and the half-space depth. Our algorithms require a few passes over the entire data set and provide tight error bounds.

The rest of the paper is organized as follows. In Section 2, we describe the important primitives used in this paper. These primitive operations are used to develop the algorithms for regression and location depth in Sections 3 and 4, respectively.

2. Primitives

External Sorting. Many algorithms have been developed to solve the external sorting problem. The merge-based algorithms[**Zhe96**, **ADADC$^+$97**] generally consist of two phases, a run formation phase and a merging phase. Each phase requires reading and writing the entire dataset. In the run formation phase, a run is read from the input file, sorted in the main memory, and written to a temporary file(s). The run size is usually a function of the main memory size. In the merging phase we merge these sorted runs. Threads and asynchronous reads are generally used to overlap IO and computation [**ADADC$^+$97**, **Aga96**].

Estimating Quantiles. The φ-quantile of an ordered sequence of data values is the element with rank $\varphi \times n$, where n is the total number of values. We are interested in deriving an estimate for the φ-quantile ($\varphi = \frac{1}{q}, \frac{2}{q}, \ldots, \frac{q-1}{q}$) for large data sets.

The following notation is used to describe the OPAQ algorithm [**ARS97**] that is used to find quantiles: Let M be the size of the main memory and n the total number of elements. Let α give the index (rank) of the quantile ($= \varphi \times n$), where φ - quantile fraction ($\varphi \in [0 \ldots 1]$) and let e_α be the value of the element corresponding to that quantile. The OPAQ algorithm consists of two phases, the sampling phase and the quantile-finding phase.

In the sampling phase, we input the whole data set as r runs. For simplicity, we assume that all runs are of equal size m ($= \frac{n}{r}$). A set of sample points $S = [s_1, \ldots, s_s]$ of size s is determined where $s_i <= s_{i+1}$, for $i < s$, for each run. The r sample lists are merged together forming a sorted sample list of size rs. The sample points are selected using the regular sampling [**LLS$^+$93**]; a sample of size s consists of the elements at relative indices $\frac{m}{s}, \ldots, s\frac{m}{s}$. [1] Each sample points thus corresponds to $\frac{m}{s}$ points less than or equal to the sample point and greater than or equal to the previous sample point. We will use the term sub-run of the sample point to denote these elements.

The sorted sample list is used in the quantile finding phase to estimate the upper and lower bounds of the true value of the φ-quantile. As a result of using a regular sampling method in deriving the sample points, it can be easily shown that the sample points have the following properties:

1. There are at least $i\frac{m}{s}$ elements less than or equal to the sample point s_i.
2. Additionally, there are at most $r - 1$ sub-runs, each with at most $\frac{m}{s} - 1$ elements less than s_i.

Thus the maximum number of elements less than s_i is given by $i\frac{m}{s} + (r - 1)(\frac{m}{s} - 1)$. These properties are used in determining $e_\alpha{}^l$ and $e_\alpha{}^u$, the lower and upper bounds on the actual value of the quantile.

---

[1] Without lose of generality, we assume that n is divisible by r and m is divisible by s.

Let *List* be the list of sorted samples. Let

$$e_\alpha{}^l = List[\lfloor \frac{s}{m}\alpha - (r-1)(1 - \frac{s}{m}) \rfloor]]$$

$$e_\alpha{}^u = List[\lceil \alpha \frac{s}{m} \rceil]]$$

It can be shown that the maximum number of elements between the true quantile and the lower bound, $e_\alpha{}^l$, is $\frac{n}{s}$. The maximum number of elements between $e_\alpha{}^u$ and the true quantile is $\frac{n}{s}$.

The total time requirement to estimate q quantiles is $O(n + rm \log s + rs \log r + q)$. This simplifies to $O(n + n \log s + \frac{n}{m}s \log \frac{n}{m} + q)$, since $r = \frac{n}{m}$. If $\frac{m \log s}{s} \geq \log \frac{n}{m}$, the total complexity of the algorithm is $O(n \log s)$. The size of the main memory M, the size of the sample s, the number of runs r, and the number of elements n are constrained by the following relation:

$$rs + \frac{n}{r} \leq M.$$

Since $s \geq 2q$ for achieving good bounds on the quantiles, this limits the maximum number of quantiles one can find using this algorithm to $O(\frac{M^2}{n})$.

3. Regression Depth

Rousseeuw and Hubert [**RH96**] presented an $O(n \lg n)$ algorithm for finding the regression depth; the storage requirements for this algorithm are $O(n)$. It first sorts Z_n on the x values. The $rdepth(Z_n, \theta)$ can then be calculated using the following:

$$rdepth(Z_n, \theta) = min(min\{L^+(x_i) + R^-(x_i), L^-(x_i) + R^+(x_i)\})$$

where:

- $L^+(t) = \#j : x_j \leq t \text{ and } r_j \geq 0$, i.e., number of positive r's on left of t.
- $L^-(t) = \#j : x_j \leq t \text{ and } r_j \leq 0$, i.e., number of negative r's on left of t.
- $R^+(t) = \#j : x_j > t \text{ and } r_j \geq 0$, i.e., number of positive r's on right of t.
- $R^-(t) = \#j : x_j > t \text{ and } r_j \leq 0$, i.e., number of negative r's on right of t.

Once the data is sorted along the x-coordinate, the L and the R values can be computed by a left-to-right scan of the entire data. A high-level description of the algorithm is given in Figure 2.

The following additional optimizations will improve the execution time. Since we are actually only interested in the sign of the r values, a boolean array could be used for storing the r values. Further, instead of actually computing the r values, we can check the following condition:

$$r_i \geq 0$$
$$\Rightarrow \quad y_i - \theta_1 \times x_i - \theta_2 \geq 0$$
(3.1)
$$\Rightarrow \quad y_i \geq \theta_1 \times x_i + \theta_2$$

The method described above requires the use of an external sorting algorithm. This may require several I/O phases. An additional I/O pass is required for calculating the L and R values; the computational requirements are proportional to $O(n)$.

In the following we describe two algorithms that estimate the regression depth with only a few passes over the entire data set. The amount of computation, as well as I/O requirements, are considerably lower. Further, we can provide tight upper and lower bounds.

ALGORITHM 1. Sorting-Based algorithm

n — Number of points
Z — Array with two fields, x and y to store the data
r — Array to store the r values
R^+ — Array to store how many positive r's on right
R^- — Array to store how many negative r's on right
Step 1. Read the data from a file into Z
Step 2. Sort Z on x values
Step 3. *for $(i = 1; i \leq n; i++)$ $r[i] = y_i - \theta_1 \times x_i - \theta_2$*
Step 4. $R^+[n] = 0$, $R^-[n] = 0$
Step 5. *for $(i = n - 1; i \geq 1; i--)$*
        Step 5.1 $R^+[i] = R^+[i+1]$, $R^-[i] = R^-[i+1]$
        Step 5.2 if $r[i+1] \geq 0$ then $R^+[i]++$
        Step 5.3 if $r[i+1] \leq 0$ then $R^-[i]++$
Step 6. $L^+ = 0$, $L^- = 0$, *rdepth* $= \infty$
Step 7. *for $(i = 1; i \leq n; i++)$*
        Step 7.1 *rdepth* $= min(rdepth, min(L^+ + R^-[i], L^- + R^+[i]))$
        Step 7.2 if $r[i] \geq 0$ then $L^+ ++$
        Step 7.3 if $r[i] \leq 0$ then $L^- ++$
Step 8. return(rdepth)

FIGURE 2. Sort-Based Algorithm for the Regression Depth

3.1. Two-Pass Algorithm. This algorithm finds the values of L^+, L^-, R^+, and R^- at a small subset of points (estimated quantiles). These values are stored in the main memory and are used to estimate the regression depth. In the first phase we estimate the quantiles (cf. Section 2) of the data points projected to the x axis. These divide the x data range into nearly equal-sized intervals. In the second phase, we initialize two counters for every interval, one for the negative r values, and one for the positive r's. For every point we find the interval that the data point lies in, using a binary search. We then calculate the r function for this point and increment either the negative or positive counter for that interval (or both, if $r = 0$).

We then perform left-to-right and right-to-left prefix sums on these counter arrays and find the values of L^+, L^-, R^+, and R^- at the estimated quantile points. This can then be used to calculate:

$$rdepth(Z_n, \theta) = min(min\{Q[i].L^+ + Q[i].R^-, Q[i].L^- + Q[i].R^+\})$$

A high-level description of this algorithm is presented in Figure 3. Assuming that q quantiles are derived in the first phase, the computational time requirements for the first phase is given by $O(n \log q)$ (Section 2). The second phase also requires $O(n \lg q)$ time. A total of two I/O passes are required on the entire dataset.

The error in the estimation is bounded by the maximum interval size, which is equal to $O(\frac{n}{q})$. The worst case occurs when an interval adjacent to the quantile that satisfies the minimum condition has all but the first element of the same r sign.

3.2. One-Pass Algorithm. This algorithm combines the two passes described to reduce the I/O requirements. The algorithm extends the OPAQ algorithm for finding quantiles (cf. Section 2). The dataset is divided into runs with each run smaller than the memory size. These runs are processed sequentially, one at a time.

ALGORITHM 2. Two-Pass algorithm
  n — Number of points
  m — Size of every run
  $runs$ — Number of runs = $\lceil \frac{n}{m} \rceil$
  s — Number of sample points from every run
  q — Number of quantiles wanted = $s \times runs$
  Q — Array of size q to store the quantiles, has four counters corresponding to L^+, L^-, R^+ and R^-
  $Buffer$ — Array of size m with two fields, x and y to store one run
  Step 1. $runs = \lceil \frac{n}{m} \rceil$
  Step 2. $q = s \times runs$
  Step 3. for $(i = 1; i \leq runs; i++)$
          Step 3.1. read m points from the input file into $Buffer$
          Step 3.2. Find s sample points, $Buffer$ is Sorted into $s + 1$ buckets
          Step 3.3. add the s quantiles to Q
  Step 4. Sort Q on x values and set all counters to zero
  Step 5. for $(i = 1; i \leq runs; i++)$
          Step 5.1. read m points from the input file into $Buffer$
          Step 5.2. for $(k = 1; k \leq m; k++)$
                  Step 5.2.1. Find which bucket b contains $Buffer[k]$ using binary search
                  Step 5.2.2. r = the r value for $Buffer[k]$
                  Step 5.2.3. if $r \geq 0$ $Q[b].L^+ ++$
                  Step 5.2.4. if $r \leq 0$ $Q[b].L^- ++$
  Step 5. Perform a right to left prefix sum to find R^+ and R^- fields of Q
  Step 6. Perform a left to right prefix sum for L^+ and L^- fields of Q
  Step 7. $rdepth = \infty$
  Step 8. for $(i = 1; i \leq q; i++)$
          Step 8.1 $rdepth = min(rdepth, min(Q[i].L^+ + Q[i].R^-, Q[i].L^- + Q[i].R^+))$
  Step 9. return(rdepth)

FIGURE 3. The Two-Pass Algorithm for the Regression Depth

Each run is read from the input file and partially sorted into $s + 1$ intervals of equal size such that every element in interval i is less than any element in bucket $i + 1$. We calculate the depth of every point at the s splitters, based only on the points in the run. This is done by calculating the r function for all elements in the current run. For each sample point (splitter), we maintain four counters corresponding to L^+, L^-, R^+, and R^-, respectively.

The sample points from all the runs are accumulated and sorted based on their x values. This is followed by a left-to-right prefix sum on the L^+ and L^- fields, and a right-to-left prefix sum on the R^+ and R^- fields. Let Q be the list of sorted sample points. The $rdepth$ can now be computed as follows:

$$rdepth(Z_n, \theta) = min(min\{Q[i].L^+ + Q[i].R^-, Q[i].L^- + Q[i].R^+\})$$

A high-level description of the algorithm is presented in Figure 4. Let n be the number of points; M be the memory size (number of points that can fit in memory); m be the size of every run; $runs$ be the number of runs $= \lceil \frac{n}{m} \rceil$. This algorithm requires $O(n \log s)$ time and $O(m)$ storage.

The error introduced in the two-pass algorithm was due to the fact that the depth was calculated at only a few points. For each of these points the depth calculation was accurate, because the r values are calculated accurately. This error is bounded by $O(\frac{n}{s})$ (cf. Section 2). In the one-pass algorithm there are additional

ALGORITHM 3. One-Pass Algorithm

n — Number of points

m — Size of every run

$runs$ — Number of runs $= \lceil \frac{n}{m} \rceil$

s — Number of sample points from every run

q — Number of quantiles wanted $= s \times runs$

Q — Array of size q to store the quantiles, it also has four counters corresponding to L^+, L^-, R^+, and R^-

$Buffer$ — Array of size m with two fields, x and y to store one run

r — Array to store the r values

Step 1. $runs = \lceil \frac{n}{m} \rceil$

Step 2. $q = s \times runs$

Step 3. $for\ (i = 1; i \leq runs; i + +)$

    Step 3.1. read m points from the input file into $Buffer$

    Step 3.2. Find s quantiles, $Buffer$ is Sorted into $s + 1$ buckets

    Step 3.3. $for\ (j = 1; j \leq m; j + +)\ r[j] = Buffer[j].y - \theta_1 \times Buffer[j].x - \theta_2$

    Step 3.9. add the s quantiles to Q and set their L^+, L^-, R^+ and R^- fields as:

        L^+ = Number of positive r's in left bucket

        L^- = Number of Negative r's in left bucket

        R^+ = Number of positive r's in right bucket

        R^- = Number of Negative r's in right bucket

Step 4. Sort Q on x values

Step 5. Do a left to right prefix sum for L^+ and L^- fields of Q

Step 6. Do a right to left prefix sum for R^+ and R^- fields of Q

Step 7. $rdepth = \infty$

Step 8. $for\ (i = 1; i \leq q; i + +)$

    Step 8.1 $rdepth = min(rdepth, min(Q[i].L^+ + Q[i].R^-, Q[i].L^- + Q[i].R^+))$

Step 9. return(rdepth)

FIGURE 4. The One-Pass Algorithm for the Regression Depth

errors due to the fact that the r values are only estimated. The estimates in the r values can also be shown to be bounded by $O(\frac{n}{s})$ (cf. Section 3.2.1). The combined error is limited by the sum of these two errors and is of the order of $O(\frac{n}{s})$.

3.2.1. *Error Analysis for the One-Pass Algorithm.* When we select the sample points, sort them, and then perform the two prefix sums (Steps 1 through 6), we know that there are at least $Q[i].L^-$ elements with negative or zero r values on the left of quantile i. The same is true for $Q[i].L^+$, $Q[i].R^-$ and $Q[i].R^+$, respectively.

Formally, let $Q[i].l^-$, $Q[i].l^+$, $Q[i].r^-$ and $Q[i].r^+$ be the exact values we are trying to estimate for $Q[i].L^-$, $Q[i].L^+$, $Q[i].R^-$ and $Q[i].R^+$, respectively. We have the following bounds on our estimation:

$$
\begin{aligned}
Q[i].L^- &\leq Q[i].l^- \\
Q[i].L^+ &\leq Q[i].l^+ \\
Q[i].R^- &\leq Q[i].r^- \\
Q[i].R^+ &\leq Q[i].r^+
\end{aligned}
$$

(3.2)

Moreover, $Q[i].L^-$ can not be greater than $Q[i].l^-$ by more than $(runs - 1) \times \frac{n}{q}$, because in the worst case all elements from the other runs are less than $Q[i].x$.

Therefore:

$$Q[i].l^- \quad \leq Q[i].L^- + (runs - 1) \times \frac{n}{q}$$

$$\Rightarrow \quad Q[i].l \quad \leq Q[i].L \quad + \frac{n^2}{mq}$$

(3.3)
$$\Rightarrow \quad Q[i].l^- \quad \leq Q[i].L^- + \frac{n}{s}$$

The same holds true for $Q[i].L^+$, $Q[i].R^-$ and $Q[i].R^+$ and we get:

$$Q[i].l^- \quad \leq \quad Q[i].L^- + \frac{n}{s}$$

$$Q[i].l^+ \quad \leq \quad Q[i].L^+ + \frac{n}{s}$$

$$Q[i].r^- \quad \leq \quad Q[i].R^- + \frac{n}{s}$$

(3.4)
$$Q[i].r^+ \quad \leq \quad Q[i].R^+ + \frac{n}{s}$$

By combining the above two relations we get:

$$Q[i].L^- \quad \leq Q[i].l^- \leq \quad Q[i].L^- + \frac{n}{s}$$

$$Q[i].L^+ \quad \leq Q[i].l^+ \leq \quad Q[i].L^+ + \frac{n}{s}$$

$$Q[i].R^- \quad \leq Q[i].r^- \leq \quad Q[i].R^- + \frac{n}{s}$$

(3.5)
$$Q[i].R^+ \quad \leq Q[i].r^+ \leq \quad Q[i].R^+ + \frac{n}{s}$$

So, the error for each of $Q[i].L^-$, $Q[i].L^+$, $Q[i].R^-$, and $Q[i].R^+$ is bounded by $\frac{n}{s}$. When we find the minimum in the equation,

$$rdepth(Z_n, \theta) = min(min\{Q[i].L^+ + Q[i].R^-, Q[i].L^- + Q[i].R^+\}),$$

the value of $error1$ is bounded by $2\frac{n}{s}$.

3.3. Experimental Results. We implemented the above algorithms for calculating the regression depth. The platform used for our implementation was a Sun *Ultra* Enterprise Server. The time requirements of our algorithms are relatively independent of the data distribution, hence we used a synthetic data set. This data set was derived by generating x and y coordinates randomly, with a uniform distribution, along each of axes. To simulate the effects of large data sets (with a limited amount of disk space available to us in our current system), we limited the memory usage available to us. In the results presented in this section we assume that the memory is limited to $64K$ data points.

Figure 5(a) plots the total time taken by different algorithms. The total number of points, n, was varied from 128k to 8M. For the one-pass and the two-pass algorithms we chose 256 equidistant samples from every run (run size was set to $64K$ points). These results show that the time taken by the one-pass algorithm is much smaller than the time taken by the other two algorithms. For the sort-based algorithm, out-of-core sorting required around 85% of the total time. For the two-pass algorithm, the second pass required around 70% of the total time; the time within the second pass was dominated by the binary search time.

The error in estimation of the regression depth was calculated for several sample points. A representative plot of the fractional error $\left(\frac{|Actual - Estimated|}{n} \right)$ is given

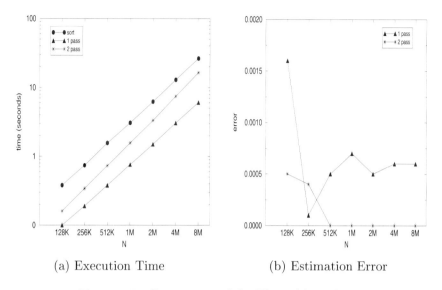

(a) Execution Time   (b) Estimation Error

FIGURE 5. Comparison of the Three Algorithms

in Figure 5(b). The actual errors derived are much smaller than the bound. Further investigation is required to study the behavior of this error for a variety of synthetic and real data sets. However, it is worth noting that the worst case error is bounded.

4. Calculating Location Depth

Rousseeuw and Ruts [**RR96**] presented an $O(n \lg n)$ algorithm for finding both the half-space depth and the simplicial depth. We use the definition and notation provided in [**RR96**] to describe this algorithm. Given three different points $(x_i, y_i), (x_j, y_j), (x_k, y_k) \in \mathcal{R}^2$, let $\triangle((x_i, y_i), (x_j, y_j), (x_k, y_k))$ be the triangle formed by the three points. Let the function I be defined as: $I(\theta) = 1$ when $\triangle((x_i, y_i), (x_j, y_j), (x_k, y_k))$ contains θ, and 0 otherwise. These depth functions increase as θ becomes centrally located in Z_n. A point with the highest depth represents a median for Z_n.

The algorithm for calculating the above depth measures has the following steps:

Step1: For each point $(x_i, y_i) \in Z_n$, compute the angle α_i formed by (x_i, y_i), θ, and (x_i, y_θ) (Figure 6(a)). The α_i's are stored in an array α of size n. This requires $O(n)$ time.

Step2: The array α is sorted. A check can be performed to determine if θ is outside Z_n by finding the largest difference between any two consecutive angles. If this difference is larger than π, then θ is outside the data cloud and its depth is 0 (unless some points in Z_n are colocated with θ). This step requires $O(n \lg n)$ time.

Step3: Rotate all angles in α around θ by α_0. The new value of $\alpha_0 = 0$. This requires $O(n)$ time.

Step4: Define h_i as the largest integer such that: $\alpha_i \leq \alpha_{i+1} \leq .... \leq \alpha_{i+h_i} < \alpha_i + \pi$; h_i gives the number of angles in the half circle starting at α_i, not including $\alpha_i + \pi$ (this corresponds to the number of angles in the shaded area in Figure 6(b)). It can easily be seen that the triangle formed by any three points in the shaded area will not include θ, thus, the number of triangles

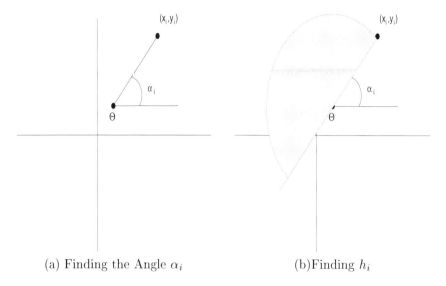

(a) Finding the Angle α_i (b)Finding h_i

FIGURE 6. Location Depth

that contain θ is given by $\sum_{1 \leq i < j < k \leq n} I(\theta \in \triangle((x_i,y_i),(x_j,y_j),(x_k,y_k))) = \binom{n}{3} - \sum_1^n \binom{h_i}{2}$. The simplicial depth is then defined as:

$$depth(\theta, Z_n) = 1 - \binom{n}{3}^{-1} \sum_1^n \binom{h_i}{2}$$

The half-space depth of θ at Z_n is given by

$$hsdepth = \frac{1}{n} \min_i \{\min(h_i + 1, n - (h_i + 1))\}$$

The naive way to compute all the h_i's requires $O(n^2)$ time. Rousseeuw and Ruts presented a way to find the h_i's in $O(n)$ time, but their method is complex. The following describes a simpler approach. Computing the h_i's can be done in one sweep of all the angles by keeping a rear and front pointer. For h_1, we move the front pointer until the half circle is completed. After this step, for computing h_2, the front pointer moves only from where it was left when computing h_1. A similar argument can be made for computing the whole sequence of h_i. It can be shown that the front pointer does not need to complete more than two circular traversals, thus a simple amortized analysis shows that the time requirements are proportional to $O(n)$.

A high-level description of this algorithm is given in Figure 7.

This algorithm for finding the simplicial depth and the half-space depth requires $O(n \lg n)$ time and $O(n)$ time. It assumes that all the data fits into the main memory. In the following sections we develop efficient algorithms for estimating the simplicial depth and the half space depth for out-of-core data.

4.1. Two-Pass Algorithm. In the first pass we find the quantiles for the α values. This divides the range 0 to 2π into q sectors such that each sector contains approximately $\frac{n}{q}$ of the α values. To select the quantiles, we use the one-pass quantiling algorithm (cf. Section 2). The values of α can be calculated for every

ALGORITHM 4. Sorting-Based Algorithm

n — Number of points

Z — Array with two fields, x and y to store the data

α — Array to store the α angles

θ — Point for which we want to find the location depth

Step 1. Read the data from a file into Z

Step 2. $NT = 0$

Step 3. *for* $(i = 1; i \leq n; i + +)$

   if $Z[i]$ is colocated with θ then $NT = NT + 1$

   else $\alpha[i - NT] =$ The angle α_i for $Z[i]$

Step 4. $NN = N - NT$

Step 5. Sort α

Step 6. $NBAD = 0$

Step 7. $NUMH = \infty$

Step 8. If θ is outside Z then goto step 11

Step 9. *for* $(i = 1; i \leq NN; i + +)$

  Step 9.1 Set h to largest integer such that: $\alpha_i \leq \alpha_{i+1} \leq .... \leq \alpha_{i+h} < \alpha_{i+\pi}$

  Step 9.2 $NBAD = NBAD + \binom{h}{2}$

  Step 9.3 $NUMH = \min(NUMH, \min(h + 1, NN - (h + 1)))$

Step 10. $NUMS = \binom{NN}{3} - NBAD$

Step 11. $NUMS = NUMS + \binom{NT}{1} * \binom{NN}{2} + \binom{NT}{2} * \binom{NN}{1} + \binom{NT}{3}$

Step 12. $NUMH = NUMH + NT$

Step 13. $SimplicialDepth = \dfrac{NUMS}{\binom{N}{3}}$

Step 14. $HalfSpaceDepth = \dfrac{NUMH}{N}$

Step 15. return(SimplicialDepth,HalfSpaceDepth)

FIGURE 7. Sort-Based Algorithm for the Location Depth

point on the fly. For every estimated quantile q_i we also add $q_i + \pi$ to this list. This helps to increase accuracy when calculating the h_i's without any change in the overall complexity.

In the second phase we count how many α values correspond to each sector. This is used to estimate the simplicial depth and the half-space depth. This can easily be done by reading the points, calculating the α value for each point, and determining the appropriate sector using a binary search. A high-level description of the algorithm is given in Figure 8.

This algorithm requires $O(n \lg s)$ and $O(n \lg q)$ time for the first and second phases, respectively. A total of two I/O passes is required on the entire data set.

The error analysis is similar to the analysis of the one-pass algorithm and is described in the next subsection. The only difference is that in the two-pass algorithm we add the antipodal angle $\beta + \pi$ for every sample point β, which allows us to derive the exact h values at sample points. This improves the accuracy of our estimation by a constant factor.

4.2. One-Pass Algorithm. This algorithm combines the two phases described in the previous section to reduce the I/O requirements.

The algorithm extends the OPAQ algorithm for finding quantiles (cf. Section 2). The dataset is divided into runs, with each run smaller than the memory size. These runs are processed sequentially one at a time.

ALGORITHM 5. Two-Pass Algorithm

   n — Number of points

   m Size of every run

   $runs$ — Number of runs = $\lceil \frac{n}{m} \rceil$

   s — Number of sample points from every run

   q — Number of quantiles wanted $= s \times runs$

   Q — Array of size q to store the quantiles, it has two fields, angle and count

   $Buffer$ — Array of size m with two fields, x and y to store one run

   α — Array of size m to store the α angles of one run

   θ — Point for which we want to find the location depth

   Step 1. $runs = \lceil \frac{n}{m} \rceil$

   Step 2. $q = s \times runs$

   Step 3. $for\ (i = 1; i \leq runs; i + +)$

        Step 3.1. read m points from the input file into $Buffer$

        Step 3.2. create the array α for points in $Buffer$

        Step 3.3. Find s sample points from α, α is Sorted into $s + 1$ buckets

        Step 3.4. add the s quantiles of α to Q

   Step 4. $for\ (i = 1; i \leq q; i + +)$

        Step 4.3. add $(Q[i].angle + \pi)\ mod\ 2\pi$ to Q

   Step 5. Sort Q on $angle$ values and set all counters to zero

   Step 6. $for\ (i = 1; i \leq runs; i + +)$

        Step 6.1. read m points from the input file into $Buffer$

        Step 6.2. create the array α for points in $Buffer$

        Step 6.3. $for\ (k = 1; k \leq m; k + +)$

             Step 6.3.1. Find which bucket b contains $\alpha[k]$ using binary search

             Step 6.3.2. $Q[b].count = Q[b].count + 1$

   Step 7. $NBAD = 0$

   Step 8. $NUMH = \infty$

   Step 9. $for\ (i = 1; i \leq (2 \times q); i + +)$

        Step 9.1 $h=$ sum of all counts in quantiles we pass to sweep π from Q[i]

        Step 9.2 $NBAD = NBAD + Q[i].count \times \binom{h}{2}$

        Step 9.3 $NUMH = \min(NUMH, \min(h + 1, n - (h + 1)))$

   Step 10. $NUMS = \binom{n}{3} - NBAD$

   Step 11. $SimplicialDepth = \frac{NUMS}{\binom{N}{3}}$

   Step 12. $HalfSpaceDepth = \frac{NUMH}{N}$

   Step 13. return(SimplicialDepth,HalfSpaceDepth)

FIGURE 8. The Two-Pass Algorithm for the Location Depth

   Each run is read from the input file and partially sorted into $s + 1$ intervals of equal size such that every element in interval i is less than any element in bucket $i + 1$. The sort is performed based on the α values.

   The sample points from all the runs are accumulated and sorted based on their α values. These sample points can then be used to select q quantiles that divide the range 0 to 2π into q sectors such that each sector will contain approximately $\frac{n}{q}$ α's (cf. Section 2).

   For each sector i, let r_i represent the minimum number of sectors that need to be swept so that a half circle is completed. We estimate h_i for all the angles belonging to the sector i by $h_i = r_i * \frac{n}{q}$. A high-level description of this algorithm is presented in Figure 9.

   We define the following terms for estimating the error:

ALGORITHM 6. One-Pass Algorithm

n — Number of points

m — Size of every run

$runs$ — Number of runs $= \lceil \frac{n}{m} \rceil$

s — Number of sample points from every run

q — Number of quantiles wanted $= s \times runs$

Q — Array of size q to store the quantiles, it has two fields, angle and count

$Buffer$ — Array of size m with two fields, x and y to store one run

α — Array of size m to store the α angles of one run

θ — Point for which we want to find the location depth

Step 1. $runs = \lceil \frac{n}{m} \rceil$

Step 2. $q = s \times runs$

Step 3. $for\ (i = 1; i \leq runs; i++)$

    Step 3.1. read m points from the input file into $Buffer$

    Step 3.2. create the array α for points in $Buffer$

    Step 3.3. Find s sample points from α, α is Sorted into $s + 1$ buckets

    Step 3.4. add the s quantiles of α to Q

Step 4. Sort Q on *angle* values and set all counters to $\frac{n}{q}$

Step 5. $NBAD = 0$

Step 6. $NUMH = \infty$

Step 7. $for\ (i = 1; i \leq (q); i++)$

    Step 7.1 $h=$ sum of all counts in quantiles we pass to sweep π from Q[i]

    Step 7.2 $NBAD = NBAD + Q[i].count \times \binom{h}{2}$

    Step 7.3 $NUMH = \min(NUMH, \min(h + 1, n - (h + 1)))$

Step 8. $NUMS = \binom{n}{3} - NBAD$

Step 9. $SimplicialDepth = \frac{NUMS}{\binom{N}{3}}$

Step 10. $HalfSpaceDepth = \frac{NUMH}{N}$

Step 11. return(SimplicialDepth,HalfSpaceDepth)

FIGURE 9. The One-Pass Algorithm for the Location Depth

- n — number of points
- q — number of quantiles
- s — sample points per run
- L_k — the set of points with α values in sector k
- W_k — the number of points in sector L_k
- r_i — the number of sectors required to sweep a half circle from quantile i

Let θ be the accurate value for the half-space depth.

$$\theta \binom{n}{3} = \binom{n}{3} - \sum_{i=1}^{n} \binom{h_i}{2}$$

(4.1)
$$= \binom{n}{3} - \sum_{k=1}^{q} \sum_{j \in L_k} \binom{h_j}{2}$$

It can easily be shown that for all $j \in L_k$:

$$h_j \leq r_k \times \frac{n}{q} + \frac{2n}{s}$$

(4.2)
$$\left(h_j - r_k \times \frac{n}{q}\right) \leq \frac{2n}{s}$$

Let

$$(4.3) \qquad \theta' \binom{n}{3} = \binom{n}{3} - \sum_{k=1}^{q} \binom{r_k \times \frac{n}{q}}{?} \times W_k$$

We will use θ' to estimate θ. The error introduced is given by:

$$(\theta - \theta') \binom{n}{3} = \binom{n}{3} - \sum_{k=1}^{q} \sum_{j \in L_k} \binom{h_j}{2} - \binom{n}{3} + \sum_{k=1}^{q} \binom{r_k \times \frac{n}{q}}{2} \times W_k$$

$$\leq \sum_{k=1}^{q} [W_k \times \{ \binom{h_k}{2} - \binom{r_k \times \frac{n}{q}}{2} \}]$$

$$= \sum_{k=1}^{q} [W_k \times \{ \frac{h_k(h_k - 1)}{2} - \frac{(r_k \times \frac{n}{q})(r_k \times \frac{n}{q} - 1)}{2} \}]$$

$$= \frac{1}{2} \sum_{k=1}^{q} [W_k \times \{ (h_k^2 - h_k) - (r_k \times \frac{n}{q})^2 + (r_k \times \frac{n}{q}) \}]$$

$$= \frac{1}{2} \sum_{k=1}^{q} [W_k \times \{ (h_k + (r_k \times \frac{n}{q})) \times (h_k - (r_k \times \frac{n}{q})) + (r_k \times \frac{n}{q}) - h_k \}]$$

$$\leq \frac{1}{2} \sum_{k=1}^{q} [W_k \times \{ (2h_k) \times \frac{2n}{s} \}]$$

$$\leq \frac{2n^2}{s} \sum_{k=1}^{q} W_k$$

$$(4.4) \qquad \leq \frac{2n^3}{s}$$

Thus

$$(4.5) \qquad \theta - \theta' \leq \frac{12}{s}$$

Thus the error in our estimation is bounded by the number of sample points chosen per run.

4.3. Experimental Results. We implemented the above algorithms for calculating the location depth and simplicial depth. Since the two depths are related we only present experimental data for location depth. The experimental setup used was the same as the one described in Section 3.3.

Figure 10(a) plots the total time taken by different algorithms. The total number of points, n, was varied from 128k to 8M. For the one-pass and the two-pass algorithm we chose 256 equidistant samples from every run (run size was set to $64K$ points). These results show that the time taken by the one-pass algorithm is much smaller than the time taken by the other two algorithms. For the sort-based algorithm, out-of-core sorting required around 77% of the total time. For the two-pass algorithm, the second pass required around 57% of the total time; the time within the second pass was dominated by the binary search time.

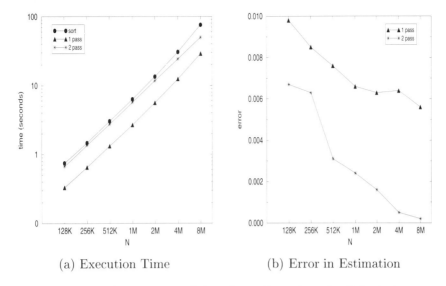

(a) Execution Time   (b) Error in Estimation

FIGURE 10. Comparison of the Three Algorithms for Calculating
the Half-Space Depth

The error in estimation of the half-space depth was calculated for several sample
points. A representative plot of the error is given in Figure 5(b). Further investigation is required to study the behavior of this error for a variety of synthetic and
real data sets. However, it is worth noting that the worst case error is bounded.

5. Conclusions

We have presented fast algorithms for finding good estimates for the location
and regression depth measures for large data sets. The data sets were assumed to
be disk resident. These methods can form a basis for finding statistics on large
data sets. An important characteristic of our methods is that they can be easily
extended to find the depth measure for several points without additional I/O requirements. This cannot be done with sorting-based methods when the sort order
changes depending on the point evaluated (as is the case with location depth).

References

[ADADC+97] Andrea C. Arpaci-Dusseau, Remzi Arpaci-Dusseau, David E. Culler, Joseph M.
            Hellerstein, and David A. Patterson, *High-performance sorting on networks of
            workstations.*, SIGMOD, ACM, 1997, pp. 243–254.
[Aga96]   Ramesh C. Agarwal, *A super scalar sort algorithm for risc processors*, ACM SIG-
            MOD (1996), 240–246.
[ARS97]   K. Alsabti, S. Ranka, and V. Singh, *A One-Pass Algorithm for Accurately Esti-
            mating Quantiles for Disk-Resident Data*, In Proceedings. of VLDB'97 Conference
            (1997), 346–355.
[HSM95]   E. Horowitz, S. Sahni, and D. Mehta, *Fundamentals of data structures in c++*,
            W. H. Freeman, 1995.
[LLS+93]   X. Li, P. Lu, J. Schaeffer, J. Shillington, P. S. Wong, and H. Shi, *On the Versatility
            of Parallel Sorting by Regular Sampling*, Parallel Computing **19(10)** (1993), 543–
            550.
[RH96]   Peter J. Rousseeuw and Mia Hubert, *Regression depth*, Technical report, University
            of Antwerp, 1996.

[RR96]   Peter J. Rousseeuw and Ida Ruts, *Bivariate location depth*, Applied Statistics **45** (1996), 516–526.
[Zhe96]   LouQuan Zheng, *Speeding up external mergesort*, IEEE Transactions on Knowledge and Data Engineering **8** (1996), no. 2, 322–332.

SYRACUSE UNIVERSITY, SYRACUSE, NY 13215
E-mail address: alfuraih@top.cis.syr.edu

AT&T RESEARCH LABS
E-mail address: johnsont@research.att.com

UNIVERSITY OF FLORIDA, GAINESVILLE, FL 32611
E-mail address: ranka@cise.ufl.edu

DIMACS Series in Discrete Mathematics
and Theoretical Computer Science
Volume **50**, 1999

Efficient Cross-Trees for External Memory

Roberto Grossi and Giuseppe F. Italiano

ABSTRACT. We describe efficient methods for organizing and maintaining large
multidimensional data sets in external memory. This is particular important as
access to external memory is currently several order of magnitudes slower than
access to main memory, and current technology advances are likely to make
this gap even wider. We focus particularly on multidimensional data sets which
must be kept simultaneously sorted under several total orderings: these orderings may be defined by the user, and may also be changed dynamically by the
user throughout the lifetime of the data structures, according to the application at hand. Besides standard insertions and deletions of data, our proposed
solution can perform efficiently *split* and *concatenate* operations on the whole
data sets according to any ordering. This allows the user: (1) to dynamically
rearrange any ordering of a segment of data, in a time that is faster than
recomputing the new ordering from scratch; (2) to efficiently answer queries
related to the data contained in a particular range of the current orderings.
Our solution fully generalizes the notion of B-trees to higher dimensions by
carefully combining space-driven and data-driven partitions. Balancing is easy
as we introduce a new multidimensional data structure, *the cross-tree*, that is
the cross product of balanced trees. As a result, the cross-tree is competitive
with other popular index data structures that require linear space (including
k-d trees, quad-trees, grid files, space filling curves, hB-trees, and R-trees).

1991 *Mathematics Subject Classification.* Primary 68P05, 68Q25.

Key words and phrases. Design and Analysis of Algorithms and Data Structures, External
Memory.

Part of this work was done while the second author was visiting the Hong Kong University
of Science & Technology. His work was supported in part by EU ESPRIT Long Term Research
Project ALCOM-IT under contract no. 20244, by a Research Grant from University "Ca' Foscari"
of Venice, and by Grants from CNR, the Italian National Research Council.

1. Introduction

Let S be a set whose items are sorted with respect to $d > 1$ total orders $\prec_1, \ldots, \prec_d$, and which is subject to dynamic operations, such as insertions and deletions of single items, split and concatenate operations performed according to any chosen order $\prec_i$ ($1 \leq i \leq d$). Let $\mathcal{P}$ be a searching problem defined on input set S with n items, and let $\mathcal{P}(x, S)$ denote its solution for a query item x. Problem $\mathcal{P}$ is *decomposable* if we can answer query $\mathcal{P}(x, S)$ by first partitioning set $S = S' \cup S''$ and computing the answers to queries $\mathcal{P}(x, S')$ and $\mathcal{P}(x, S'')$ recursively, and then combining them through a suitable associative operator $\Diamond$. Formally, $\mathcal{P}$ is said to be $f(n)$-*decomposable* if and only if $\mathcal{P}(x, S) = \Diamond(\mathcal{P}(x, S'), \mathcal{P}(x, S''))$ for any partition $S = S' \cup S''$ and any query item x, where $\Diamond$ is an operator whose computation requires $O(f(n))$ time. Throughout, we assume that $f(n)$ is a nondecreasing and smooth function, i.e., $f(\Theta(n)) = \Theta(f(n))$, and that $f(n) = o(n)$. Some simple examples of $O(1)$-decomposable searching problems include: membership queries, where $\Diamond$ is the logical or function; closest point queries, where $\Diamond$ is the minimal distance; range queries, where $\Diamond$ is the list append operation. Convex hull searching is not decomposable as the fact that a point $x \in S$ belongs to the convex hull of S' or S'' does not necessarily imply that x belongs to the convex hull of $S = S' \cup S''$. The definition of decomposable *search* problems can be extended also to the decomposable *set* problems in which the query item is not specified (e.g., finding the minimum weight item, where $\Diamond$ is the minimum). Thus, we shall denote a generic solution to a decomposable problem $\mathcal{P}$ by $\mathcal{P}(S)$.

In this paper, we consider *order decomposable problems*. A problem $\mathcal{P}$ is $f(n)$-*order* decomposable with respect to total order $\prec_i$ if $\mathcal{P}(S) = \Diamond(\mathcal{P}(S'), \mathcal{P}(S''))$ for any *ordered partition* $S = S' \cup S''$ (i.e., $x' \prec_i x''$ for all $x' \in S'$ and $x'' \in S''$), where operator $\Diamond$ takes $O(f(n))$ time. Problem $\mathcal{P}$ is $f(n)$-*order decomposable* if it is $f(n)$-order decomposable with respect to any total order $\prec_i$, $1 \leq i \leq d$. For example, performing multidimensional range queries is $O(1)$-order decomposable, and convex hull searching is $O(\log_2 n)$-order decomposable. Many other examples of order decomposable problems can be found in basic data structures, computational geometry, database applications and statistics [9, 23, 27].

In the *static* case, there is an obvious divide-and-conquer algorithm for these problems on a sorted set S: split S into two (equally sized) subsets S' and S'', solve the problem recursively on S' and S'', and combine their solutions in $O(f(n))$ time. We aim at maintaining a *dynamic* set S with d total orders, for constant $d > 1$, under insertions of a single item, deletions of a single item, and re-arrangements of any of the total orders $\prec_1, \ldots, \prec_d$ on S by means of split and concatenate operations. In several cases, executing splits and concatenates is faster than performing a batch of insertions and deletions to achieve the same goal. Our queries involve finding the solution $\mathcal{P}(R)$ for the items contained in the subset $R \subseteq S$ defined by some ranges in the orders $\prec_1, \ldots, \prec_d$.

In this paper, we deal with the following *multiordered set splitting and merging problem* introduced in [16]:

> *split*($S, z, \prec_i$): Split S into S' and S'' according to item z and the specified total order $\prec_i$ ($1 \leq i \leq d$). That is, $x' \prec_i z$ and $z \prec_i x''$ for all $x' \in S'$ and $x'' \in S''$. S is no longer available after this operation.
>
> *concatenate*($S', S'', \prec_i', \prec_i''$): Combine S' and S'' together according to their respective i-th total orders $\prec_i'$ and $\prec_i''$ ($1 \leq i \leq d$) into a new set $S = S' \cup S''$.

After this operation, S' and S'' are no longer available. The items in the resulting set S undergo the new order $\prec_i$ obtained by concatenating $\prec_i'$ and $\prec_i''$. That is, $x \prec_i y$ in S if and only if one of the following three conditions holds:

1. $x \prec_i' y$ and $x, y \in S'$;
2. $x \prec_i'' y$ and $x, y \in S''$;
3. $x \in S'$ and $y \in S''$.

insert(z, S): Insert item z into set S according to all orders $\prec_1, \dots, \prec_d$.

delete(z, S): Delete item z from set S.

range$(\langle a_1, b_1 \rangle, \dots, \langle a_d, b_d \rangle, S)$: Let $R = \{z \in S : a_i \prec_i z \prec_i b_i, \text{ for } 1 \leq i \leq d\}$. Find solution $\mathcal{P}(R)$ to problem $\mathcal{P}$ restricted to the items in region R only.

For $d = 1$, the recursive nature of order decomposable problems gives an immediate tree structure, and each of the above operations can be simply implemented with $O(f(n) \log_2 n)$ time (with $O(f(n))$ time per tree node) by using a 2-3-tree [2] or a B-tree [6, 11]. However, the multidimensional version of this problem ($d > 1$) is much more complicated: indeed, maintaining $d > 1$ total orders on the same set S, while splitting or merging each order independently of the others, makes things highly non-trivial due to the interplay among different orders.

In previous work [16] we introduced efficient techniques and data structures for this problem in main memory, and showed how these data structures could efficiently implement multidimensional priority queues, multidimensional search trees, and concatenable interval trees. These applications improved many previously known results on decomposable problems under split and concatenate operations, including membership query, minimum weight item, range query, and convex hulls.

OUR RESULTS. The main result of this paper is to make the main memory technique of [16] suitable for efficient external memory implementation. To achieve this goal, we have to prove and exploit other features and properties of the technique introduced in [16], and implement it differently. This produces an external memory technique for solving order decomposable problems on S under insertions, deletions, splits, concatenates and range queries, yielding new and efficient concatenable multidimensional data structures. Differently from other approaches, our technique is based more on simple geometric properties rather than on underlying sophisticated data structures, and exploits the fact that some data structures can be built on (pre)sorted items more efficiently. Balancing is easy as we introduce a new multidimensional data structure, *the cross-tree*, that is the cross product of balanced trees.

The model we consider for external memory is quite standard [1, 37]: it consists of a fast memory of size M, and a number D of slow independent disks. Data transfers between disks and internal memory occur in *blocks* of size B, each single data transfer accessing a single block per disk in parallel. It is assumed that the fast memory is large enough to accommodate items from the disks, e.g., $M \geq 2BD$. The objective is to minimize the number of transferred blocks, called *inputs/outputs or I/Os*, performed by an algorithm or required to access a data structure. A single I/O handles BD items, one block of B items per disk. The I/O complexity is typically expressed in terms of the input size and the "machine" parameters B, M and D. For example, the I/O complexity of sorting n constant size keys in external memory is $Sort(n) = \Theta(\frac{n}{BD} \log_{M/B} \frac{n}{B})$ I/Os [1, 25]. The general problem of sorting variable length keys in external memory is discussed in [3]. For the sake of presentation, we will assume here the the number of disks is $D = 1$; however, our results smoothly extend to $D > 1$ parallel disks by using the disk striping technique [31] with a block size of $B' = BD$ and $D' = 1$.

In the simple case $d = 1$, B-trees [6, 11] are very popular data structures that can be successfully employed in decomposable search problems analogously to concatenable 2-3-trees [2]. For higher dimensions ($d > 1$), no provably good external memory data structures for splitting and concatenating along any dimension were previously known in the literature. We show how to maintain S with the following I/O bounds: $O(\log_B n)$ for the insertion or the deletion of a single item, where n is the number of items currently in S; $O((n/B)^{1-1/d})$ for splits and concatenates along any order; $O((n/B)^{1-1/d}))$ plus an optimal output sensitive cost for rectangular range queries, i.e., an $O(r/B)$ cost for retrieving r *pointers* to the items. This is helpful when processing the output lists subsequently, e.g., by answering to a boolean query. Furthermore, we get an optimal cost when actually scanning the pointed items as there is *no replication of data*. Namely, if each item occupies b bytes, we require optimal $O(br/B)$ I/Os to scan the content of all retrieved items. This is not always possible with the known multidimensional data structures as pointed out in [19].

We require that our external memory algorithms must use optimal storage. In our case, the space required is linear, i.e., $O(n/B)$ disk blocks. In general, we have to face some problems that do not appear in internal memory, especially when we have to maintain dynamically data structures with optimal storage. Consider for instance the simple case of moving an item u of indegree h from its block to a different block. The h incoming pointers could be in different blocks and, when no specific block strategy is adopted, moving u might cause the explicit update of all these pointers. If h is not bounded a priori, this single update becomes costly as we have to pay $\Theta(h)$ I/Os. More details on this problem can be found in [12].

RELATED WORK. There is a great deal of work on decomposable searching problems. They were introduced by Bentley [8] for dynamizing static data structures. The initial goal was to support insertions with low amortized times, without affecting much of the query efficiency. Other dynamization techniques were subsequently given (see [10, 27] for many references to the literature). The main idea behind these techniques is to partition a big data structure into a collection of small data structures, called *blocks*, and to tune properly the number of blocks in order to obtain a good tradeoff between queries and updates. Later on, van Kreveld and Overmars [20, 21] introduced two new operations to this dynamic setting: split and concatenate. Namely, they showed how to maintain a set of multidimensional points subject to the following repertoire of operations: range queries, insertion of a point, deletion of a point, split and concatenate along one coordinate.

In external memory, rectangular range queries on ordered decomposable problems can be done with quad-trees [13], k-d trees [7] and many other data structures, such as grid files [24], space filling curves, e.g. [26], hB-trees [22], and R-trees [17] just to cite a few (see the survey [14] for a more complete scenario and paper [19] for a discussion of their worst-case complexity). These data structures were originally designed to support some operations for windowing problems in computer graphics and databases, and in several cases it was rather difficult to keep them balanced [29, 34]. In contrast, our cross-tree can be easily kept balanced. Many other powerful data structures for range queries were devised subsequently and we refer the reader to [10] for a comprehensive survey on this topic and a list of references. Recently, some elegant data structures [5, 19, 30, 32, 35, 36] were devised to support fast range queries in external memory, and Arge *et al.* [4] have dealt with some decomposable problems in external memory. However, none of these data structures seems to be able to support efficiently split and concatenate along any coordinate.

Ravi and Singh [33], following up on a previous lower bound of Hellerstein *et al.* [18], considered the complexity of range searching in a simple index tree model. In this model, they proved a lower bound of $\Omega((n/B)^{1-1/d} + r/B)$ for a range query retrieving the pointers to r out of n items, where B is the capacity of index nodes. This lower bound holds on pointer machines and for non-replicating index structures, which occupy linear space. Our range query matches this lower bound, and uses simply the list append operator as operator $\Diamond$: consequently, it can be easily implemented on a pointer machine. As cross-trees use some auxiliary pointers for supporting efficiently split and concatenate operations, they are not strictly considered non-replicating in the terminology of Ravi and Singh: however, we do not believe this to be a substantial infringement of the non-replicating assumption. Ravi and Singh [33] introduced also the O-tree, a pointer-based strictly non-replicating data structure which also matches their lower bound. Once again, O-trees support only queries, insertions and deletions but are not able to support efficiently split and concatenate operations.

The remainder of this paper consists of seven sections. In Section 2 we present our data structure, which consists of two levels. The first level operates a space-driven partition of the points into basic squares and is described in more details in Section 3. The second level, considered in Section 4, is based upon a data-driven multilevel hierarchical partition. The two levels are next combined in Section 5. Section 6 illustrates how the overall data structure can be used to implement the operations, while Section 7 mentions some further applications of our techniques. Finally, Section 8 contains some concluding remarks.

2. Our Data Structure

Similarly to [16], we define a two-level data structure: the *lower* level data is only locally kept sorted, while the *upper* level data follows a hierarchical and global order. More specifically, the lower level organizes the embedding space of data items according to their "spatial" coordinates so as to facilitate their location search. This is done by partitioning the space into a collection of disjoint *basic squares*, forming a flexible grid. Any row or column formed by basic squares needs only local changes to preserve a global notion of balancedness. Order inside a basic square is strictly not a requirement of our data structures, but can be maintained efficiently. The upper level consists of a multidimensional tree, called the *cross-tree*, whose leaves are in one-to-one correspondence with the basic squares of the lower level. This upper level organizes the data items according to their "relative" coordinates in the orderings, by exploiting their value comparisons. Contrarily to k-d trees [7], quad-trees [13] and their variants in external memory, cross-trees can be easily and dynamically kept balanced by means of node splits and merges as they are "cross product" of one-dimensional balanced trees. In other words, cross-trees extend the natural notion of balance to higher dimensions: *in order to update a*

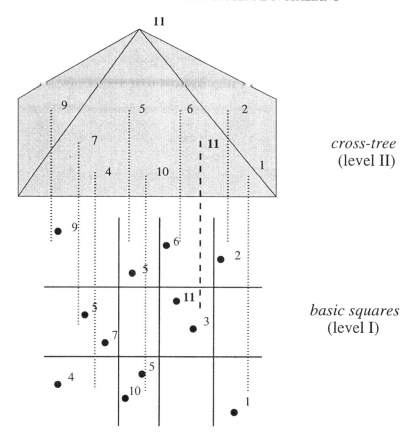

cross-tree
(level II)

basic squares
(level I)

FIGURE 1. An example of the two-level data structure for maintaining the solution $\mathcal{P}(S)$ to the decomposable problem "find the maximum weight element in S" with $\Diamond = \max$ and $d = 2$. The cross-tree is shown in Figures 2–3.

cross-tree, it is sufficient to work on one of the (one-dimensional) balanced trees and broaden its updates to the whole cross-tree.

The power of our data structure for external memory lies in this two-level organization: we can run a fast search in the upper level at the price of an increased update cost, while we can run a fast update in the lower level at the price of an increased search cost. The resulting good performance is obtained by a proper trade-off between total search and update costs and by carefully tuning some parameters that bound the size of the two levels. Differently from other approaches, our technique is based more on simple geometric properties rather than on underlying sophisticated data structures, and exploits the fact that some data structures can be built on (pre)sorted items more efficiently. As a result, our data structure may be particularly attractive as a multidimensional index tool for many kinds of terabyte applications. An example is shown in Figure 1.

For the sake of simplicity, we start out by describing our data structure in the case of $d = 2$ total orders, which we denote by $\prec_X$ and $\prec_Y$. We will see later on (Section 7) how to extend this to the more general case. Let n be the number of items in S, where $n > M$. Each item $z \in S$ is associated with a point $(X(z), Y(z))$

in the Cartesian plane, such that $X(z)$ is the rank of z in S with respect to order $\prec_X$ and $Y(z)$ is the rank of z in S with respect to $\prec_Y$. Note that $X(z)$ and $Y(z)$ are affected by our dynamic operations and so the mapping from z onto point $(X(z), Y(z))$ changes dynamically throughout the sequence of operations. Starting from n items in S, we obtain n points in the Cartesian plane, which can be stored in the form of a $n \times n$ *sparse* and *dynamic* matrix $\mathcal{M}$. We use $\mathcal{M}$ as it is more suitable for external memory implementation. The operations in S can be simulated by a certain number of operations in $\mathcal{M}$ by suitably using the ranks of the involved items. For any integers h_1, h_2, v_1, v_2 (with $1 \le h_1 \le h_2 \le n$ and $1 \le v_1 \le v_2 \le n$), we use $\mathcal{M}[h_1, h_2; v_1, v_2]$ to denote the submatrix of $\mathcal{M}$ that contains entries $\mathcal{M}[i, j]$ with $h_1 \le i \le h_2$ and $v_1 \le j \le v_2$. We call this submatrix a *region*. The operations defined on the matrix $\mathcal{M}$ are:

  h_split$(\mathcal{M}, i)$: Split $\mathcal{M}$ horizontally at row i and obtain two new matrices $\mathcal{M}_1$ and $\mathcal{M}_2$, such that $\mathcal{M}_1 = \mathcal{M}[1, i; 1, n]$ and $\mathcal{M}_2 = \mathcal{M}[i + 1, n; 1, n]$. In other words, $\mathcal{M}_1$ is given by the first i rows of $\mathcal{M}$ and $\mathcal{M}_2$ is given by the last $(n - i)$ rows of $\mathcal{M}$. $\mathcal{M}$ is no longer available after the operation.

  h_concatenate$(\mathcal{M}_1, \mathcal{M}_2)$: Let $\mathcal{M}_1$ have size $m_1 \times n$ and $\mathcal{M}_2$ have size $m_2 \times n$. We meld $\mathcal{M}_1$ and $\mathcal{M}_2$ horizontally and produce a matrix $\mathcal{M}$ of size $(m_1 + m_2) \times n$, such that $\mathcal{M}[1, m_1; 1, n] = \mathcal{M}_1$ and $\mathcal{M}[m_1 + 1, m_1 + m_2; 1, n] = \mathcal{M}_2$. In other words, the first m_1 rows of $\mathcal{M}$ are given by $\mathcal{M}_1$ and the last m_2 rows of $\mathcal{M}$ are given by $\mathcal{M}_2$. This operation assumes that $\mathcal{M}_1$ and $\mathcal{M}_2$ have the same number of columns. $\mathcal{M}_1$ and $\mathcal{M}_2$ are no longer available after the operation.

  v_split$(\mathcal{M}, j)$ and *v_concatenate*$(\mathcal{M}_1, \mathcal{M}_2)$: they are similarly defined.

  set$(i, j, w, \mathcal{M})$: Update $\mathcal{M}$ by setting $\mathcal{M}[i, j] = w$. This corresponds either to an insertion (if w is nonempty) or to a deletion (if w is empty). It causes the implicit renumbering of the horizontal and vertical rankings.

  range$(h_1, h_2, v_1, v_2, \mathcal{M})$: Find the solution $\mathcal{P}(R)$ to problem $\mathcal{P}$ restricted to the nonempty entries contained in region $R = \mathcal{M}[h_1, h_2; v_1, v_2]$.

Our technique works for a general matrix $\mathcal{M}$. However, in the remainder of this paper we restrict ourselves to the special case where each row or column of $\mathcal{M}$ contains a constant number of points. This is without loss of generality, as a row or a column with s points can be represented by a sequence of $\Theta(s)$ columns or rows with a constant number of points: this transformation does not affect the achieved bounds and can be easily maintained throughout our sequence of operations.

Before describing the two levels of our data structures for the splitting and merging problem, we need some preliminary definitions. Let $X = \{x_1, x_2, \ldots, x_q\}$ be a sorted sequence of q elements, according to some total order $\prec$: $x_1 \prec x_2 \prec \cdots \prec x_q$. Let $I_1, \ldots, I_s$ be a partition of X into adjacent intervals, so that for $1 \le i \le s - 1$ all the elements in I_i precedes all the elements in I_{i+1}. For $1 \le i \le s$, let $|I_i|$ denote the size of interval I_i, defined as the number of elements in I_i.

DEFINITION 2.1. **(Size Invariant)** Let $k \ge 1$ be a positive integer. The adjacent intervals $I_1, \ldots, I_s$ satisfy the *size invariant of order k* if the following two conditions are met: (a) $|I_i| \le k$, $1 \le i \le s$; and (b) $|I_i| + |I_{i+1}| > k$, $1 \le i \le s - 1$.

The size invariant of order k in Definition 2.1 implies that the number s of intervals is $O(q/k)$. We remark that the size invariant can be easily maintained when an element is deleted from X or a new element is inserted into X. This can be done as follows. When an element x is inserted into interval I_i, we check whether the size of I_i still satisfies condition (a) of Definition 2.1. If condition (a) is violated, $|I_i| = k + 1$, and we split I_i into two intervals, say I'_i and I''_i such that all the elements of I'_i precedes all the elements of I''_i. Next, if condition (b) is violated for I_{i-1} and I'_i (resp. I''_i and I_{i+1}), we merge I'_i (resp. I''_i) with I_{i-1} (resp. I_{i+1}). If an element x is deleted from interval I_i, then we have still to check that condition (b) is met for I_{i-1} and I_i and for I_i and I_{i+1}. In the worst case, this implies again a constant number of interval merges and splits.

3. Level I: The Basic Squares

We now describe the first level of our data structure for a matrix $\mathcal{M}$, which operates a space-driven partition of the points. We refer to the n nonempty entries of $\mathcal{M}$ as the *points* of $\mathcal{M}$ and let k be a slack parameter, where k is an integer with $1 \leq k \leq n$. We handle the *sparse* $n \times n$ matrix $\mathcal{M}$ *as if it were a dense matrix* of size $\Theta(n/k + k) \times \Theta(n/k + k)$. We then tune k according to the chosen problem $\mathcal{P}$. We group adjacent rows and columns of matrix $\mathcal{M}$ into respectively *horizontal and vertical stripes* so as to satisfy the size invariant of order k: namely, any horizontal (resp. vertical) stripe contains at most k rows (resp. columns), and no two adjacent horizontal (resp. vertical) contains more than k rows (resp. columns). This guarantees that each stripe contains at most $O(k)$ points and that the total number of horizontal and vertical stripes is $O(n/k)$. The partition into horizontal and vertical stripes induces a partition of $\mathcal{M}$ into $O(n^2/k^2)$ squares, such that each square intersects no more than k rows and k columns. These are the *basic squares* in $\mathcal{M}$. For each such basic square, we maintain a solution to problem $\mathcal{P}$ for the points in the square.

We store these points in external memory blocks of size B according to this partition scheme so as to guarantee a fast access to the basic squares in a stripe along the two dimensions, i.e., both horizontally and vertically. Ideally, we would like to get an $O(\sqrt{n/B})$ cost by organizing all the n points into a "dense" ($\sqrt{n} \times \sqrt{n}$) square grid and by covering the grid with a number of ($\sqrt{B} \times \sqrt{B}$)-shaped squares, such that the union of these (disjoint) squares is the grid and each square is stored into one block. However, this organization seems quite difficult to maintain as the matrix is both sparse and dynamic: we insert and delete points in a sparse order and perform split and concatenate operations as well. We will instead follow a different approach whose cost is $O(n/k + k/B)$, which is still $O(\sqrt{n/B})$ if we choose $k = \lceil \sqrt{nB} \rceil \leq n$.

We store points in external memory by vertical stripes, as follows. Let σ be a vertical stripe, and assume without loss of generality that σ does not fit into one block, i.e., $k = \Omega(B)$ (otherwise, we can always stuff a number of adjacent vertical stripes into the same block and then access a stripe with a single I/O). We group *adjacent* basic squares of σ into $O(k/B)$ disjoint groups. These groups satisfy the size invariant of order B on the total size of their basic squares, with the notable exception of basic squares with size larger than B. Each basic square of this kind contains $\Omega(B)$ points, and so it forms a singleton group on its own and is stored in a list of blocks. We implement this list by means of a B$^+$-tree [11] that can be updated, split and concatenated in $O(\log_B k)$ I/Os. To this end, we maintain a $\prec_B$-order on the points in the basic square at hand. Note that the size of the basic square remains proportional to the number of points in the square, as the size of this B$^+$-tree is proportional to the number of points divided by B. The remaining groups all contain the basic squares with size bounded by B and each group can be stored into a single block. Although unnecessary, we also keep the points in each single-block group sorted by $\prec_B$, i.e., a single-node B-tree. Note that a vertical stripe can be stored in at most $O(k/B)$ different blocks.

LEMMA 3.1. *The points contained in the basic squares belonging to either a horizontal or a vertical stripe in $\mathcal{M}$ can be retrieved in the correct order (i.e., left to right and bottom-up) with $O(n/k + k/B)$ I/Os.*

PROOF. A vertical stripe can be easily accessed with $O(k/B)$ I/Os. Consider now a horizontal stripe σ. As σ contains $O(k)$ points, at most $O(k/B)$ blocks can have all of their points completely inside σ. Furthermore, at most $O(n/k)$ blocks can contain points both from σ and from either of its neighbor horizontal stripe, as there are no more than two such blocks per basic square. Hence, the total number of different blocks required to access a horizontal stripe in our vertically oriented organization is $O(n/k + k/B)$.   □

We conclude this section by observing that the total number of blocks needed to store the matrix $\mathcal{M}$ is optimal, i.e., $\Theta(n/B)$. Furthermore, we can maintain efficiently the size invariant for any two adjacent stripes and their blocks with a constant number of scans.

4. Level II: The Cross-Tree

In the second level, the basic squares are uniquely associated with the leaves of a *cross-tree* [16], a multidimensional hierarchical data structure that describes recursively the partition of $\mathcal{M}$ into its basic squares and that supports efficiently *split and concatenate operations*.

A cross-tree [16] describes a balanced decomposition of a bidimensional set, and it is based upon two balanced trees (such as 1-2-trees, 2-3-trees [2] or B+-trees [11]). Let T_H and T_V be two balanced trees, whose leaves are on the same level and correspond to the horizontal and vertical stripes respectively, and which recursively describe a multilevel horizontal and vertical partition into stripes. Then a *cross-tree* $CT(T_H \times T_V)$ for dimension $d = 2$ is the cross product of T_H and T_V. In particular:

- For each pair of nodes $u \in T_H$ and $v \in T_V$ on the same level, there is a node α_{uv} in $CT(T_H \times T_V)$ with no more than b^2 children, where b is the maximum degree of u and v.
- For each pair of edges $(u, \hat{u}) \in T_H$ and $(v, \hat{v}) \in T_V$, such that u and v are on the same level, there is an edge $(\alpha_{uv}, \alpha_{\hat{u}\hat{v}})$ in $CT(T_H \times T_V)$.

An example illustrating a cross-tree is shown in Figure 2. One important property of a cross tree is the following. Let π be a leaf-to-root path in either balanced tree (say T_H). Then, all the cross tree nodes that are composed with nodes from π, induce a subtree of $CT(T_H \times T_V)$ which is isomorphic to T_V. For instance in Figure 2, the cross-tree part identified by nodes b1, b2, b3, b4, d5, d6, d7, f8, f9, g10, is isomorphic to T_V and corresponds to leaf-to-root path $\pi = \{b, d, f, g\}$ in T_H. Analogously, the part identified by a3, b3, c3, d6, e6, f8, g10, is isomorphic to T_H and corresponds to path $\pi = \{3, 6, 8, 10\}$ in T_V. This property is clearly important if we have to split (or concatenate) cross-trees in response to splits (or concatenates) in their trees T_H and T_V. We refer the reader to [16] for further properties and details on cross-trees.

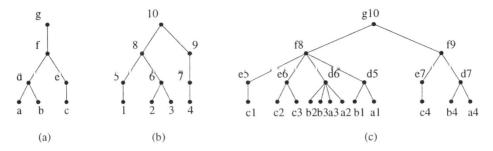

FIGURE 2. An example of cross-tree $CT(T_H \times T_V)$ for the two trees T_H and T_V respectively shown in (a)–(b) and corresponding to the horizontal and vertical stripes of the basic squares illustrated in Figure 1. Here, T_H and T_V have maximum degree $b = 2$ and so the cross-tree has maximum degree $b^2 = 4$. The actual representation of $CT(T_H \times T_V)$ is shown in (c).

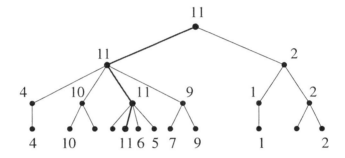

FIGURE 3. The cross-tree of Figure 2 with nonempty solutions stored in its nodes for the problem treated in Figure 1.

As problem $\mathcal{P}$ is decomposable, we store the solutions of the basic squares in the corresponding leaves of the cross-tree and percolate these solutions up in a heap-like fashion by using operator $\diamond$ as shown in Figure 3. We are assuming that each solution takes constant space, and we treat the case of solutions requiring more space later on.

We store the cross-tree $CT(T_H \times T_V)$, built according to the partition of $\mathcal{M}$ into basic squares, in an optimal number $\Theta(n/B)$ of external memory blocks. As we want to support efficient split and concatenate operations, we choose to use two *weight-balanced B-trees* T_H and T_V in our cross tree: they are a weight-balanced variant of B-trees and 2-3-trees, introduced by Arge and Vitter [5]. We bound their maximum number of children by $b = \sqrt{B}$, so that the cross-tree has branching factor $\Theta(B)$. The weight-balanced B-trees have several interesting features, including leaf insertions, leaf deletions, splits and concatenates in $O(\log_B |T|)$ I/Os per operation. Each operation only involves the nodes in a leaf-to-root path and their children. If the two trees have different heights, we replace the root of the lower tree with a chain of $O(1)$ unary nodes so that they both have the same height. Consequently, the cross-tree $CT(T_H \times T_V)$ has height $O(\log_B(n/k))$ and has a total number of $\Theta(n/B)$ nodes (i.e., blocks). The following lemma is easy to derive.

LEMMA 4.1. *The cross-tree leaves corresponding to the basic squares in either a horizontal or a vertical stripe in $\mathcal{M}$ can be retrieved (together with their ancestors) in bottom-up order, with $O(n/k)$ I/Os.*

5. Putting the Two Levels Together

We now have all the ingredients for a general view of our data structure. The basic squares in the lower level are associated with the leaves of the cross-tree in the upper level. The recursive nature of decomposable problem $\mathcal{P}$ suggests us that its solution $\mathcal{P}(S) = \Diamond(\mathcal{P}(S'), \mathcal{P}(S''))$ can be stored in the root of the cross-tree and that $\mathcal{P}(S'), \mathcal{P}(S'')$ can be stored recursively in its two children. Each internal node therefore stores the solution to $\mathcal{P}$ for the items in S corresponding to its descendent leaves. Our basic data structure comprises matrix $\mathcal{M}$ and cross-tree $CT(T_H \times T_V)$. Trees T_H and T_V have $O(n/k)$ leaves, one for each stripe of $\mathcal{M}$, and a total of $O(n/k)$ nodes. Consequently, the resulting cross-tree $CT(T_H \times T_V)$ has $O(n^2/k^2)$ leaves, one for each basic square of $\mathcal{M}$, and a total of $O(n^2/k^2)$ nodes. Our data structure has the following features:

1. For each nonempty basic square of $\mathcal{M}$, we keep its points sorted according to an order $\prec_B$ (which is not necessarily equal to $\prec_X$ or $\prec_Y$) by means of a B$^+$-tree. Searching, inserting and deleting a point takes $O(\log_B k)$ I/Os. We introduce order $\prec_B$ because some data structures can be (re)built more efficiently on a sorted set of points.

2. Each cross-tree node corresponds to a region of matrix $\mathcal{M}$. If a node is empty then its corresponding region is empty. A leaf corresponds to a basic square. An internal node α corresponds to a *region* $\mathcal{M}[h_1, h_2; v_1, v_2]$, and has no more than B children $\alpha_1, \ldots, \alpha_j$ corresponding to an orthogonal partition of $\mathcal{M}[h_1, h_2; v_1, v_2]$ into smaller regions. For example, if α has $j = 4$ children, they correspond to the regions: $\mathcal{M}[h_3, h_2; v_1, v_3]$, $\mathcal{M}[h_3, h_2; v_3, v_2]$, $\mathcal{M}[h_1, h_3; v_3, v_2]$, and $\mathcal{M}[h_1, h_3; v_1, v_3]$, for some $h_1 \leq h_3 \leq h_2$ and $v_1 \leq v_3 \leq v_2$. In other words, the smaller (non-overlapping) regions corresponding to $\alpha_1, \ldots, \alpha_j$ can be united to produce the region corresponding to α.

3. For each nonempty basic square of $\mathcal{M}$, we examine its points and store their solution to problem $\mathcal{P}$ into their corresponding cross-tree leaf. We then percolate this information up in the cross-tree: let α be an internal node, and let $s_1, \ldots, s_j$ be the solutions stored in the $j \leq B$ children of α. As $\mathcal{P}$ is a decomposable problem, the internal node α stores its solution $\Diamond(s_1, \ldots, s_j)$ for the points in its corresponding region in $\mathcal{M}$. (Recall that $\Diamond$ is associative and so is well-defined also for an arbitrary number of arguments.)

We assumed so far that the solutions take constant space each. When this is not the case, an internal node can store *implicitly* its corresponding solution $\mathcal{P}(R)$: namely, the node stores the $O(f(n))$ actions taken to compute $\mathcal{P}(R) = \Diamond(\mathcal{P}(R'), \mathcal{P}(R''))$ and its (say, two) children store only the leftover pieces of $\mathcal{P}(R')$ and $\mathcal{P}(R'')$ that are not employed to build $\mathcal{P}(R)$. A full solution is available only at the root. In order to obtain the solution in the other nodes, we can use a simple top-down procedure to obtain $\mathcal{P}(R'), \mathcal{P}(R'')$ from $\mathcal{P}(R)$ recursively along a downward path, and another simple bottom-up procedure to obtain $\mathcal{P}(R)$ from $\mathcal{P}(R'), \mathcal{P}(R'')$ along an upward path. We therefore say that $\Diamond$ is *invertible* if we can keep $O(f(n))$ bookkeeping information associated with any solution $\mathcal{P}(R)$ so that we can compute $\Diamond^{-1}(\mathcal{P}(R))$ with $O(f(n))$ I/Os. Operator $\Diamond$ corresponds to executing the bottom-up procedure while the inverse operator $\Diamond^{-1}$ corresponds to the top-down procedure. At this point a comment is in order. As cross-tree degree is upper bounded by B, we could require $O(f(n)\,B)$ I/Os to (re)compute the solution in a node. While this does not change the I/O complexity of the split and concatenate operations, it may change the one of insertions and deletions. In this case, we say that $\Diamond$ is *strongly invertible* if we can obtain the solution $\mathcal{P}(R')$ in any children of a node, along with what remains $\mathcal{P}(R - R')$ from its solution $\mathcal{P}(R)$, with $O(f(n))$ I/Os. For example, if $\Diamond$ is the destructive list append with cost $f(n)$, we can simply keep a pointer to the last item in the appended lists to "de-append" and "re-append" any of them with $O(f(n))$ I/Os by means of $\Diamond^{-1}$. In summary, when dealing with point 3 above, if $\Diamond$ is strongly invertible and solution $\Diamond(s_1, \ldots, s_j)$ is not of constant size, we use $\Diamond^{-1}$ to store the $O(f(n)\,B)$ actions needed to compute $\Diamond(s_1, \ldots, s_j)$ into α itself and store only the leftover of solutions $s_1, \ldots, s_j$ into the children of α. We can indeed recover any of $s_1, \ldots, s_j$ in $O(f(n))$ I/Os by applying $\Diamond^{-1}$ to the solution in α, which is itself recursively computed from the parent of α.

Since we have all the details for our data structure, we can now analyze its performance.

THEOREM 5.1. *We can insert in, delete from, concatenate and split T_H or T_V in order to update cross-tree $CT(T_H \times T_V)$ with $O(f(n)(n/k))$ I/Os, where $f(n) = o(n)$.*

PROOF. We only analyze the complexity of splitting as the other operations have an analogous analysis. Without loss of generality, assume that T_H is split into T_1 and T_2 (horizontal split).

Splitting T_H takes $O(\log_B(n/k))$ time, and the cost of splitting the cross-tree $CT(T_H \times T_V)$ accordingly consists of two tasks: (1) updating the cross-tree topology and (2) recomputing the solutions to $\mathcal{P}$ by means of operators $\Diamond$ and $\Diamond^{-1}$.

Let n_ℓ be the number of nodes of T_V on level ℓ, where $0 \le \ell \le h = O(\log_B(n/k))$. In task (a), we examine $O(n_\ell)$ nodes per level in the cross-tree and execute $O(n_\ell)$ work on them. By summing over all the levels ℓ, we obtain a total cost of

$$\sum_{\ell=0}^{h} O(n_\ell) = O(|T_V|) = O(n/k),$$

since the path that is splitting T_H induces a subtree of $CT(T_H \times T_V)$ which is isomorphic to T_V.

In order to analyze task (2), we proceed as follows. We call a cross-tree node α nonempty if its corresponding region in $\mathcal{M}$ contains at least one point, and we call it empty otherwise. Since we traverse $O(n/k)$ empty nodes by the analysis of task (1), we have to bound the cost for the nonempty nodes and take their cost $f(n)$ into account. We define I_ℓ to be the set of nonempty nodes involved, on level ℓ, by the splitting operation. Their total contribution can be expressed as

$$O\left(\sum_{\ell=0}^{h} \sum_{\alpha \in I_\ell} f(r_\alpha)\right),$$

where r_α denotes the number of points in the region corresponding to α, with $1 \le r_\alpha \le n$. As $f(r_\alpha) \le f(n)$, we have that:

$$O(\sum_{\ell=0}^{h} \sum_{\alpha \in I_\ell} f(n)) = O(\sum_{\ell=0}^{h} |I_\ell| f(n)) = O(f(n) \sum_{\ell=0}^{h} |I_\ell|) = O(f(n)(n/k)),$$

where $|I_\ell| = O(n_\ell)$ by the split operation.

Combining the analysis of tasks (1) and (2) and adding the $O(\log_B(n/k))$ splitting cost of T_H yields the claimed bounds.   □

By an analogous argument, we can prove:

THEOREM 5.2. *Building the cross-tree* $CT(T_H \times T_V)$ *requires* $O(f(n)(n^2/k^2))$ *I/Os and* $O(f(n)(n^2/(k^2B)))$ *space, where* $f(n) = o(n)$.

6. Implementing the Operations

We now show how to use our data structure for solving an order decomposable problem $\mathcal{P}$. To keep our presentation general, we introduce a few functions that help us to state some I/O and space bounds.

$P(k)$: The number of I/Os required to preprocess an $O(k)$-point stripe and solve problem $\mathcal{P}$ for every basic square in the stripe. We will exploit the fact that "large" basic squares are already $\prec_B$-ordered to determine $P(k)$, and assume that $P(k) \ge k/B$ as we need to scan the k points, and that $P(k)/k$ is a smooth nondecreasing function.

$U(k)$: The number of I/Os required to update the solution to problem $\mathcal{P}$ for a basic square in an $O(k)$-point stripe after its preprocessing. We assume that $U(k) \ge \log_B k$, since we have to update at least the B$^+$-tree in the basic square.

$S(k)$: The number of blocks occupied by an $O(k)$-point stripe. We also assume that $S(k) \ge k/B$ is a smooth nondecreasing function.

In most of our applications, we will have $P(k) = O(k/B)$, $S(k) = O(k/B)$, $U(k) = O(\log_B k)$ (and $k = \lceil \sqrt{nB} \rceil \le n$).

Preprocessing n points requires their external memory sorting in lexicographic order with $Sort(n)$ I/Os and an external memory sorting of all the points in every basic square in $\prec_B$-order. Next, the B$^+$-trees are built in a total of $\Theta(n/B)$ I/Os. It takes further $O(P(n))$ I/Os to scan the stripes, determine the solutions in their basic squares and to group the basic squares into blocks. Further $O(f(n)(n^2/k^2))$ I/Os are needed to build the cross-tree and recompute the solution in its nodes by means of operator $\Diamond$ according to Theorem 5.2. In summary, the preprocessing step takes a total $O(Sort(n) + P(n) + f(n)(n^2/k^2))$ I/Os, where $Sort(n)$ is the I/O complexity of sorting n points.

We now describe how to perform the dynamic operations on $\mathcal{M}$.

OPERATIONS h_split, v_split. We perform $v\_split(\mathcal{M}, j)$ as follows. Column j might fall inside a vertical stripe σ, which must necessarily be split. We examine the basic squares of σ. Given a basic square, we scan its points according to their $\prec_B$-order and produce two $\prec_B$-*ordered* lists with a simple scan: one list contains all the points whose second coordinate is smaller than or equal to j and the other list contains the remaining points. We split the basic square into two squares and, if necessary, build two B$^+$-trees by using the two $\prec_B$-ordered lists. By repeating this for all the $O(n/k)$ basic squares in σ, which contains $O(k)$ points, we split it into new stripes σ_1 and σ_2, with $O(k/B + n/k)$ I/Os by Lemma 3.1. We check to see if we can combine σ_1 and σ_2 with their neighbor stripes to maintain the size invariant of order k. For any two such stripes to be merged, we examine their basic squares in pairs (a square per stripe), such that the two squares are on the same horizontal stripe. We take their two $\prec_B$-ordered lists of points and merge them to build a B^+-tree on the resulting list with a linear number of I/Os. Again, this requires $O(k/B + n/k)$ I/Os. It is worth noting that splitting and merging stripes preserve the order of their presorted points. We maintain the size invariant of order B to adjust the number of blocks storing these basic squares, with further $O(n/k + k/B)$ I/Os. After that, we recompute the solutions in the $O(n/k)$ basic squares contained in a constant number of stripes with $O(P(k))$ I/Os. It remains to split cross-tree $CT(T_H \times T_V)$ to reflect the split operation on σ. We split tree T_V at the leaf w corresponding to σ. We split w into two new leaves w_1 and w_2, corresponding to the split of σ into σ_1 and σ_2. If σ_1 or σ_2 are combined with their neighbor stripes, we do the same on w_1 and w_2 and their neighbor leaves. Globally, we create no more than $O(n/k)$ leaves corresponding to the new basic squares in $O(1)$ stripes, and we update their ancestors by a bottom-up traversal of the cross-tree, with $O(n/k)$ I/Os by Lemma 4.1. We then take $O(f(n)(n/k))$ I/Os to update the solutions in the internal nodes by means of $\Diamond$ (and possibly of $\Diamond^{-1}$). We therefore spend a total of $O(f(n)(n/k) + k/B + P(k)) = O(n/k + P(k))$ I/Os, as $P(k) \geq k/B$. The implementation of h_split is completely analogous.

OPERATIONS $h\_concatenate$, $v\_concatenate$. They are the inverse of h_split and v_split respectively, and can be implemented in a total of $O(f(n)(n/k) + P(k))$ I/Os.

OPERATION $range(h_1, h_2, v_1, v_2, \mathcal{M})$. We perform it by executing an h_split at row $(h_1 - 1)$ followed by another h_split at row h_2. We let r be the number of rows of $\mathcal{M}$. These two horizontal splits produce three matrices $\mathcal{M}_1, \mathcal{M}_2$, and $\mathcal{M}_3$, such that $\mathcal{M}_1$ consists of the first $(h_1 - 1)$ rows of $\mathcal{M}$, $\mathcal{M}_2$ contains the rows of $\mathcal{M}$ between h_1 and h_2, and $\mathcal{M}_3$ contains the last $(r - h_2 + 1)$ rows of $\mathcal{M}$. We produce submatrix $\mathcal{M}[h_1, h_2; v_1, v_2]$ by carrying out two additional vertical splits on $\mathcal{M}_2$ at columns $(v_1 - 1)$ and v_2, respectively. We give the solution that is stored in the root of the resulting cross-tree, which corresponds to region $\mathcal{M}[h_1, h_2; v_1, v_2]$, and then assemble back the pieces by a proper sequence of four $concatenate$ operations. Since it can easily be mimicked by a proper sequence of $O(1)$ $split$ and $concatenate$ operations, a $range$ operation takes $O(f(n)(n/k) + P(k))$ I/Os.

OPERATION $set(i, j, w, \mathcal{M})$. We use it to perform either an insertion (if w is nonempty) or a deletion (if w is empty). Let us assume first that we have only insertions. We locate the basic square corresponding to position (i, j) by traversing a path π of length $O(\log_B(n/k))$ from the cross-tree root to the proper leaf. If necessary, we insert (i, j) into the B^+-tree in the basic square. Since we use the $\prec_B$-order, this takes $O(\log_B k)$ I/Os. We then update the solution in the basic square with $U(k)$ I/Os and link it to its relative cross-tree leaf if the basic square becomes nonempty. We propagate the new solution upwards along path π by applying operator $\Diamond$ to its nodes, with $O(f(n))$ I/Os per node (and possibly with the help of $\Diamond^{-1}$). The total cost is $O(f(n)\log_B(n/k))$ I/Os. As setting $\mathcal{M}[i, j] = w$ can cause the insertion of a new row at position i and a new column at position j, the horizontal stripe and the vertical stripe whose intersection gives the updated basic square can violate the size invariant of order k. Since we are treating insertions, this means that each such stripe σ has increased its number of rows (columns) to $k + 1$. Consequently, we split σ into two stripes σ_1

and σ_2 of at most $\lceil (k+1)/2 \rceil$ rows (columns) each in order to maintain the size invariant, with $O(P(k) + f(n)(n/k))$ I/Os (as in the split operation). This gives a total amortized cost of $O(f(n)\log_B(n/k) + U(k) + P(k)/k + f(n)(n/k^2))$ per insertion. In order to obtain good worst-case bounds for the single insertion, we spread this cost among the subsequent operations on the items in σ. To this end, we use Overmars' global rebuilding technique [27] by adding an extra worst-case cost $O(P(k)/k + f(n)(n/k^2))$ to each insertion, such that $\lceil (k+1)/2 \rceil - 1$ insertions are sufficient to cover the $O(P(k) + f(n)(n/k))$ splitting cost for σ. To perform deletions, we simply mark the given item as logically deleted and update its basic square with $U(k)$ I/Os and the solutions along the path π to its cross-tree leaf in a total of $O(\log_B(n/k) + U(k))$ I/Os. We apply again global rebuilding to bound the number of deleted items. In summary, set can be supported in $O(f(n)\log_B(n/k) + U(k) + P(k)/k + f(n)(n/k^2))$ I/Os.

This is summarized in the following theorem.

THEOREM 6.1. *The splitting and merging problem on $n > M$ points can be solved with the following number of I/Os, for a parameter k, and for $f(n) = o(n)$:*

- *range, h_split, v_split, h_concatenate, v_concatenate in $O(f(n)(n/k) + P(k) + P(n)/n)$, with $P(k) \geq k/B$;*
- *set in $O(U(k) + f(n)(\log_B(n/k) + (n/k^2)) + P(n)/n)$, with $U(k) \geq \log_B k$.*

The number of occupied blocks is $O(S(n) + f(n)(n^2/(k^2 B)))$ and the preprocessing cost is $O(Sort(n) + P(n) + f(n)(n^2/k^2))$ I/Os, where $Sort(n)$ is the I/O complexity of sorting n points.

In most of our applications $P(k) = O(k/B)$, $U(k) = O(\log_B k)$ and $S(k) = O(k/B)$:

COROLLARY 6.2. *When $P(k) = O(k/B)$, $U(k) = O(\log_B k)$, $S(k) = O(k/B)$, and $f(n) = O(1)$, the splitting and merging problem on n items can be solved with the following I/O bounds:*

- *h_split, v_split, h_concatenate, v_concatenate in $O(\sqrt{n/B})$;*
- *range in $O(\sqrt{n/B} + r/B)$, where r is the number of points reported by the search;*
- *set in $O(\log_B n)$.*

The space required is $O(n/B)$ blocks and the preprocessing cost is $O(Sort(n))$ I/Os.

PROOF. Fix $k = \lceil \sqrt{n B} \rceil \leq n$. The corollary is now a direct consequence of Theorem 6.1. □

7. Some Applications

We can apply Theorem 6.1 and Corollary 6.2 to a large set of order decomposable problems for basic data structures (e.g., member searching, predecessor, ranking, priority queues), in computational geometry (e.g., neighbor queries, union and intersection queries, interval queries, convex hull), and database applications (e.g., partial match queries, range queries). The reader is referred to [16] for more details on this.

We now sketch an alternative approach to solve our splitting and merging problem. We illustrate it by implementing a two-dimensional priority queue in external memory, in which split, concatenate, insertions and deletions have all

the same cost: $O(Sort(\sqrt{n \log_2 B})) = O\left(\frac{\sqrt{n \log_2 B}}{B} \log_{M/B} \frac{\sqrt{n \log_2 B}}{B}\right)$ I/Os. This
bound is asymptotically greater than the one given in Corollary 6.2. However, due
to the typical values of B, we expect that it could be factor in practice.

We describe the main ideas by examining again the "ideal" memory organiza-
tion in which the n points in matrix $\mathcal{M}$ are organized in a "dense" $(\sqrt{n} \times \sqrt{n})$
square grid. The result in Goodrich et al. [15] implies that an arbitrary $\sqrt{n}$ long
path in the grid must require $\Theta(\sqrt{n/B})$ I/Os. Although $\mathcal{M}$ is *sparse* and *dynamic*,
the approach leading to Corollary 6.2 obtains this bound in a simple and practical
way by fixing the slack parameter k so as to minimize the term $n/k + k/B$. If $\mathcal{M}$ is
static, we can improve n/k and obtain a bound better: we duplicate $\mathcal{M}$ and store
one copy in column order and the other copy in row order (in a sequence of adjacent
blocks). Consequently, we can access a row or a column in the grid by retrieving
$\sqrt{n}/B$ blocks from the proper copy. This is not in contrast with the $\Theta(\sqrt{n/B})$
bound given by Goodrich *et al.*, as they allow for an arbitrary path in the square
grid while here we only deal with some axis parallel paths. Basically, we would like
to follow a similar approach when $\mathcal{M}$ is dynamic.

THEOREM 7.1. *A two-dimensional priority queue for a set of n items can be
maintained in external memory with the following I/O bounds:*

- *an item insertion or deletion, a split or concatenate of one order, a minimum-
  weight query in a region: $O(Sort(\sqrt{n \log_2 B}))$.*

The space required is $O(n/B)$ blocks and the preprocessing cost is $O(Sort(n))$ I/Os.

PROOF. We replace $\mathcal{M}$ by a *hierarchy* of $\log_2 B$ smaller and smaller matrices
that can be handled in a "loosely" fashion. In this way, the cross-tree is pruned in
its $\log_2 B$ lower levels, and the matrix on level $\ell < \log_2 B$ basically stores the same
information as the pruned cross-tree nodes that originally were on level ℓ. Each
such matrix is stored in two copies according to its horizontal and vertical stripes,
respectively. The splits and concatenates cause some swaps on its columns and its
rows while producing some item deletions and insertions due to the recomputation
of the solutions (previously kept in the pruned cross-tree nodes). We delay and
buffer these swaps on the several levels of the matrix hierarchy. For a given level,
the matrix copy stored according to its horizontal stripes uses a buffer to store the
pending vertical swaps while the horizontal swaps are promptly executed. Analogously, the matrix copy with the vertical stripes uses a buffer to store the pending
horizontal swaps while the vertical swaps are promptly executed. We apply the
pending swaps to a given stripe with s items in lazy fashion, namely soon before
accessing it. We do it by carefully sorting the swaps and the deleted/inserted items
pending on the stripe, with $Sort(s)$ worst-case I/Os. Because of this buffering, our
insertions and deletions have to collect the points scattered inside each stripe but
we can pay $Sort(s)$ for handling that. As a result, each operation costs $Sort(s)$
per level, where $s = O(n/k + k)$ for the lowest level and $s = O(n/k)$ for the
other levels. Since the number of cross-tree nodes is $O(n^2/(k^2/B^2))$ as a result of
pruning, the cost of accessing the cross-tree is $O(n/(kB))$. This gives a total of
$O(Sort(n/k + k) + Sort(n/k) \log_2 B + n/(kB))$ I/Os, which is

$$O\left(\frac{k}{B} \log_{M/B} \frac{k}{B} + \left(\frac{n/k}{B} \log_{M/B} \frac{n/k}{B}\right) \log_2 B\right)$$

as $Sort(n) = \Theta(\frac{n}{B} \log_{M/B} \frac{n}{B})$. Choosing $k = \lceil \sqrt{n \log_2 B} \rceil$ yields

$$O\left(\frac{\sqrt{n \log_2 B}}{B} \log_{M/B} \frac{\sqrt{n \log_2 B}}{B} + \left(\frac{\sqrt{n}}{B\sqrt{\log_2 B}} \log_{M/B} \frac{\sqrt{n}}{B\sqrt{\log_2 B}} \right) \log_2 B \right)$$

which is $O(Sort(\sqrt{n \log_2 B}))$. $\square$

We conclude this section by mentioning that our technique can be easily extended to the multidimensional case, i.e., when we have $d > 2$ total orders $\prec_1$, ... , $\prec_d$. In this case, we reduce our problem to a matrix problem in which $\mathcal{M}$ is a d-dimensional matrix and item $x \in S$ is a d-dimensional point $(x_1, \ldots, x_d)$, such that x_i is the rank of x in S according to order $\prec_i$, $i = 1, \ldots, d$. We begin by defining the cross-trees for dimension $d > 2$. A cross-tree $CT(T_1 \times T_2 \times \cdots \times T_d)$ of d weight-balanced B-trees $T_1, T_2, \ldots, T_d$ having the same height and their leaves on the same level, is defined as follows:

- For each choice of nodes $u_1, u_2, \ldots, u_d$ on the same level, such that $u_i \in T_i$ for $1 \leq i \leq d$, there is a node $\alpha_{u_1 u_2 .. u_d}$ in $CT(T_1 \times T_2 \times \cdots \times T_d)$.
- For each choice of edges $(u_1, \hat{u}_1), (u_2, \hat{u}_2), \ldots, (u_d, \hat{u}_d)$, where $u_1, u_2, \ldots, u_d$ are on the same level and $(u_i, \hat{u}_i) \in T_i$ for $1 \leq i \leq d$, there is an edge $(\alpha_{u_1 u_2 .. u_d}, \alpha_{\hat{u}_1 \hat{u}_2 .. \hat{u}_d})$ in $CT(T_1 \times T_2 \times \cdots \times T_d)$.

We define the stripes in $\mathcal{M}$ by following the approach described before. In dimension two, a stripe is composed of $O(k)$ rows or columns, which can be seen as matrices of dimension one. We therefore define the stripes as the groups of the $O(k)$ adjacent matrices of dimension $(d-1)$, each such matrix containing $O(1)$ items. There are $O(n/k)$ stripes defined for each coordinate of $\mathcal{M}$ (i.e., we have a total of d sets of $O(n/k)$ stripes each) and satisfy the size invariant of order k (Definition 2.1) on their number of matrices of dimension $(d-1)$. In this way, a stripe contains still $O(k)$ points. By intersecting these d sets of stripes, we obtain a partition of $\mathcal{M}$ into $O((n/k)^d)$ basic squares, such that each basic square overlaps no more than k matrices of dimension $(d-1)$ in each direction. We keep the items in a basic square sorted by a $\prec_B$-order to be specified and build a cross-tree $CT(T_1 \times T_2 \times \cdots \times T_d)$ on the top of these basic squares, where T_i is a weight-balanced B-tree whose leaves are in one-to-one correspondence to the stripes for the i-th coordinate $(1 \leq i \leq d)$. We then store the solutions in the cross-tree nodes exactly as before. As $|T_i| = O(n/k)$, we can use an argument analogous to the one presented in Theorem 5.1 to obtain the following lemma.

LEMMA 7.2. *We can insert in, delete from, concatenate and split any tree T_i in order to update cross-tree $CT(T_1 \times T_2 \times \cdots \times T_d)$ in $O(f(n)(n/k)^{d-1})$ I/Os, where $f(n) = o(n)$.*

Theorem 6.1 can be generalized to dimension d:

THEOREM 7.3. *The splitting and merging problem on n items undergoing $d > 1$ total orders can be solved with the following number of I/Os, for a parameter k $(1 \leq k \leq n)$, and for an operator cost $f(n) = o(n)$.*

- *$O(f(n)(n/k)^{d-1} + P(k) + P(n)/n)$ for operations range, split, and concatenate along any coordinate, with $P(k) \geq k/B$;*
- *$O(U(k) + f(n)(\log_B(n/k) + n^{d-1}/k^d) + P(n)/n)$ for inserting and deleting a point with operation set, with $U(k) \geq \log_B k$;*

The total number of blocks required is $O(S(n) + f(n)(n^d/k^d B))$ and the total pre-processing time is $O(Sort(n) + P(n) + f(n)(n^d/k^d))$.

PROOF. By Lemma 7.2, splitting and merging the cross-tree along one coordinate take $O(f(n)(n/k)^{d-1})$ I/Os. Scanning a stripe takes $O(k/D)$ I/Os for retrieving its $O(k)$ items plus $O((n/k)^{d-1})$ I/Os for traversing the cross-tree. The total cost of processing a stripe is therefore $O(f(n) \ (n/k)^{d-1} + P(k))$ I/Os. With this difference in mind, the operations can be implemented and analyzed as before. For example, the term $(n^{d-1}/k^d)f(n)$ in the complexity for *set* is due to a stripe overflow or underflow which requires splitting and merging the cross-tree every $\Theta(k)$ insert operations. Hence, the analysis of the d-dimensional case follows directly from the two-dimensional case. □

Theorem 7.3 finds applications to the d-dimensional versions of some order decomposable problems. For example, we can tune properly k (i.e., choose $k = \lceil B^{1/d} n^{1-1/d} \rceil$) to obtain external-memory multidimensional 2-3-trees:

THEOREM 7.4. *A d-dimensional 2-3-tree storing n d-dimensional points can be maintained with the following I/O bounds:*

- *a point insertion or deletion in $O(\log_B n)$;*
- *a split or a concatenate along any coordinate in $O((n/B)^{1-1/d})$;*
- *a range search in $O((n/B)^{1-1/d} + r/B)$, where r is the number of points reported by the search.*

The space required is $O(n/B)$ blocks and the preprocessing cost is $O(Sort(n))$ I/Os.

8. Conclusions

We described some efficient methods for organizing large multidimensional data sets in external memory, and for maintaining them simultaneously sorted under several total orderings. Besides standard insertions and deletions of single items, our proposed solutions perform efficiently *split* and *concatenate* operations on the whole data sets according to any ordering. We made some techniques devised for main memory [16] suitable for efficient external memory implementation. Differently from other approaches, our techniques are based more on simple geometric properties rather than on underlying sophisticated data structures, and exploit the fact that some data structures can be built on (pre)sorted items more efficiently. Balancing is easy as we introduce a new multidimensional data structure, *the cross-tree*, that is the cross product of balanced trees and generalizes the notion of B-trees to higher dimensions by carefully combining space-driven and data-driven partitions. As a result, the cross-tree is competitive with other popular index data structures that require linear space, such as quad-trees [13], k-d trees [7], grid files [24], space filling curves, e.g. [26], hB-trees [22], and R-trees [17].

ACKNOWLEDGMENTS. We are indebted to Lars Arge and to Paolo Ferragina for helpful discussions.

References

[1] A. Aggarwal and J.S. Vitter, The Input/Output Complexity of Sorting and Related Problems, *Comm. ACM* 31 (1988), 1116–1127.

[2] A.V. Aho, J.E. Hopcroft, and J.D. Ullman. *The Design and Analysis of Computer Algorithms.* Addison–Wesley, Reading, MA, 1974.

[3] L. Arge, P. Ferragina, R. Grossi and J.S. Vitter. On Sorting Strings in External Memory. In 29th ACM Symposium on the Theory of Computing (1997), pp.540–548.

[4] L. Arge, O. Procopiuc, S. Ramaswamy, T. Suel, and J.S. Vitter. Theory and Practice of I/O-Efficient Algorithms for Multidimensional Batched Searching Problems. In Proc. *9th Annual ACM-SIAM Symp. on Discrete Algorithms* (1998).

[5] L. Arge and J.S. Vitter, Optimal Dynamic Interval Management in External Memory. 37th IEEE Symp. on Foundations of Computer Science (1996).

[6] R. Bayer and C. McCreight, Organization and Maintenance of Large Ordered Indexes. *Acta Informatica 1*, 3 (1972), 173–189.

[7] J.L. Bentley, Multidimensional Binary Search Trees Used for Associated Searching. *Comm. ACM*, 19 (1975), 509–517.

[8] J.L. Bentley, Decomposable Searching Problems. *Information Processing Letters*, 8 (1979), 244–251.

[9] J.L. Bentley and J.B. Saxe, Decomposable Searching Problems I. Static–to–Dynamic Transformation. *J. of Algorithms*, 1 (1980), 301–358.

[10] Y.-J. Chiang and R. Tamassia, Dynamic Algorithms in Computational Geometry, *Proceedings of the IEEE*, Special issue on Computational Geometry, G. Toussaint, ed., 80 (1992) 1412–1434.

[11] D. Comer, The Ubiquitous B-Tree. *Computing Surveys 11* (1979), 121–137.

[12] P. Ferragina and R. Grossi. A Fully Dynamic Data Structure for External Substring Search, 27th ACM Symp. on Theory of Computing (1995). Full version in: An External-Memory Indexing Data Structure and its Applications, submitted to journal.

[13] R.A. Finkel and J.L. Bentley, Quad–trees: A Data Structure for Retrieval of Composite Keys. *Acta Inform.*, 4 (1974), 1–9.

[14] V. Gaede and O. Günther. Multidimensional Access Methods. To appear in *ACM Computing Surveys* (1998).

[15] M.T. Goodrich, M.H. Nodine, and J.S. Vitter, Blocking for External Graph Searching, *Algorithmica*, 16 (1996), 181–214.

[16] R. Grossi and G.F. Italiano. Efficient Splitting and Merging Algorithms for Order Decomposable Problems. In Proc. 24th International Colloquium on Automata, Languages and Programming, LNCS (1997) Springer-Verlag, 605–615. A full version to appear in *Information and Computation* (the revised paper is available as http://www.dsi.unive.it/~italiano/Papers/multiord.ps.gz).

[17] A. Guttman. R-trees: A Dynamic Index Structure for Spatial Searching. In Proc. ACM SIGMOD International Conference on the Management of Data (1984) 47–57.

[18] J.M. Hellerstein, E. Koutsopias and C.H. Papadimitriou. On the Analysis of Indexing Schemes. In Proc. ACM Symp. on Principles of Database Systems (1997), 249–256.

[19] P.C. Kanellakis, S. Ramaswamy, D.E. Vengroff and J.S. Vitter. Indexing for Data Models with Constraints and Classes. *Journal of Computer and System Sciences*, 52 (1996), 589–612.

[20] M.J. van Kreveld and M.H. Overmars, Divided k–d Trees, *Algorithmica*, 6 (1991), 840–858.

[21] M.J. van Kreveld and M.H. Overmars, Concatenable Structures for Decomposable Problems, *Information and Computation*, 110 (1994), 130–148.

[22] D.B. Lomet and B. Salzberg, The hB-tree: A Multiattribute Indexing Method with Good Guaranteed Performance. *ACM Transactions on Databases Systems*, 15 (1990), 625–658.

[23] K. Mehlhorn, *Multi-Dimensional Searching and Computational Geometry* EATCS Monographs on Theoretical Computer Science, vol. 3, Springer–Verlag, Berlin/New York, 1984.

[24] J. Nievergelt, H. Hinterberger and K.C. Sevcik. The Grid File: An Adaptable, Symmetric Multikey File Structure. *ACM Transactions on Databases Systems*, 9 (1984), 38–71.

[25] M.H. Nodine and J.S. Vitter. Greed Sort: Optimal Deterministic Sorting on Parallel Disks. *J. ACM*, 42 (1995) 919–933.

[26] J.A. Orenstein, Spatial Query Processing in an Object-Oriented Database Sytem. In Proc. ACM SIGMOD International Conference on the Management of Data (1986) 326–336.

[27] M.H. Overmars, *The Design of Dynamic Data Structures*, LNCS 156, Springer–Verlag, Berlin/New York, 1983.

[28] M.H. Overmars and J. van Leeuwen, Worst-Case Optimal Insertion and Deletion Methods for Decomposable Searching Problems. *Information Processing Letters*, 12 (1981), 168–173.

[29] M.H. Overmars and J. van Leeuwen, Dynamic Multi-Dimensional Data Structures Based on Quad– and k–d Trees. *Acta Inform.*, 17 (1982), 267–285.

[30] M.H. Overmars, M. Smid, M.T. de berg, and M.J. van Kreveld. Maintaining Range Trees in Secondary Storage. part I: partitions. *Acta Inform.*, 27 (1990), 423–452.

[31] N. P. Patt (ed.), Special Issue on "The I/O Subsystem: A Candidate for Improvement", *IEEE Computer*, (1994).

[32] C. Ramaswamy and S. Subramanian. Path Caching: A Technique for Optimal External Searching. In Proc. ACM Symp. on Principles of Database Systems (1994).

[33] K.V. Ravi Kanth and A.K. Sing, Optimal Dynamic Range Searching in Non-Replicating Index Structures. UCSB Technical Report TRCS97-13, July 1997.

[34] H. Samet, Bibliography on Quad–Trees and Related Hierarchical Data Structures. In *Data Structures for Raster Graphics*, L. Kessenaar, F. Peters, and M. van Lierop eds., Springer-Verlag, Berlin, (1986), 181–201.

[35] S. Subramanian and S. Ramaswamy. The P-Range Tree: A New Data Structure for Range Searching in Secondary Memory. In Proc. 5th ACM–SIAM Annual Symp. on Discrete Algorithms (1995), 378–387.

[36] D.E. Vengroff and J.S. Vitter. Efficient 3-D Range Searching in External Memory. In Proc. 28th ACM Symp. on Theory of Computing (1996) 192–201.

[37] J.S. Vitter and E.A.M. Shriver, Algorithms for Parallel Memory, I: Two–Level Memories, *Algorithmica*, 12 (1994), 110–147. Also in Proc. 22nd ACM Symp. on Theory of Computing (1990) 159–169.

DIPARTIMENTO DI SISTEMI E INFORMATICA, UNIVERSITÀ DI FIRENZE, VIA LOMBROSO 6/17, 50134 FIRENZE, ITALY.
  E-mail address: `grossi@dsi.unifi.it`

DIPARTIMENTO DI MATEMATICA APPLICATA ED INFORMATICA, UNIVERSITÀ "CA' FOSCARI" DI VENEZIA, VIA TORINO 155, 30173 MESTRE (VE), ITALY.
  E-mail address: `italiano@dsi.unive.it`
  URL: `http://www.dsi.unive.it/~italiano`

DIMACS Series in Discrete Mathematics
and Theoretical Computer Science
Volume **50**, 1999

Computing on data streams

Monika R. Henzinger, Prabhakar Raghavan, and Sridhar Rajagopalan

ABSTRACT. In this paper we study the space requirement of algorithms that make only one (or a small number of) pass(es) over the input data. We study such algorithms under a model of *data streams* that we introduce here. We give a number of upper and lower bounds for problems stemming from query-processing, invoking in the process tools from the area of *communication complexity*.

1. Overview

In this paper we study the space requirement of algorithms that make only one (or a small number of) pass(es) over the input data. We study such algorithms under a model of *data streams* that we introduce here. We develop an intimate connection between this setting and the classical theory of *communication complexity* [**AMS96, NK, Yao79**].

1.1. Motivation. A data stream, intuitively, is a sequence of data items that can be read once by an algorithm, in the prescribed sequence. A number of technological factors motivate our study of data streams.

The most economical way to store and move large volumes of data is on secondary and tertiary storage; these devices naturally produce data streams. Moreover, multiple passes are prohibitive due to the volumes of data on these media; for example internet archives [**Ale**] sell (a large fraction of) the web on tape. This problem is exacerbated by the growing disparity between the costs of secondary/tertiary storage and the cost of memory/processor speed. Thus to sustain performance for basic systems operations, core utilities are restricted to read the input only once. For example, storage managers (such as IBM's ADSM [**ADSM**]) use one-pass differential backup [**ABFLS**].

Networks are bringing to the desktop ever-increasing quantities of data in the form of data streams. For data in networked storage, each pass over the data results in an additional, expensive network access.

The SELECT/PROJECT model of data access common in database systems give a "one-pass like" access to data through system calls independent of the physical storage layout. The interface between the storage manager and the application

---

1991 *Mathematics Subject Classification.* Primary 68Q05; Secondary 68Q25.

layer in a modern database system is well-modeled as a data stream. This data could be a filtered version of the stored data, and thus might not be contiguous in physical storage.

In a large multithreaded database system, the available main memory is partitioned between various computational threads. Moreover, operators such as the hash-based GROUP BY operator compute multiple aggregation results concurrently using only a single scan over the data. Thus, the amount of memory used in each thread influences efficiency in two ways: (1) It limits the number of concurrent accesses to the database system. (2) It limits the number of different computations that can be performed simultaneously with one scan over the data. Thus effectively, even a 1Gb machine will have provide under 1Mb to each thread when supporting a thousand concurrent threads, especially after the operating system and the DBMS take their share of memory.   There is thus a need for studying algorithms that operate on data streams.

1.2. Scope of the present work. A *data stream* is a sequence of data items $x_1, \ldots, x_i, \ldots, x_n$ such that the items are read once in increasing order of the indices i. Our model of computation can be described by two parameters: The number P of *passes* over the data stream and the *workspace S* (in bits) required in main memory, measured as function of the input size n. We seek algorithms for various problems that use one or a small number of passes and require a workspace that is smaller than the size of the input. Our model does not require a bound on the computation time, though for all the algorithms we present, the time required is small. In query settings, the size of the output is typically much smaller than the size of the workspace.

For example, for graph problems (e.g., [**MAGQW, MM**]), we view the input as a sequence of m (possibly directed) edges between n nodes. Our goal is to find algorithms where space requirements can be bounded as a function of n (rather than m), or in establishing that the space must grow with m.

Our goal is to expose dichotomies in the space requirements along the different axis: (i) between one-pass and multi-pass algorithms, (ii) between deterministic and randomized algorithms, and (iii) between exact and approximation algorithms.

We first describe some classes of problems which can be described in this context.

(1) Systems such as LORE[**MAGQW**] and WEBSQL [**AMM, MM, MMM**] view a database as a graph/hypergraph. For instance, a directed edge might represent a hyperlink on the web, or a citation between scientific papers, or a pair of cities connected by a flight; in a database of airline passengers a hyperedge may relate a passenger, the airports he uses, and the airline he uses in a flight. Some typical queries might be: From which airport in Africa can one reach the most distinct airports within 3 hops? (The *MAXTOTAL* problem below.) Of the papers citing the most referenced graphics paper, which has the largest bibliography? (The *MAX* problem below.)

We propose four problems that model queries such as those described here: Consider a directed multigraph with node set $V_1 \cup V_2 \cdots \cup V_k$, all of whose edges are directed from a node in V_i to a node in V_{i+1}. Let $n = \max_i |V_i|$. The degree of a vertex is its indegree unless specified otherwise.

The *MAX* **problem.:** Let u_1 be the node of largest outdegree in V_1. Let $u_i \in V_i$ be a node of largest degree among those incident to u_{i-1}. Find u_k.

The *MAXNEIGHBOR* **problem.:** Let u_1 have largest outdegree in V_1. Let $u_i \in V_i$ have the largest number of edges to u_{i-1}, determine u_k.

The *MAXTOTAL* **problem.:** Find a node $u_1 \in V_1$ which is connected to the largest number of nodes of V_k.

The *MAXPATH* **problem.:** Find nodes $u_1 \in V_1$, $u_k \in V_k$ such that they are connected by the largest possible number of paths.

(2) A second problem class is verifying consistency in databases. For instance, check if each customer in a database has a unique address, or if each employee has a unique manager/salary. We model these problems as consistency verification problems of relations. Let a k-ary relation R over $\{0, 1, \cdots n\}$ be given. Let $\phi = \forall u_1, u_2, \ldots ! \exists (v_1, v_2, \ldots) : f(u_1, \ldots, v_1, \ldots) \text{for} (u_1, u_2, \ldots, v_1, v_2, \ldots) \in R$.

The Consistency Verification problem.: Verify that R satisfies ϕ.

(3) More traditional graph problems like connectivity arise [**BGMZ97**], while analyzing various properties of the web. In database query optimization estimating the size of the transitive closure is important [**LN89**]. This motivates our study of various traditional graph properties.

(4) As pointed out in [**SALP79, AMS96**] estimates of the frequency moments of a data set can be used effectively for database query optimization. This motivates our study of approximate frequency estimation problems and approximate selection problems (e.g., find a product whose sales are within 10% of the most popular product).

1.3. Definitions. *Las-Vegas and Monte-Carlo algorithms.* A *randomized* algorithm is an algorithm that flips coins, i.e., uses random bits: no probabilistic assumption is made of the distribution of inputs. A randomized algorithm is called *Las-Vegas* if it gives the correct answer on all input sequences; its running time or workspace could be a random variable depending on the coin tosses. A randomized algorithm is called *Monte-Carlo with error probability* ϵ if on every input sequence it gives the right answer with probability at least $1 - \epsilon$. If no ϵ is specified, it is assumed to be 2/3.

Our principal tool for showing lower bounds on the workspace of limited-pass algorithms is drawn from the area of *communication complexity.*

Communication complexity. Let X, Y, and Z be finite sets and let $f : X \times Y \to Z$ be a function. The *(2-party) communication model* consist of two *players, A* and *B* such that A is given an $x \in X$ and B is given an $y \in Y$ and they want to compute $f(x, y)$. The problem is that A does not know y and B does not know x. Thus, they need to communicate, i.e., exchange bits according to an agreed-upon *protocol.* The *communication complexity of a function* f is the minimum over all communication protocols of the maximum over all $x \in X$ and all $y \in Y$ of the number of bits that need to be exchanged to compute $f(x, y)$. The protocol can be deterministic, Las Vegas or Monte Carlo. Finally, if the communication is restricted to one player transmitting and the other receiving, then this is termed *one-way communication complexity.* In a one-way protocol, it is critical to specify which player is the transmitter and which the receiver. Only the receiver needs to be able to compute f.

1.4. Related previous work. Estimation of order statistics and outliers [**ARS97, AS95, JC85, RML97, Olk93**] has received much attention in the context of sorting [**DNS91**], selectivity estimation [**PIHS96**], query optimization [**SALP79**] and in providing online user feedback [**Hel**]. The survey by Yannakakis [**Yan90**] is a comprehensive account of graph-theoretic methods in database theory.

Classical work on time-space tradeoffs [**Cob66, Tom80**] may be interpreted as lower bounds on workspace for problems such as verifying palindromes, perfect squares and undirected st connectivity. Paterson and Munro [**MP80**] studied the space required in selecting the kth largest out of n elements using at most P passes over the data. They showed an upper bound of $n^{1/P} \log n$ and an almost matching lower bound of $n^{1/P}$ for large enough k. Alon, Matias and Szegedy [**AMS96**] studied the space complexity of estimating the *frequency moments* of a sequence of elements in one-pass. In this context, they show (almost) tight upper and lower bounds for a large number of frequency moments and show how communication complexity techniques can be used to prove lower bounds on the space requirements.

Our model appears at first sight to be closely related to papers on I/O complexity [**HK81**], hierarchical memory [**AACS87**], paging [**ST85**] and competitive analysis [**KMRS88**], as well as external memory algorithms [**VV96**]. However, our model is considerably more stringent: whereas in these papers on memory management one can bring back (into fast memory) a data item that was previously evicted (and is required again), in our model we cannot retrieve items that are discarded.

1.5. Our main results. We expose the following dichotomies in our model. (i) Some problems require large space in one pass but small space in two. (ii) We show that there can be an exponential gap in space bounds between Monte Carlo and Las Vegas algorithms. (iii) We show that if we settle for an approximate solution, we can reduce the space requirement substantially. Our tight lower bounds for the approximate solution apply communication complexity techniques to approximation algorithms.

THEOREM 1.1. *In one pass, the* MAX *problem requires* $\Omega(kn^2)$ *space and has an* $O(kn^2 \log n)$ *space solution. In* $P > 1$ *passes it requires* $\Omega(kn/P)$ *space and can be solved in* $O((kn \log n)/P)$ *space.*

THEOREM 1.2. *In one pass, the* MAXNEIGHBOR, MAXTOTAL, *and* MAX-PATH *problem require* $\Omega(kn^2)$ *space and have* $O(kn^2 \log n)$ *space solutions.*

Notice however, that unlike the *MAX* problem, the other three do not seem to admit efficient two pass solutions. Resolving this remains an open issue. We believe that no constant number of passes will result in substantial savings.

Let R be a $k = (k_1 + k_2)$-ary relation over $\{1, \ldots n\}$. Consider the formula

$$\phi = \forall u_1 \ldots u_{k_1}, !\exists (v_1 \ldots v_{k_2}) : f(u_1 \ldots u_{k_1}, v_1 \ldots v_{k_2}) \text{ for } (u_1, \ldots v_1 \ldots) \in R$$

where f is a function assumed to be provided via an oracle. Also suppose that we are presented the relation R one tuple at a time. Then, we have the following:

THEOREM 1.3. *Verifying that* R *satisfies* ϕ *can be done by an* $O(\log \frac{1}{\delta} \log n)$ *space Monte Carlo algorithm that outputs the correct answer with probability* $1 - \delta$. *Any Las Vegas algorithm that verifies that* R *satisfies* ϕ *requires at least* $\Omega(n^2)$ *space.*

Theorem 1.3 shows an exponential gap between Las Vegas and Monte Carlo algorithms. In Section 4, we describe an algorithm and its analysis; these are easily modified to yield Theorem 1.3 through a *completeness* property described further in Section 4. We also have the following open problem: Let R be a binary relation. Let $\phi = \forall x \exists u : (x, u) \in R$. Is there any sub-linear space Monte Carlo algorithm that verifies that R satisfies ϕ?

THEOREM 1.4. *Given a sequence of m numbers in $\{1, \ldots, n\}$ with multiple occurrences finding the k most frequent items requires $\Omega(n/k)$ space. Random sampling yields an upper bound of $O(n(\log m + \log n)/k)$.*

The proof of Theorem 1.4 is in Section 5.

The *approximate median problem* requires finding a number whose rank is in the interval $[m/2 - \epsilon m, m/2 + \epsilon m]$. It can be solved by a one-pass Monte Carlo algorithm with error probability $1/10$ and $O(\log n(\log 1/\epsilon)^2/\epsilon)$ space [**RML97**]. We give a corresponding lower bound in Section 5.

THEOREM 1.5. *Any 1-pass Las Vegas algorithm for the approximate median problem requires $\Omega(1/\epsilon)$ space.*

Easy one-pass reductions from the communication complexity of the DISJOINTNESS function [**NK**] yields:

THEOREM 1.6. *In P passes, the following graph problems on an n-node graph all require $\Omega(n/P)$ space: computing the connected components, k-edge connected components with $1 < k < n$, k-vertex connected components with $1 < k < n$, testing graph planarity. Finding the sinks in a directed graph requires $\Theta(n/P)$ space.*

Incremental graph algorithms give one-pass algorithms for all the problems of Theorem 1.6. Thus, there are one-pass algorithms for connected components, k-edge and k-vertex connectivity with $k \leq 3$, and planarity testing that use $O(n \log n)$ space.

THEOREM 1.7. *For any $1 > \epsilon > 0$, estimating in one pass the size of the transitive closure to within a factor of ϵ requires space $\Omega(m)$.*

We prove this theorem in Section 5. Computing the exact size of the transitive closure requires $O(m \log n)$ space.

The lower bounds of Theorems 1.1, 1.2, 1.6 and 1.7 hold even for Monte Carlo algorithms that are correct with error probability ϵ for a sufficiently small ϵ.

All our lower bounds are information-theoretic, placing no bounds on the computational power of the algorithms. The upper bounds, on the other hand, are all "efficient": in all cases, the running time is about the same as the space usage.

2. Three lower bounds from communication complexity

Many of the lower bounds in our model build on three lower bounds in communication complexity. We review these lower bounds in this section.

Bit-Vector Probing.: Let A have a bit-vector x of length m. Let B have an index $0 < i \leq m$. B needs to know x_i, the ith input bit. The only communication allowed is from A to B.

There is no better method for A to communicate x_i to B than to send the entire string x. More precisely, any algorithm that succeeds in B guessing x_i correctly with probability better than $(1 + \epsilon)/2$, requires at least ϵm bits of communication [**NK**].

Bit-Vector Comparison.: Let A and B both have bit-vectors x and y respectively, each of length m. B wishes to verify that $x = y$.

Any deterministic or Las Vegas algorithm that successfully solves this problem must essentially send the entire string x from A to B, or vice versa. More precisely, any algorithm that outputs the correct answer with probability at least ϵ and never outputs the wrong answer must communicate ϵm bits [**NK**].

Bit-Vector Disjointness.: Let A and B both have bit-vectors x and y respectively, each of length m. B wishes to find an index i such that $x_i = 1$ and $y_i = 1$.

There is no better protocol than to essentially send the entire string x from A to B, or vice versa. More precisely, any algorithm that outputs the correct answer with probability at least $1 - \epsilon$ (for some small enough ϵ) must communicate $\Omega(m)$ bits [**NK**].

Notice that the second theorem is weaker than the first and the third in some respects: it does not apply to Monte Carlo algorithms. There is a good reason: there is a Monte Carlo algorithm that does much better, i.e. communicates only $O(\log n)$ bits. On the other hand, the first theorem is weaker than the second and third in some respects: it insists that there be no communication in one of the two directions. This too is for good reason: B could send A the index, and then A could respond with the bit. For a description of these and other issues in this area, see [**NK**].

3. One pass versus many passes

Our goal in this section is to outline the proof of Theorem 1.1 showing that some problems require large space in one pass but small space in two. We give here a lower bound of $\Omega(n^2)$ on the space used by any Monte Carlo one-pass algorithm for the 2-layer *MAX* problem; a somewhat more elaborate construction (omitted here) yields a lower bound of $\Omega(kn^2)$ for the k-layer version.

PROOF OF THEOREM 1.1. We provide a reduction from the bit-vector probing problem. Denote the node-set on the left of the bipartite graph by U, and the node-set on the right by V, where $|U| = |V| = n$. We further partition U into U_1, U_2 where $|U_1| = n/3 = \sqrt{m}$. Likewise we partition V into V_1, V_2 where $|V_1| = n/3$. The bit-string x is interpreted as specifying the edges of a bipartite graph on $U_1 \times V_1$ in the natural way, i.e. the edge (u, v) corresponds to index $u\sqrt{m} + v$. On getting the query i, we translate it to an edge $(u, v), u \in U_1, v \in V_1$ and augment the graph with edges (u, v') and (u', v) for each $u' \in U_2$ and $v' \in V_2$. The answer to the *MAX* problem on this graph is v if and only if the edge (u, v) is in the bipartite graph, i.e. the ith input bit is set.

The *MAX* problem is solved in 2 passes with space $O(kn \log n)$, even on k layered graphs: In the first pass find the degree of each vertex. In the second pass determine the highest degree neighbor in V_i for each vertex in V_{i-1}. Then compute u_1 and repeatedly find the highest degree neighbor of the current node until u_k is determined. This algorithm can be modified to use only space $O(kn \log n/P)$ in P passes. $\qquad\square$

Note that the lower bound proof that we provide above applies to approximate versions of the MAX problem as well, namely, it requires $\Omega(n^2)$ space to compute a near max degree neighbor of a vertex with near max degree in U.

PROOF OF THEOREM 1.2. The proofs for all three problems are reductions from the bit-vector probing problem similar to the proof of Theorem 1.1.

To show the bound for the $MAXNEIGHBOR$ problem we construct the same initial graph as in the proof of Theorem 1.1, but double each edge from V_1 to U_1. On getting the query i, we translate it into an edge (u,v), $u \in U_1, v \in V_1$, and add this edge to the graph. Additionally we augment the graph with two edges (u,v') for each $v' \in V_2$. Then there are three edges between u and v iff the ith input bit is set; otherwise there is only one edge between u and v. Thus, v is returned iff the ith input bit is set.

For the $MAXTOTAL$ problem construct a tripartite graph with node set $U \cup V \cup W$, where $|U| = |V| = |W| = \sqrt{m}+1$. The nodes in set U are numbered from 1 to $\sqrt{m}+1$. The same holds for set V and set W. As in the proof of Theorem 1.1, the bit-string x is translated into edges from U to V as follows: edge (u,v) exists in the graph iff index $u\sqrt{m}+v$ of x is set. Additionally there is an edge from node $\sqrt{m}+1$ in U to node $\sqrt{m}+1$ in V and from the latter node to node $\sqrt{m}+1$ in W. On getting a query i we translate it into an edge (u,v) with $u \in U$ and $v \in V$ and augment the graph by edges (v,w) for each $w \in W$. Then u reaches the most nodes in W iff the ith input bit was set; otherwise node $\sqrt{m}+1$ of U reaches the most nodes of W.

For the $MAXPATH$ problem augment the graph for the $MAXTOTAL$ problem with a fourth node set X and connect every node of W with an edge to the same node x of X. Using the same reduction for a query as for problem $MAXTOTAL$ shows that u and x are connected by the largest number of paths iff the ith input bit was set; otherwise node $\sqrt{m}+1$ and x are connected by the largest number of paths. $\qquad\square$

4. Las Vegas versus Monte Carlo

In this section we present an exponential gap between Las Vegas and Monte Carlo one-pass algorithms. The *symmetry* property given a sequence of ordered pairs over $\{1,\dots,n\}$, is that (u,v) is in the sequence if and only if there is a unique (v,u) in it. In this section, we show a $O(\log n)$ space Monte Carlo algorithm to verify the symmetry property. By contrast, any one pass Las Vegas algorithm requires $\Omega(m)$ space, where m is the size of the relation.

ALGORITHM 4.1. Choose p, a random prime smaller than n^3. Let $l_{u,v} = n^{2((nu)+v)}$ if $u < v$ and $l_{u,v} = -(n^{2((nu)+v)})$ if $u > v$ and 0 otherwise. Compute the sum $s = \sum_{(u,v)\in R} l_{u,v}$ modulo p. Check if $s = 0 \mod p$ at the end. Storing $s \mod p$ requires only $\log p < 3\log n$ space. Also check that there are no more than n^2 edges in all.

THEOREM 4.1. *Algorithm 4.1 will output a correct response with probability at least* $1 - (2\log^2 n/n)$*. Moreover, any one pass Las Vegas algorithm that outputs the correct response with probability* $2/3$ *or more uses* $\Omega(n^2)$ *space.*

PROOF. It is easily seen that $\sum_{(u,v)\in R} l_{u,v}$ is 0 if and only if R is symmetric. On the other hand, it follows from the Chinese Remainder Theorem and the fact that there are at least $n^3/\log n$ primes smaller than n^3, that the probability that a non zero sum evaluates to 0 modulo a random prime is smaller than $2\log^2 n/n$. This follows, since s could be 0 modulo p for at most $2n^2 \log n$ of them since $s < 2^{2n^2 \log n}$.

The lower bound follows from a reduction from the bit-vector comparison problem. Let as specified earlier, player A and B have strings x and y each of length $n' = \frac{n(n-1)}{2}$ srespectively. Interpret each string as an upper triangular matrix specifying undirected graphs G_x and G_y each containing n vertices respectively. Clearly, $G_x = G_y$ if and only if $x = y$. Now construct the sequence $\sigma = \sigma_x + \sigma_y = \{(u,v): \{u,v\} \in G_x \text{and} u < v\} + \{(v',u'): \{u',v'\} \in G_y \text{and} u' < v'\}$. Again, it is easily verified that this sequence is symmetric if and only if $G_x = G_y$, i.e. $x = y$. Also, the size of the sequence σ is the sum of the number of edges in G_x and G_y.

Let S be any algorithm which uses space s and tests symmetry. Then, S can be turned into a Las Vegas communication protocol using $s + O(1)$ bits for bit-vector comparison as follows: A and B interpret x and y as sequences σ_x and σ_y as above. A runs S on σ_x and then communicates $s + O(1)$ bits of state to B, who continues the execution of S on σ_y. Since the sequence $\sigma_x + \sigma_y$ represents a symmetric relation if and only if $x = y$, B will be able to judge if x indeed matches y. Consequently, $s \in \Omega(n') = \Omega(n^2)$. □

More interesting, however, is a completeness property that arises here. Namely, that every problem of the form in Theorem 1.3 can be reduced to the symmetry problem.

PROOF OF THEOREM 1.3. The reduction works as follows: we encode each tuple $(u_1, \ldots u_{k_1})$ as an index i using a standard Gödel encoding, g, let this encoding range from 1 through m. Then we output the ordered pair $(i, 0)$ for each $1 \leq i \leq m$. Then for each tuple $(u_1 \ldots, v_1, \ldots)$ we output $(0, g(u_1, \ldots u_{k_1}))$ if and only if $f(u_1, \ldots, v_1 \ldots)$. The resultant graph is symmetric if and only if the relation R satisfies ϕ. □

5. Exact versus approximate computation

In this section, we show that if we settle for an approximate solution, we can reduce the space requirement substantially. Our matching lower bounds for the approximate solution require a generalization of communication complexity techniques to approximation algorithms.

PROOF OF THEOREM 1.4. Alon *et al.* show that finding the mode (i.e., the most frequently-occurring number) of a sequence of m numbers in the range $\{1, \ldots, n\}$ requires space $\Omega(n)$. By a simple reduction (replace each number i in the original sequence by a sequence of k numbers $ki+1, ki+2, \ldots ki+k$), it follows that that finding one of the k most frequent items in one pass requires space $\Omega(n/k)$

The almost matching upper bound is given by the following Monte-Carlo algorithm that succeeds with constant probability: before the start of the sequence sample each number in the range with probability $1/k$ and then only keep a counter for the successfully sampled numbers. Output the successfully sampled number with largest count. With constant probability one of the k-th most frequent numbers has been sampled successfully. This needs $O(n(\log m + \log n)/k)$ space. $\qquad\square$

PROOF OF THEOREM 1.5. We show that any algorithm that solves the ϵ - approximate median problem requires $\Omega(1/\epsilon)$ space. The proof follows from a reduction from the bit-vector probing problem. Let $b_1, b_2 \cdots b_n$ be a bit vector followed by a query index i. This is translated to a sequence of numbers as follows: First output $2j + b_j$ for each j. Then on getting the query, output $n - i - 1$ copies of 0 and $i + 1$ copies of $2(n + 1)$. It is easily verified that the least significant bit of the exact median of this sequence is the value of b_i. Choose $\epsilon = \frac{1}{2n}$. Thus, the ϵ approximate median is the exact median. Thus, any one pass algorithm that requires fewer than $\frac{1}{2\epsilon} = n$ bits of memory can be used to derive a communication protocol that requires fewer than n bits to be communicated from A to B in solving bit vector probing. Since every protocol that solves the bit-vector probing problem must communicate n bits, this is a contradiction. $\qquad\square$

We next prove Theorem 1.6.

PROOF OF THEOREM 1.6. We reduce the bit-vector disjointness problem to the graph connectivity problem. Construct a graph whose node-set is $\{a, b, 1, 2, \ldots , n\}$. Insert an edge (a, i) if bit i is set in A's vector, and insert an edge (b, i) if bit i is set in B's vector. Now, a and b are connected in the graph if and only if there exists a bit that is set in both A's vector and B's vector. By the lower bound for the bit-vector disjointness problem, every protocol must exchange $\Omega(n)$ bits between A and B. Thus, if there are P passes over the data, one of the passes must use at least $\Omega(n/P)$ space. The reduction for k-edge or k-vertex connectivity follows by adding $k - 1$ nodes $c_1, \ldots, c_{k-1}$ and an edge from each c_j, $1 \leq j \leq k - 1$ to both a and b.

To reduce to planarity testing we add four nodes c_1, c_2, c_3, c_4 and connect them pairwise. Additionally we add the edges (c_1, a), (c_2, a), (c_3, a), and (c_4, b). Then the graph contains K_5 as a minor if and only if a and b are connected.

We also reduce the bit-vector disjointness problem to the problem of deciding whether the graph contains a sink. Construct a graph whose node-set is $\{a, b, 1, 2, \ldots, n\}$. Insert edges (a, b) and (b, a) to guarantee that neither of them is a sink. If bit i is set in A's vector, insert an edge (a, i), otherwise insert an edge (i, a). Similarly, if bit i is set in B's vector, insert an edge (b, i), otherwise insert an edge (i, b). Now node i is a sink if and only if bit i is set in both A's and B's vector. It follows that the graph contains a sink if and only if there exists a bit that is set in both A's vector and B's vector. By the lower bound for the bit-vector disjointness problem, every protocol must exchange $\Omega(n)$ bits between A and B. Thus, if there are P passes over the data, one of the passes must use at least $\Omega(n/P)$ space.

A P-pass algorithm that keeps a bit for node $(i-1)n/P, (i-1)n/P+1, \ldots, in/P- 1$ in pass i indicating whether an edge leaving the node was read gives the desired upper bound. $\qquad\square$

We also provide a lower bound here for the transitive closure problem.

PROOF OF THEOREM 1.7. We reduce the bit-vector probing problem to the transitive closure estimation problem. Let $d \geq 1$ be a constant. Given a bit-vector of length m, we construct a graph G on $2(dm + \sqrt{m})$ vertices, V_i, $1 \leq i \leq 4$, with $|V_2| = |V_3| = \sqrt{m}$ and $|V_1| = |V_4| = dm$, such that edge (i, j) with $i \in V_2$ and $j \in V_3$ exists iff entry $i\sqrt{m} + j$ is set in the vector. To test whether entry $i\sqrt{m} + j$ is set, add edges from each vertex in V_1 to $i \in V_2$ and from $j \in V_3$ to each vertex in V_4. The size of the transitive closure is larger than m if and only if the edge (i, j) is in the graph. Furthermore, for $\epsilon < 1 - 2/(d^2 + 1)$, any ϵ-approximation algorithm for the transitive closure can answer a query correctly. Thus, any ϵ-approximation algorithm must use $\Omega(m)$ space. □

6. Further work

Our work raises a number of directions for further work; we list some here:

1. We need more general techniques for both lower and upper bounds when multiple passes can be performed over the data. They might also imply interesting new results about communication complexity. From a practical perspective, algorithms are needed for a wider class of problems than the selection problem that has been extensively studied [**ARS97, AS95, JC85, RML97, Olk93**].

2. Can we design algorithms that minimize the number of passes performed over the data given the amount of memory available? This would be useful when, for instance, the number of active concurrent threads governs the memory available at runtime

3. How can we arrange the data physically in a linear order with the express goal of optimizing the memory required to process some set of queries? Recall that the results of a query may not necessarily be physically contiguous (e.g., in the database of airports, the subset from Africa may not be together; more generally, we will have to cope with the results of some class of SELECT and GROUPBY operations). Can we model the class of "likely" queries and use it to drive the data layout?

4. From a theoretical perspective, we have highlighted the importance of studying the communication complexity of approximation problems (as in our bounds for the approximate solutions of selection and transitive closure); existing work only treats computations that yield exact answers.

References

[ADSM] http://www.storage.ibm.com/software/adsm/addoc.htm

[AACS87] A. Aggarwal, B. Alpern, A.K. Chandra and M. Snir. A model for hierarchical memory. *Proc. ACM STOC*, 305–314, 1987.

[ABFLS] M. Ajtai, R. Burns, R. Fagin, D. Long, and L. Stockmeyer, Compactly encoding arbitrary files with differential compression. In preparation.

[Ale] http://www.alexa.com

[AMS96] N. Alon, Y. Matias and M. Szegedy. The space complexity of approximating frequency moments. *Proc. ACM STOC*, 20–29, 1996.

[ARS97] K. Alsabti, S. Ranka, and V. Singh. A One-Pass Algorithm for Accurately Estimating Quantiles for Disk-Resident Data. In *Proc. 23rd VLDB Conference*, 346–355, Athens, Greece, 1997.

[AS95] R. Agrawal and A. Swami. A One-Pass Space-Efficient Algorithm for Finding Quantiles. In *Proc. 7th Intl. Conf. Management of Data (COMAD-95)*, Pune, India, 1995.

[AMM] G. Arocena, A. Mendelzon, G. Mihaila. Applications of a Web Query language. *Proceedings of the 6th International WWW Conference*, 1997.

[AAFL96] B. Awerbuch, Y. Azar, A. Fiat, T. Leighton. Making commitments in the face of uncertainty: how to pick a winner almost every time In *Proc. of 28th STOC*, 519–530, 1996.

[BGMZ97] A. Broder, S. Glassman, M. Manasse, and G. Zweig. Syntactic clustering of the Web. In *Proc. of the 6th International WWW Conference*, 391–404, April 1997.

[Cob66] A. Cobham. The recognition problem for the set of perfect squares. IBM Research Report RC 1704, T.J. Watson Research Center, 1966.

[DNS91] D. DeWitt, J. Naughton, and D. Schneider. Parallel Sorting on a Shared-Nothing Architecture using Probabilistic Splitting. In *Proc. Intl. Conf. on Parallel and Distributed Inf. Sys.*, 280–291, Miami Beach, 1991.

[Hel] J. M. Hellerstein. Online Processing Redux. *IEEE Data Engineering Bulletin*, 1997.

[HK81] J-W Hong and H.T. Kung. I/O Complexity: the red-blue pebble game. *Proc. ACM STOC*, 326–333, 1981.

[JC85] R. Jain and I. Chlamtac. The P^2 Algorithm for Dynamic Calculation for Quantiles and Histograms without Storing Observations. *CACM*, 28(10):1076–1085, 1985.

[KMRS88] A. R. Karlin, M. S. Manasse, L. Rudolph, and D.D. Sleator. Competitive snoopy caching. *Algorithmica*, 3(1):70–119, 1988.

[MAGQW] J. McHugh, S. Abiteboul, R. Goldman, D. Quass, and J. Widom. Lore: A database management system for semistructured data. SIGMOD Record, 26(3), 54–66, 1997.

[MP80] J. I. Munro and M. S. Paterson. Selection and Sorting with Limited Storage. *Theoretical Computer Science*, 12:315–323, 1980.

[LN89] R.J. Lipton and J.F. Naughton. Estimating the size of generalized transitive closure. *Proc. 15th VLDB*, 165–172, 1989.

[MM] A. Mendelzon and T. Milo. Formal Models of the Web. *Proceedings of ACM PODS Conference*, 1997.

[MMM] A. Mendelzon, G. Mihaila, T. Milo. Querying the World Wide Web. *Journal of Digital Libraries* 1(1), 68–88, 1997.

[NK] N. Nisan and E. Kushilevitz. Communication Complexity. Cambridge University Press, 1997.

[Olk93] F. Olken. *Random Sampling from Databases*. PhD thesis, University of California Berkeley, 1993.

[PIHS96] V. Poosala, Y. E. Ioannidis, P. J. Haas, and E. J. Shekita. Improved Histograms for Selectivity Estimation of Range Predicates. In *ACM SIGMOD 96*, pages 294–305, Montreal, June 1996.

[RML97] G. Manku, S. Rajagopalan, and B. Lindsay. Approximate medians and other order statistics in one pass and with limited memory: Theory and Database applications. *ACM SIGMOD 98.*, Seattle, June 1998.

[SD77] B. W. Schmeiser and S. J. Deutsch. Quantile Estimation from Grouped Data: The Cell MidPoint. *Communications in Statistics: Simulation and Computation*, B6(3):221–234, 1977.

[SALP79] P. G. Selinger, M. M. Astrahan, R. A. Lories, and T. G. Price. Access Path Selection in a Relational Database Management System. In *ACM SIGMOD 79*, June 1979.

[ST85] D.D. Sleator and R.E. Tarjan. Amortized efficiency of list update and paging rules. *Communications of the ACM*, 28:202–208, February 1985.

[VV96] D.E. Vengroff and J.S. Vitter. I/O-efficient algorithms and environments. *Computing Surveys* 28, 212, 1996.

[Tom80] M. Tompa. Time-Space Tradeoffs for Computing Functions Using Connectivity Properties of their Circuits. *JCSS* 20, 118–132, 1980.

[Yan90] M. Yannakakis. Graph-theoretic methods in database theory. *Proc. PODS*, 230–242, 1990.

[Yao79] A. C-C. Yao. Some complexity questions related to distributed computing. *Proc. 11th ACM STOC*, 209–213, 1979.

DIGITAL EQUIPMENT CORPORATION, SYSTEMS RESEARCH CENTER, 130 LYTTON AVE., PALO ALTO, CA 94301

E-mail address: monika@pa.dec.com

IBM ALMADEN RESEARCH CENTER, 650 HARRY ROAD, SAN JOSE CA 95120
E-mail address: pragh@almaden.ibm.com

IBM ALMADEN RESEARCH CENTER, 650 HARRY ROAD, SAN JOSE CA 95120
E-mail address: sridhar@almaden.ibm.com

DIMACS Series in Discrete Mathematics
and Theoretical Computer Science
Volume **50**, 1999

On maximum clique problems in very large graphs

J. Abello, P. M. Pardalos, and M. G. C. Resende

ABSTRACT. We present an approach for clique and quasi-clique computations
in very large multi-digraphs. We discuss graph decomposition schemes used to
break up the problem into several pieces of manageable dimensions. A semi-
external greedy randomized adaptive search procedure (GRASP) for finding
approximate solutions to the maximum clique problem and maximum quasi-
clique problem in very large sparse graphs is presented. We experiment with
this heuristic on real data sets collected in the telecommunications industry.
These graphs contain on the order of millions of vertices and edges.

1. Introduction

The proliferation of massive data sets brings with it a series of special computational challenges. Many of these data sets can be modeled as very large multi-
digraphs M with a special set of edge attributes that represent special characteristics of the application at hand [**1**]. Understanding the structure of the underlying
digraph $D(M)$ is essential for storage organization and information retrieval. In
this paper we present a new approach for finding large cliques in large sparse multi-
digraphs with millions of vertices and edges.

Let $G = (V, E)$ be an undirected graph where $V = \{v_1, v_2, ..., v_n\}$ is the set of
vertices and E is the set of edges in G. For a subset $S \subseteq V$, we let $G(S)$ denote
the subgraph induced by S.

A graph $G = (V, E)$ is *complete* if its vertices are pairwise adjacent, i.e.
$\forall \; i, j \in V, \{i, j\} \in E$. A *clique* C is a subset of V such that the induced graph
$G(C)$ is complete. The maximum clique problem is to find a clique of maximum
cardinality in a graph G. A multigraph M is just an undirected graph with an integer multiplicity associated with every edge. A multi-digraph is defined similarly.

The maximum clique problem is known to be *NP*-complete. Furthermore, the
complexity of its approximation remained an open question until recently. In [**16**],
Papadimitriou and Yannakakis introduced the complexity class *MAX SNP* and
showed that many natural problems are complete in this class, relative to a re-
ducibility that preserves the quality of approximation. For example, the vertex
cover problem (for constant degree graphs), minimum cut problem, dominating set
problem, and the MAX 3-SAT problem are such complete problems [**19**].

1991 *Mathematics Subject Classification.* Primary 68R10; Secondary 68U30.

Key words and phrases. maximum clique problem, GRASP, heuristics, experimentation,
large data sets, very large sparse graphs, applications.

If the solution to any of these complete problems can be approximated to arbitrary small constant factors, then the optimal solution to any problem in the class can be approximated to arbitrarily small constant factors. The question of whether such approximation schemes can be found for the complete problems in this class was left unresolved. In [5], Berman and Schnitger show that if one of the MAX SNP problems does not have polynomial time approximation schemes, then there is an $\epsilon > 0$ such that the maximum clique problem cannot be approximated in polynomial time with performance ratio (size of maximum clique to size of approximate clique) of $O(n^\epsilon)$, where n is the number of vertices in the graph (see also Feige et al. [6], where a connection between approximation complexity and interactive proof systems is discussed).

A breakthrough in approximation complexity is the recent result by Arora et al. [3, 4]. It is shown that the maximum number of satisfiable clauses in a 3-SAT formula (MAX 3-SAT) cannot be approximated to arbitrary small constants (unless $P = NP$), thus resolving the open question in [16]. This immediately shows the difficulty of finding good approximate solutions to all the above listed problems. In particular, it is shown that no polynomial time algorithm can approximate the maximum clique size within a factor of n^ϵ ($\epsilon > 0$), unless $P = NP$ (by using the results of Feige et al. [6]).

Although these complexity results characterize worst case instances, they nevertheless indicate that the maximum clique problem is indeed a very difficult problem to solve. The maximum clique problem can be approached in two different ways: by exact solutions or heuristic approximation. Since the problem is NP-complete, one can expect exact solution methods to have limited performance on large dense problems. On the other hand, with a heuristic one can never know how close to the actual maximum clique the solution is.

The main difficulty with heuristic approximation of the maximum clique problem is that a local optimum can be far from a global optimum. This is handled in many heuristics by devices that allow escape from poor local optimal solutions. Such devices are present in heuristics, such as simulated annealing [15], tabu search [10, 11], and genetic algorithms [13], that move from one solution to another, as well as in the multistart heuristic GRASP [7, 8], which samples different regions of the solution space, finding a local minimum each time. For further information on various algorithms and heuristics see [14, 17].

The maximum clique problem has many practical applications in science and engineering. Some of them are project selection, classification theory, fault tolerance, coding theory, computer vision, economics, information retrieval, signal transmission theory and aligning DNA with protein sequences [17, 12].

The paper is organized as follows. In Section 2 we consider several graph decomposition schemes used in the experiment. Section 3 proposes a GRASP for finding large cliques in very large graphs. Preliminary computational results are described in Section 4 and concluding remarks are made in Section 5.

2. Graph decomposition

In this section, we discuss some decomposition schemes that make very large sparse graphs suitable for processing by graph optimization algorithms. We discuss two schemes.

First, we take the approach proposed by Abello [1] to partition the edge set of a multi-digraph M. Consider the underlying directed graph

$$D(M) = \{(x, y) \mid (x, y) \text{ is an edge in } M\}$$

and the corresponding underlying undirected graph

$$U(M) = \{\{x, y\} \mid (x, y) \text{ is an edge in } D(M) \}.$$

For a vertex $u \in M$, let

$$\text{out}(u) = \{x \mid (u, x) \in D(M)\} \text{ and } \text{in}(u) = \{y \mid (y, u) \in D(M)\}.$$

Furthermore, define the out-degree $\text{outdeg}(u) = |\text{out}(u)|$, the in-degree $\text{indeg}(u) = |\text{in}(u)|$, and for C, a subset of vertices in M, let the neighborhood of u with respect to C be $N_C(u) = \text{out}(u) \cup \text{in}(u) \cap C$. The degree of u with respect to C is $\deg_C(u) = | N_C(u) |$.

In a preprocessing phase, we use efficient external memory algorithms for computing the connected components of $U(M)$ [2]. For each connected component, consider the sub-digraph of $D(M)$ induced by its vertex set and classify its vertices as sources (indeg = 0), sinks (outdeg = 0) and transmitters (indeg and outdeg > 0). We then partition the directed edge set of each connected component by computing directed Depth First Search Trees (DFST) from each source vertex (or from high degree transmitters). These DFSTs impose an ordering of the edges which is used in a greedy randomized adaptive search procedure (GRASP) tailored for the local maximum clique problem that is described in the next section. The collection of DFSTs defines a neighborhood structure which is used to search for an improvement to a local clique or it can be taken itself as a directed weighted digraph where each DFST now becomes a vertex and the edges running between two DFSTs are collected into a weighted directed edge. It is worth to notice that this digraph DAG(M) is acyclic, suggesting a topological order of processing. Undirected cliques on DAG(M) give global structural information about M. The product of the maximum clique size in DAG(M) times the maximum local clique size in the DFSTs gives an upper bound on the maximum clique size of M. We are currently experimenting with heuristics that use DAG(M) as the neighborhood structure to move from one DFST to another comparing their local max cliques and gluing them together to search for an improvement.

Another simple decomposition scheme works on large connected components of the graph $U(M)$. The approach is repeatedly applied until no further reduction in the size of the graph is possible. In this reduction scheme, we assume that cliques of size k or smaller are of no interest. All edges having at least one vertex of degree less than k are deleted. This, of course, affects the degrees of other vertices, hence, further simplification is possible by reapplying the reduction scheme. In Section 4 we illustrate these reduction schemes on a graph that arises from real telecommunications data.

3. GRASP for maximum clique and maximum quasi-clique

Feo, Resende, and Smith [9, 18] proposed a greedy randomized adaptive search procedure (GRASP) for the maximum independent set problem. In this paper, we propose a GRASP, tailored for the maximum clique problem and the maximum quasi-clique problem on very large sparse graphs, based on the procedure described in [9].

```
procedure construct(V,E,α,Q)
1      Set initial clique Q = ∅;
2      Set C = V;
3      while |C| > 0 do
4          Let G(C) be the subgraph induced by the vertices in C;
5          Let deg_{G(C)}(u) be the degree of u ∈ C with respect to G(C);
6          d = min{deg_{G(C)}(u) | u ∈ C};
7          d̄ = max{deg_{G(C)}(u) | u ∈ C};
8          RCL = {u ∈ C | deg_{G(C)}(u) ≥ d + α(d̄ − d)};
9          Select u at random from the RCL;
10         Q = Q ∪ {u};
11         C = N_C(u);
12     end while;
end construct;
```

FIGURE 1. GRASP construction procedure

GRASP [7, 8] is an iterative method that, at each iteration, constructs a randomized, greedily biased, solution and then finds a locally optimal solution in the neighborhood of the constructed solution. The role of randomization is to generate different solutions from which to initiate local search. Greediness leads to constructed solutions of good quality. Local search applied from these greedily biased solutions usually converges quickly to a local minimum and consequently many probes of different, potentially good, neighborhoods can be carried out in a limited amount of time. We initially describe the GRASP for maximum clique. To describe a GRASP, one needs to specify a construction mechanism and a local search procedure.

The construction phase of the GRASP for maximum clique proposed here builds a clique, one vertex at a time. It uses vertex degrees as a guide for construction. Assume that a clique is being constructed. At each step of the construction phase we have a clique on hand. We call the vertices of this clique the *selected* vertices. A vertex is a *candidate* to be included in the clique if it is adjacent to all previously selected clique vertices. Let C denote the set of candidate vertices. Initially, all vertices are candidates, i.e. $C = V$. A *restricted candidate list* (RCL) contains all candidate vertices of high degree in the subgraph $G(C)$ induced by the candidate vertices. A vertex $u \in C$ is said to have high degree in the graph induced by the candidate vertices if its degree $\deg_{G(C)}(u)$ with respect to the subgraph induced by the candidate nodes is at least $d + \alpha(\bar{d} - d)\}$, where $d = \min\{\deg_{G(C)}(u) \mid u \in C\}$ and $\bar{d} = \max\{\deg_{G(C)}(u) \mid u \in C\}$ and α is a real parameter in the interval [0,1]. One vertex, among those in the RCL, is selected at random and is added to the clique under construction. To take this selection into consideration, the newly chosen vertex and all other candidate vertices not adjacent to it are eliminated from the set of candidate vertices, i.e. the new candidate set is simply the neighborhood of the newly selected vertex with respect to the old candidate set. This process is repeated until the set of candidate vertices is empty. Figure 1 shows pseudo-code for the construction phase of GRASP.

```
procedure local(V,E,Q)
1       H = {(v, u, w) | v, u, w ∈ V, (v, u) ∈ E, w ∈ Q,
             and v and u are adjacent to all vertices of Q except w};
2       while |H| > 0 do
3           Select (u, v, w) ∈ H;
4           Q = Q ∪ {u, v} \ {w};
5           H = {(v, u, w) | v, u, w ∈ V, (v, u) ∈ E, w ∈ Q,
                 and v and u are adjacent to all vertices of Q except w};
6       end while;
end local;
```

FIGURE 2. GRASP $(2,1)$-exchange local search procedure

```
procedure grasp(V,E,maxitr,Q*)
1       CliqueSize = −∞;
2       for k = 1, 2, . . . , maxitr do
3           Select α, at random, from interval [0,1];
4           construct(V,E,α,Q);
5           local(V,E,Q);
6           if |Q| > CliqueSize do
7               CliqueSize = |Q|;
8               Q* = Q;
9           end do;
10      end for;
end grasp;
```

FIGURE 3. GRASP for maximum clique problem

Local search can be implemented in many ways. A simple $(2,1)$-exchange approach seeks a vertex in the clique whose removal allows two adjacent vertices not in the clique to be included in the clique, thus increasing the clique size by one. Likewise, one could look for pairs of vertices in the clique whose removal would permit three vertices not in the clique to be placed in the clique. Figure 2 shows pseudo-code for the $(2,1)$-exchange local search used in our implementation.

Figure 3 shows pseudo-code indicating how the two procedures described above make up the GRASP for maximum clique.

The GRASP described in this section requires access to the edges and vertices of the graph. This limits its use to graphs small enough to fit in memory. We next propose a semi-external procedure that works only with vertex degrees and a subset of the edges in-memory, while most of the edges can be kept in secondary disk storage. Besides enabling its use on smaller memory machines, the procedure we describe below also speeds up the computation of the GRASP.

To describe this procedure, we first define the peel operation on a graph. Given a parameter q, peel(G, q) recursively deletes from G all vertices having degree less than q along with their incident edges.

```
procedure clique(V,E,maxitr_s, maxitr_l,T_G,Q)
1      Let 𝒢 = (𝒱,ℰ) be a subgraph of G = (V,E) such that | ℰ | ≤ T_G;
2      while 𝒢 ≠ G do
3            𝒢' = 𝒢;
4            grasp(𝒱,ℰ,maxitr_s,Q);
5            q = |Q|;
6            peel(V,E,q);
7            Let 𝒢 = (𝒱,ℰ) be a subgraph of G = (V,E) such that | ℰ | ≤ T_G;
8            if 𝒢⁺ == 𝒢 break;
9      end while;
10     Partition E into E_1,...,E_k such that | E_j | ≤ T_G, for j = 1,...,k;
11     for j = 1,...,k do
12           Let V_j be the set of vertices in E_j;
13           grasp(V_j,E_j,maxitr_l,Q_j);
14     end for;
15     𝒢⁺ = 𝒢;
16     q = max {|Q_1|,|Q_2|,...,|Q_k|};
17     peel(V,E,q);
18     if 𝒢⁺ ≠ G then
19           clique(V,E,maxitr_s,maxitr_l, T_G,Q);
20     end if;
end clique;
```

FIGURE 4. Semi-external approach for maximum clique

The semi-external procedure clique first samples a subset ℰ of edges of E such that | ℰ | < T_G, where T_G is a threshhold function of the graph G. The subgraph corresponding to ℰ is denoted by 𝒢 = (𝒱,ℰ), where 𝒱 is the vertex set of 𝒢. The procedure grasp is applied to 𝒢 to produce a clique Q. Let the size of the clique found be q = |Q|. Since q is a lower bound on the largest clique in G, any vertex in G with degree less than q cannot possibly be in a maximum clique and can be therefore discarded from further consideration. This is done by applying peel to G with parameter q. Procedures grasp and peel are reapplied until no further reduction is possible. The aim is to delete irrelevant vertices and edges, allowing grasp to focus on the subgraph of interest. Reducing the size of the graph allows GRASP to explore portions of the solution space at greater depth, since GRASP iterations are faster on smaller graphs. If the reduction results in a subgraph smaller than the specified threshhold, GRASP can be made to explore the solution space in more detail by increasing the number of iterations to maxitr_l. This is what usually occurs, in practice, when the graph is very sparse. However, it may be possible that the repeated applications of grasp and peel do not reduce the graph to the desired size. In this case, we propose partitioning the edges that remain into sets that are smaller than the threshhold and applying grasp to each resulting subgraph. The size of the largest clique found is used as parameter q in a peel operation and if a reduction is achieved the procedure clique is recursively called. Figure 4 shows pseudo-code for this semi-external approach.

In procedure `clique`, edges of the graph are sampled. As discussed earlier, we seek to find a clique in a connected component, examining one component at a time. Within each component, we wish to maintain edges that share vertices close together so that when they are sampled in `clique` those edges are likely to be selected together. To do this, we compute Depth First Search Trees on the directed subgraph in the component and store the edges for sampling in the order determined by the trees.

A related problem of interest is to find quasi-cliques in very large sparse graphs. A *quasi-clique* is defined to be a dense subgraph. A clique is a completely dense quasi-clique. A quasi-clique $\mathcal{Q}(\gamma, q)$ is defined to be a subgraph having q vertices and edge density γ, i.e. the number of edges in $\mathcal{Q}(\gamma, q)$ is $\lfloor \gamma q(q-1)/2 \rfloor$. One can define several optimization problems for quasi-cliques. Three examples are:

- max γq,
- fix q and max γ,
- fix γ and max q.

The approach described above for finding cliques can be extended to address the third problem as follows.

The GRASP for quasi-clique constructs a clique to serve as a seed to grow the quasi-clique from, using a modified local search procedure. The construction procedure is identical to the one used in the case of cliques. The local search procedure looks for vertices (v, u, w) such that $(v, u) \in E$, and $w \in \mathcal{Q}$ and v and u are adjacent to at least $\gamma(q-1)$ vertices of $\mathcal{Q} \setminus \{w\}$. If such vertices exist, then w is removed from the quasi-clique and v and u are added to the quasi-clique, increasing its size by one. Note that if $\gamma = 1$, then this local search procedure is identical to the one used for cliques.

A semi-external procedure similar to the one proposed for finding large cliques can be derived for quasi-cliques. As before, the procedure samples edges from the original graph such that the subgraph induced by the vertices of those edges is of a specified size. The GRASP for quasi-cliques is applied to the subgraph producing a quasi-clique $\mathcal{Q}$ of size q. In the original graph, any vertex having degree not greater than γq can be discarded from further consideration, since it cannot possibly be in a quasi-clique of size at least q.

4. Experiments with a very large graph

In this section, we outline the preliminary exploratory experiments done with a sample dataset. The experiments were done on a Silicon Graphics Challenge computer (20 MIPS 196MHz R10000 processors with 6.144 Gbytes of main memory). A substantial amount of disk space was also used.

Our current data comes from telecommunications traffic. The corresponding multi-graph has 53,767,087 vertices and over 170 million edges. We found 3,667,448 connected components out of which only 302,468 were components of size greater than 3 (there were 255 self-loops, 2,766,206 pairs and 598,519 triplets).

A giant component with 44,989,297 vertices was detected. It is tantalizing to suggest then that we may be witnessing here a behavior similar to the one predicted by random graph theory even though our graphs are certainly not random. The giant component has 13,799,430 Depth First Search Trees (DFSTs) and one of them is a giant DFST (it has 10,355,749 vertices and 19,072,448 edges). Most of the DFSTs have no more than 5 vertices. The interesting trees have sizes between

TABLE 1. Cliques found by `construct` and `local`

size	cliques found by construct	local	distinct cliques
2	63	62	
3	473	320	
4	95	176	
5	73	103	14
6	116	95	11
7	59	38	25
8	54	63	28
9	22	33	14
10	17	10	9
11	15	38	35
12	10	32	22
13	1	26	18
14	0	3	3
15	0	1	1

5 and 100. Their corresponding induced subgraphs are most of the time very sparse ($|E| < |V| \log |V|$), except for some occasional dense subgraphs ($|E| > |V| \sqrt{|V|}$) with 11 to 32 vertices.

We argue that the largest clique in this component has size not greater than 32. Cliques are either within a subgraph induced by the vertices of a DFST, or distributed among the different DFSTs. We expect the former to occur. There are several large DFSTs, the largest having about 19 million edges. By counting the edges in the trees, one observes that there remain very few edges to go between trees and consequently it is more likely that cliques are within the graphs induced by the nodes of a tree. Since the largest dense subgraph induced by the vertices of a tree has 32 vertices, we should not expect many cliques larger that 32 to be found.

In this exploratory phase of our work, we did not integrate the `grasp` and `peel` procedures into procedure `clique`, but rather applied them manually. To begin our experimentation, we considered 10% of the edges in the large component from which we recursively removed all vertices of degree one by applying `peel`$(V, E, 1)$. This resulted in a graph with 2,438,911 vertices and 5,856,224 edges, which fits in memory. In this graph we search for large cliques. Our first motivation is to identify a lower bound on the size of the maximum clique so that we can delete higher-degree vertices on larger portions of the graph to possibly identify larger cliques. The GRASP was repeated 1000 times, with each iteration producing a locally optimal clique. Though applying local search on every constructed solution may not be efficient from a running time point of view, we applied local search to all constructed solutions to explore its effect in improving clique sizes. Because of the independent nature of the GRASP iterations and since our computer is configured with 20 processors, we created 10 threads, each independently running GRASP starting from a different random number generator seed.

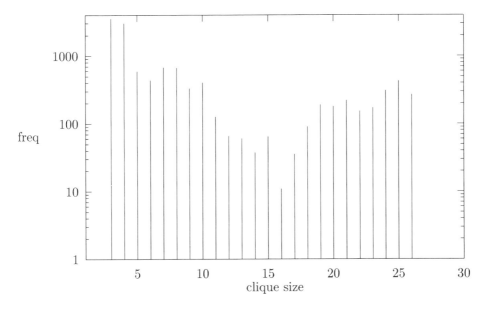

FIGURE 5. Frequency of clique sizes found on 25% of data

Table 1 summarizes the first part of the experimental results. It shows, for each clique size found, the number of GRASP iterations that constructed or improved such solution, and from sizes 5 to 15, the number of distinct cliques that were found by the GRASP iterations. It is interesting to observe that these cliques, even though distinct, share a large number of vertices. Applying a greedy procedure to these cliques to identify a disjoint set of cliques produced one clique of size 15, 12, 9, and 7, and 4 cliques of size 6, and 5 of size 5.

Next, we considered 25% of the edges in the large component from which we recursively removed all vertices of degree 10 or less. The resulting graph had 291,944 vertices and 2,184,751 edges. 12,188 iterations of GRASP produced cliques of size 26. In 271 iterations, a clique of size 26 was produced. In 431 iterations, a clique of size 25 was produced. In 313 iterations, a clique of size 24 was produced. Figure 5 shows the frequencies of cliques of different sizes found by the algorithm.

Having found cliques of size 26 in a quarter of the graph, we next intensified our search on the entire huge connected component. In this component, we recursively removed all vertices of degree 20 or less. The resulting graph has 27,019 vertices and 757,876 edges. Figure 6 shows the frequencies of cliques of different sizes found by the algorithm. Figure 7 shows the statistics of the improvement attained by local search.

Over 20,000 GRASP iterations were carried out on the 27,019 vertex – 757,876 edge graph. Cliques of 30 vertices were found. These cliques are very likely to be optimal because we do not expect cliques larger than 32 vertices to be found. The local search can be seen to improve the constructed solution not only for large constructed cliques, but also for small cliques. In fact, in 26 iterations, constructed cliques of size 3 were improved by the local search to size 30.

Finally to increase our confidence that the cliques of size 30 found are maximum, we applied peel($V, E, 30$), resulting in a graph with 8724 vertices and about 320

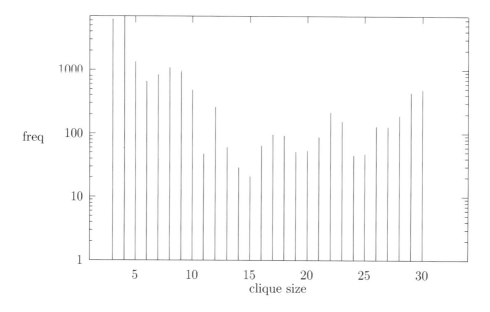

FIGURE 6. Frequency of clique sizes found on entire dataset

thousand edges. We ran 100,000 GRASP iterations on the graph taking 10 parallel processors about one and a half days to finish. The largest clique found had 30 vertices. Of the 100,000 cliques generated, 14,141 were distinct, although many of them share one or more vertices.

Finally, to compute quasi-cliques on this test data, we looked for large quasi-cliques with density parameters $\gamma = .9, .8, .7,$ and $.5$. Quasi-cliques of sizes 44, 57, 65, and 98, respectively, were found.

5. Concluding remarks

We presented an approach for clique and quasi-clique computations in very large multi-digraphs. We discussed graph decomposition schemes used to break up the problem into several pieces of manageable dimensions. A greedy randomized adaptive search procedure (GRASP) for finding approximate solutions to the maximum clique (quasi-clique) problem in very large sparse graphs was presented. We experimented with this heuristic on real data sets collected in the telecommunications industry. These graphs contain on the order of millions of vertices and edges.

In a variety of industrial applications it is necessary to deal explicitly with weights, edge directions, and multiplicities. In this regard, we are currently extending the methods presented here to find directed weighted quasi-cliques. These are subgraphs that satisfy special local density conditions which make them amenable to local search based techniques. Our findings in this direction will be reported in a forthcoming paper.

References

[1] J. Abello. A dynamic multigraph model for massive data sets. Technical report, AT&T Labs Research, Florham Park, NJ, December 1997.

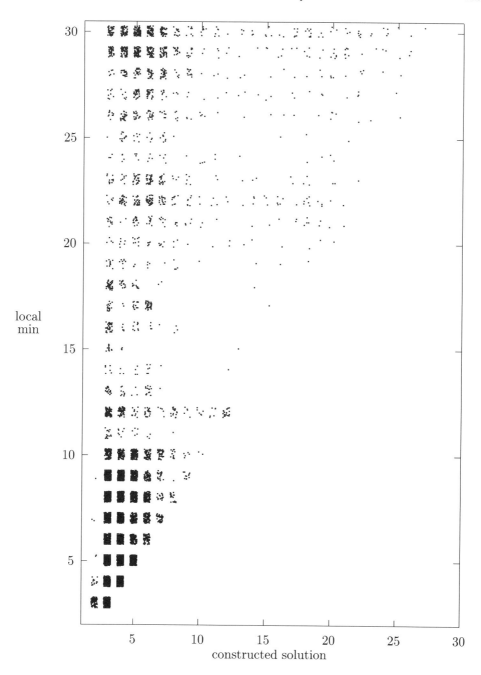

FIGURE 7. Local search improvement

[2] J. Abello, A. Bushbaum, and J. Westbrook. A functional aproach to external memory algorithms. In *European Symposium on Algorithms*, volume 1461 of *Lecture Notes in Computer
Science*, pages 332–343. Springer-Verlag, 1998.

[3] S. Arora, C. Lund, R. Motwani, M. Sudan, and M. Szegedy. On the intractability of approximation problems. Technical report, University of Rochester, 1992.

[4] S. Arora and S. Safra. Approximating the maximum clique is *NP*-complete. Technical report, University of Rochester, 1992.

[5] P. Berman and G. Schnitger. On the complexity of approximating the independent set problem. *Lecture Notes in Computer Science*, 349:256–267, 1989.

[6] U. Feige, S. Goldwasser, L. Lovász, S. Safra, and M. Szegedy. Approximating the maximum clique is almost *NP*-complete. In *Proc. 32nd IEEE Symp. on Foundations of Computer Science*, pages 2–12, 1991.

[7] T.A. Feo and M.G.C. Resende. A probabilistic heuristic for a computationally difficult set covering problem. *Operations Research Letters*, 8:67–71, 1989.

[8] T.A. Feo and M.G.C. Resende. Greedy randomized adaptive search procedures. *Journal of Global Optimization*, 6:109–133, 1995.

[9] T.A. Feo, M.G.C. Resende, and S.H. Smith. A greedy randomized adaptive search procedure for maximum independent set. *Operations Research*, 42:860–878, 1994.

[10] F. Glover. Tabu search. Part I. *ORSA J. Comput.*, 1:190–206, 1989.

[11] F. Glover. Tabu search. Part II. *ORSA J. Comput.*, 2:4–32, 1990.

[12] J. Hasselberg, P. M. Pardalos, and G. Vairaktarakis. Test case generators and computational results for the maximum clique problem. *Journal of Global Optimization*, 3:463–482, 1993.

[13] J. H. Holland. *Adaptation in Natural and Artificial Systems*. University of Michigan Press, Ann Arbor, MI, 1975.

[14] D. S. Johnson and M. A. Trick, editors. *Cliques, Graph Coloring, and Satisfiability: Second DIMACS Implementation Challenge*, volume 26 of *DIMACS Series on Discrete Mathematics and Theoretical Computer Science*. American Mathematical Society, 1996.

[15] S. Kirkpatrick, C. D. Gellat Jr., and M. P. Vecchi. Optimization by simulated annealing. *Science*, 220:671–680, 1983.

[16] C. Papadimitriou and M. Yannakakis. Optimization, approximation and complexity classes. In *Proc. of the Twentieth Annual ACM STOC*, pages 229–234, 1988.

[17] P. M. Pardalos and J. Xue. The maximum clique problem. *Journal of Global Optimization*, pages 301–328, 1994.

[18] M.G.C. Resende, T.A. Feo, and S. H. Smith. FORTRAN subroutines for approximate solution of maximum independent set problems using grasp. Technical report, AT&T Labs Research, Florham Park, NJ, 1997. To appear in *ACM Trans. Math. Software*.

[19] M. Yannakakis. On the approximation of maximum satisfiability. In *Proc. 3rd Annual ACM-SIAM Symp. on Discrete Algorithms*, 1992.

(J. Abello) NETWORK SERVICES RESEARCH CENTER, AT&T LABS – RESEARCH, SHANNON LABORATORY, FLORHAM PARK, NJ 07932 USA
    E-mail address, J. Abello: `abello@research.att.com`

(P. M. Pardalos) CENTER FOR APPLIED OPTIMIZATION, DEPARTMENT OF INDUSTRIAL AND SYSTEMS ENGINEERING, UNIVERSITY OF FLORIDA, GAINESVILLE, FL 32611 USA
    E-mail address, P. M. Pardalos: `pardalos@ufl.edu`

(M. G. C. Resende) INFORMATION SCIENCES RESEARCH CENTER, AT&T LABS – RESEARCH, SHANNON LABORATORY, FLORHAM PARK, NJ 07932 USA
    E-mail address, M. G. C. Resende: `mgcr@research.att.com`

DIMACS Series in Discrete Mathematics
and Theoretical Computer Science
Volume 50, 1999

I/O-Optimal Computation of Segment Intersections

A. Crauser, P. Ferragina, K. Mehlhorn, U. Meyer, and E.A. Ramos

ABSTRACT. We investigate the I/O-complexity of computing the trapezoidal decomposition defined by a set of N line segments in the plane. We present a randomized algorithm which solves optimally this problem requiring $O(\frac{N}{B} \log_{M/B} \frac{N}{B} + \frac{K}{B})$ expected I/O operations, where K is the number of pairwise intersections, M is the size of available internal memory and B is the size of the block transfer. The proposed algorithm requires an optimal expected number of internal operations. As a by-product, the algorithm also solves the segment intersections problem requiring the same number of I/Os and internal operations.

1. Introduction

Computer graphics [19] as well as Geographic Information Systems [12, 14] are nowadays rich sources of large-scale computational problems. Therefore, they need the design of appropriate external-memory techniques and data structures to efficiently cope with the enormous amount of spatial data which have to be searched, stored and manipulated. In those applications, most of the subproblems require the processing of line segments, so that Computational Geometry turns out to be a hot topic of research in the external memory context. In this paper, we study the I/O-complexity of the problem of computing the trapezoidal decomposition of a set of line segments [18]. We give a *randomized* algorithm and analyze its I/O-complexity in the external memory model [20], in which M is the available internal memory size and B is the size of the block transfer (where $1 \leq B \leq M/2$). As a by-product, the algorithm also solves the segment intersections problem.

Let S be a set of N segments in the plane with a set $\mathcal{K}(S)$ of pairwise intersection points. The *trapezoidal decomposition* of S is obtained by extending vertically each segment endpoint, and each intersection point, upward and downward until it hits another segment. The resulting connected regions of the plane are called *trapezoids* (see Fig. 1).

The set of these trapezoids is denoted by $\mathcal{T}(S)$. Note that $|\mathcal{T}(S)|$ is $\Theta(N + K)$ where $K = |\mathcal{K}(S)|$. Given S, the *segment intersections problem* consists in computing $\mathcal{K}(S)$, whereas the *trapezoidal decomposition problem* consists in computing $\mathcal{T}(S)$. Clearly, by solving the second problem we also get a solution for the first one.

1991 *Mathematics Subject Classification.* Primary 68U05; Secondary 68Q20.

Max-Planck-Institut für Informatik, Im Stadtwald, 66123 Saarbrücken, Germany {crauser, paolo, mehlhorn, umeyer, ramos}@mpi-sb.mpg.de. Supported in part by EU ES-PRIT LTR Project N. 20244 (ALCOM-IT).

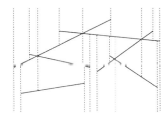

FIGURE 1. Trapezoidal decomposition

The internal memory version of the line segments intersections problem has been extensively studied. A *deterministic* algorithm is known that finds the intersections in $O(N \log N + K)$ optimal time [7]. There are also much simpler *randomized* algorithms [10, 16]. All of these algorithms actually compute $\mathcal{T}(S)$ (or some variation of it). There are also algorithms that use space $O(N)$ and output $\mathcal{K}(S)$ in time $O(N \log N + K)$ [2, 4, 10].

In the external memory model, *optimality* means that the total number of I/Os required to compute the K intersections or the trapezoidal decomposition of S, is $\Theta(n \log_m n + k)$, where $n = N/B$, $m = M/B$ and $k = K/B$. In fact, the first term derives from the optimal I/O-cost for sorting the N segments of S [1], whereas the latter term derives from the cost due to the storage of the produced output on the disk. No I/O-optimal algorithm is known. The best known algorithms are due to Arge *et al.* [3]. They solve the segment intersection problem (their algorithm does *not* compute the trapezoidal decomposition) in sub-optimal $O((n+k) \log_m n)$ I/Os, and they show how to compute the trapezoidal decomposition induced by a set of N *non-intersecting* segments in optimal $O(n \log_m n)$ I/Os.

In this paper, we present a *randomized* algorithm that computes $\mathcal{T}(S)$ for an arbitrary set of line segments S using an optimal expected number of I/Os. It uses a variation of the *randomized incremental construction* approach (RIC) [10] that has been successfully employed in the design of geometric algorithms. Our algorithm also requires an optimal expected number of internal operations. As a by-product, the algorithm also computes $\mathcal{K}(S)$.

The content of this extended abstract is as follows. In Sec. 2, we review the sampling concepts and results that are needed in the design and analysis of our algorithm. In Sec. 3, we present the basic RIC approach via gradations. Then, in Secs. 4 and 5, we present its implementation in external memory. Complete details can be found in the full version of the paper [11].

2. Preliminaries from Geometric Sampling

Let $R \subseteq S$. For $\sigma \in \mathcal{T}(R)$, S_σ denotes the set of segments in S that intersect the interior of σ, and is called the *conflict list*. Let $N_\sigma = |S_\sigma|$. A *p-sample* R from S is obtained by choosing every $s \in S$ into R independently with probability p. Note that for a p-sample R the expected value of $|\mathcal{T}(R)|$ is $\Theta(f(p, S))$ where $f(p, S) = p|S| + p^2 K(S)$. There are two main properties of this sampling process that are relevant for the analysis of the algorithms [9, 10, 17]. First, *the average* *conflict list size is at most* $1/p$. More precisely, for a constant $C > 0$:

$$(1) \qquad \mathbf{E}\left[\sum_{\sigma \in \mathcal{T}(R)} N_\sigma\right] \leq C \frac{f(p, S)}{p}.$$

Second, *the deviation of the conflict list size is* $O(\log s)$ *with high probability*. More precisely, for $s \geq pN$, given $c > 0$ there is $C > 0$, such that with probability at least $1 - 1/s^c$:

$$(2) \qquad \max_{\sigma \in \mathcal{T}(R)} N_\sigma \leq C \; \frac{\log s}{p}.$$

For the optimal algorithms under general conditions, the concept of a $(1/r)$-*cutting* for S is needed [8, 15]: *A decomposition of the plane into a set T of disjoint trapezoids such that* $\max_{\sigma \in T} N_\sigma \leq N/r$. In particular, the following fact is used:

FACT 1. *There are constants c and d such that for any set S of N segments, a* $(1/r)$-*cutting for S of size $O(r^c)$ can be computed using $O(r^d N)$ expected operations.*

3. RIC via Gradations

To facilitate the presentation we use the following additional notation: Given a set of segments X and a set of trapezoids T, we write $T[X]$ to denote the set of conflict lists X_σ for $\sigma \in T$, and write $|T[X]|$ to denote $\sum_{\sigma \in T} |X_\sigma|$.

3.1. Gradation. The algorithm is a variant of the RIC approach [10] and follows [6, 5]. Given parameters μ and μ', it chooses a sequence of subsets of S, called a *gradation*:

$$\emptyset = S_0 \subseteq S_1 \subseteq \ldots \subseteq S_{l-1} \subseteq S_l = S$$

where S_{i-1} is a $(1/\mu)$-sample from S_i for $i < l$ and S_{l-1} is a $(1/\mu')$-sample from $S_l = S$ (the need for two parameters will become clear when discussing the external-memory implementation). Then, it iteratively constructs the decomposition $\mathcal{T}(S_i)$, for all i, $1 \leq i \leq l$. At the beginning of the i-th *round*, the decomposition $T_{i-1} = \mathcal{T}(S_{i-1})$ and its conflict lists $T_{i-1}[S]$ are available (initially, $S_0 = \emptyset$ and thus T_0 consists of a single "unbounded" trapezoid having the whole S as conflict list), then the algorithm constructs the new decomposition $T_i = \mathcal{T}(S_i)$ by using T_{i-1} and $T_{i-1}[S]$. We refer to the l-th round as the *last* round, and to all the others as *early* rounds. Let R_i be $S_i - S_{i-1}$, the set of segments added in the i-th round, and for each trapezoid $\sigma \in T_{i-1}$, let $R_{i,\sigma}$ be the subset of the segments in R_i which are conflicting with σ. Because of the random sampling, the sets $R_{i,\sigma}$ will be well balanced on the average. More precisely, in each early round, the average is at most μ, and in the last round the average is at most μ' (recall Eqn. (1)).

3.2. The Generic Round. The i-th round computes T_i and $T_i[S]$ in three steps:

1. *Intermediate decomposition:* For each $\sigma \in T_{i-1}$, identify $R_{i,\sigma}$ by scanning S_σ and taking from it the segments which belong to R_i. Then compute the restriction of $\mathcal{T}(R_{i,\sigma})$ to σ, denoted T_σ. This results in an intermediate decomposition $T_i^I = \bigcup_{\sigma \in T_{i-1}} T_\sigma$.

2. *Intermediate conflict lists:* For each $\sigma \in T_{i-1}$, compute the conflict lists $T_\sigma[S]$ by taking each $s \in S_\sigma$ at a time and, after an *initial search* in T_σ that locates an endpoint of s, determine the conflicts of s with the trapezoids in T_σ by appropriately walking through it. These are the intermediate conflict lists $T_i^I[S]$.

3. *Clean-up:* Obtain T_i and its conflict lists $T_i[S]$ from the intermediate decomposition T_i^I and its conflict lists $T_i^I[S]$. Observe that $\tau \in T_i$ can be *chopped* into pieces $\tau \cap \sigma$, for $\sigma \in T_{i-1}$. So we need to stitch together τ from its pieces $\tau \cap \sigma$ and also build its conflict list from the conflict lists of its pieces.

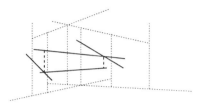

FIGURE 2. A chopped trapezoid

Remarks. In Step 1, if $|R_{i,\sigma}|$ is $O(1)$ then a non-optimal polynomial algorithm suffices. In our external-memory implementation, $|R_{i,\sigma}|$ is relatively large and hence, we require the use of an optimal (internal memory) algorithm. In Step 3, we need to recover $\mathcal{T}(S_i)$ so that we can apply the sampling results for S_i with respect to S. If the clean-up is not performed, then we could only use the sampling results *locally* in each trapezoid for S_i with respect to S_{i-1}.

The expected running time of the algorithm can be estimated using Eqn. (1) and other more general bounds. The result is that the algorithm performs an optimal expected number of operations $O(N \log N + K)$.

4. The Algorithm for External Memory (Ignoring Deviations)

We now present an I/O-efficient implementation of the previous algorithm (recall the notation $m = M/B$, $n = N/B$ and $k = K/B$). First, as a technical point, we assume that the gradation is constructed before the algorithm starts, so that for each segment $s \in S$ there is an associated *tag* that indicates the round in which s is *inserted*. The tag is carried by each copy of s (in each conflict list) so that, in the i-th round, the sets $R_{i,\sigma}$ can be easily determined by scanning the conflict lists S_σ. This is important in an efficient external memory implementation where we cannot assume random access to data.

The choice of parameters μ and μ' is done as follows: $\mu = m^{1/2}$ and $\mu' = \max\{B, M^{1/2}\}$. Therefore, the expected number of levels in the gradation is $l = O(\log_\mu(N/\mu')) = O(\log_m n)$. The main observations that lead to an I/O efficient implementation are:

(i) The choice of μ' implies that $f(1/\mu', S) = O(f(S)/B)$. As a result, in each early i-th round, the algorithm does not need to handle T_{i-1} in an I/O-efficient manner: even incurring *one* I/O per operation of the internal memory algorithm is acceptable. Only the last round must handle T_l in an I/O-efficient manner, but then this is aided by the knowledge of T_{l-1}.

(ii) The choice of μ implies that in each early round the average size of T_σ is at most $\mu^2 = M/B$ (since the average size of $R_{i,\sigma}$ is at most μ). Therefore, ignoring deviations, T_σ can be computed in internal memory and one block of memory allocated for each trapezoid of T_σ, allowing the conflict lists to be written in an I/O-efficient manner. Similarly, in the last round, the choice of μ' implies that, also ignoring deviations, the average size of $R_{l,\sigma} = S_\sigma$ is at most μ', for each $\sigma \in T_{l-1}$. If $B^2 \leq M$ then $\mu' = M^{1/2}$ and an optimal internal memory algorithm can be used to construct T_σ, because the decomposition fits in internal memory. If $B^2 > M$ then $\mu' = B$ and T_σ can be larger than M, so that an I/O-optimal algorithm must be devised to handle small inputs of size less than B.

We discuss the I/O operations needed to implement the basic algorithm of Section 3.2 in external memory under the hypothesis that $|R_{i,\sigma}| \leq \mu$ for $i < l$ and $|R_{l,\sigma}| \leq \mu'$. We will relax these assumptions in Section 5 where we will cope with the deviations of the $R_{i,\sigma}$'s.

4.1. The Early Rounds. We detail below the external-memory implementation of the three steps forming the i-th early round in the basic algorithm.

Step 1. Each trapezoid $\sigma \in T_{i-1}$ in turn, and its conflict list S_σ, are loaded in *internal memory* so that $R_{i,\sigma}$ is determined by checking the tags and $T_\sigma = \mathcal{T}(R_{i,\sigma})$ is computed. Since the size of T_σ is at most $\mu^2 \leq M/B$, then T_σ can be constructed by an optimal internal-memory algorithm without performing any I/Os. Thus, the number of I/Os required by this step is proportional to:

$$\sum_{\sigma \in T_{i-1}} \left(\left\lceil \frac{N_\sigma}{B} \right\rceil + \left\lceil \frac{|R_{i,\sigma}|}{B} \right\rceil + \left\lceil \frac{|T_\sigma|}{B} \right\rceil \right) = O\left(|T_{i-1}| + \frac{|T_{i-1}[S]|}{B} + \frac{|T_i^I|}{B} \right).$$

Step 2. By the hypothesis $|T_\sigma| \leq \mu^2 = M/B$, hence we can reserve in internal memory a buffer of size B for each trapezoid $\tau \in T_\sigma$. The conflict lists $T_\sigma[S]$, for $\sigma \in T_{i-1}$, are computed by scanning S_σ and walking through T_σ in internal memory. As a conflict of s with some τ is determined, it is written into the buffer associated with τ. As a buffer becomes full, it is written to external memory. In this way, the number of I/Os is proportional to the size of the scanned and returned conflict lists, divided by B. Therefore, the overall number of I/Os required by this step is proportional to:

$$\sum_{\sigma \in T_{i-1}} \left\lceil \frac{N_\sigma}{B} \right\rceil + \sum_{\sigma \in T_{i-1}} \sum_{\tau \in T_\sigma} \left\lceil \frac{N_\tau}{B} \right\rceil = O\left(\frac{|T_i^I[S]|}{B} + |T_i^I| \right).$$

Step 3. We have the intermediate decomposition T_i^I and we need to determine the decomposition T_i. As noted earlier, a trapezoid $\tau \in T_i$ may occur in T_i^I chopped into pieces $\tau \cap \sigma$, for $\sigma \in T_{i-1}$. To achieve optimal bounds, the stitching of pieces $\tau \cap \sigma$ has to be performed using a number of I/Os proportional to the total number of blocks of size B required to hold the conflict lists $T_i^I[S]$, and the resulting conflict lists $T_i[S]$. Achieving this goal presents some difficulties because we cannot afford to use sorting since this would lead to a suboptimal algorithm paying an extra-factor $\log_m n$ in the final I/O-complexity. Our approach is to first consider the partial ordering between trapezoids induced by their vertical adjacencies; then, compute a linear order by *topologically sorting* that partial order; and finally, traverse the decomposition T_{i-1} by following this linear order. A difficulty is that we do not know how to topologically sort with a "linear" number of I/Os, that is, $|T_{i-1}|/B$. Fortunately, $|T_{i-1}|$ is "small" so that one can afford to use an optimal internal-memory algorithm that even performs one I/O per internal memory operation. As T_{i-1} is traversed, the chopped trapezoids of T_i are put together by maintaining, for each trapezoid in T_{i-1} already visited but whose right neighbor is yet to be visited, a list of the chopped trapezoids in T_i that cross its right vertical boundary. This list is matched to the list of chopped trapezoids in T_i that cross the left vertical boundary of the adjacent trapezoid in T_{i-1} when it comes under consideration.

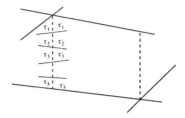

FIGURE 3. Trapezoids crossing a vertical edge

The conclusion is that Step 3 can be performed using a number of I/Os which is proportional to the total number of blocks required to hold the conflict lists $T_i^I[S]$ and the resulting conflict lists $T_i[S]$, plus the I/Os needed for the topological sort (i.e., $|T_i^I|$):

$$\sum_{\sigma \in T_{i-1}} \sum_{\tau \in T_\sigma} \left\lceil \frac{N_\tau}{B} \right\rceil + \sum_{\sigma \in T_i} \left\lceil \frac{N_\sigma}{B} \right\rceil \;=\; O\left(\frac{|T_i^I[S]|}{B} + |T_i^I| \right).$$

Essentially, Steps 1–3 require an *I/O-efficient* handling of the conflict lists $T_{i-1}[S]$, $T_i^I[S]$ (a number of I/Os proportional to their sizes divided by the block size B). On the other hand, Steps 1–3 allow us to handle in an *I/O-inefficient* way the decompositions T_{i-1}, T_i^I and T_i (a number of I/Os proportional to their whole size).

4.2. The Last Round. The goal is now to compute the final decomposition $T_l = \mathcal{T}(S)$ from T_{l-1} in a reduced number of I/Os. We recall that S_{l-1} is a $(1/\mu')$-sample from S, and we assumed that $|R_{l,\sigma}| \leq \mu'$. We need an optimal algorithm that handles the small case, that is, it computes $\mathcal{T}(R_{l,\sigma})$ for $|R_{l,\sigma}| \leq \mu' \leq M$ requiring $O((N_\sigma/B)\log_m(N_\sigma/B) + K_\sigma/B)$ I/Os, where $N_\sigma = |R_{l,\sigma}|$ and $K_\sigma = |\mathcal{K}(R_{l,\sigma}) \cap \sigma|$ (that is, K_σ is the number of pairwise intersections of $R_{l,\sigma}$ inside σ). This is trivial if $K_\sigma = 0$ because we can use an internal-memory algorithm that requires linear space. But when this is not the case $\mathcal{T}(R_{l,\sigma})$ might not fit in internal memory at once and a proper *optimal external-memory* algorithm is needed. The algorithm for such a "small case" can be obtained again using sampling: Take a sample of size $\sqrt{M}$, compute its decomposition using an internal memory algorithm, compute its conflict lists, compute each of the resulting subproblems again in internal memory, and finally put together the result. It can be shown that for this case, the clean-up step is possible by using sorting (details in the full version [**11**]).

We note that in the discussion above we ignored the deviations, so that we are actually assuming that the internal memory can always accommodate $R_{i,\sigma}$, its decomposition and corresponding buffers. In this situation we can prove the following result:

THEOREM 1. *Assuming that the internal memory can always accommodate $R_{i,\sigma}$, its decomposition T_σ and the corresponding $|T_\sigma|$ buffers of size B, the trapezoidal decomposition problem on a set of N line segments with K pairwise intersections can be optimally solved in $O(n \log_m n + k)$ expected I/Os.*

The algorithm can be modified to handle degeneracies and still achieve optimal complexity: $O(n \log_m n + i)$ where $i = I/B$ and I is the number of intersection points (which can be much smaller than the number of pairwise intersections).

5. Handling Deviations

We describe how to remove the conditions imposed on $|R_{i,\sigma}|$, and thus cope with the *deviation* of these random variables. This is achieved by using a refinement approach of Chazelle and Friedman [8, 15]. For $i < l$, the idea is to construct using Fact 1, for each $\sigma \in T_{i-1}$, a $(1/t_\sigma)$-cutting $\hat{T}_\sigma$ for $R_{i,\sigma}$ restricted to σ, where $t_\sigma = |R_{i,\sigma}|/\mu$. Since each trapezoid $\tau \in \hat{T}_\sigma$ satisfies the desired constraint on the size of its conflict list, we can apply to it the Steps 1 and 2 of the i-th round in the algorithm of the previous section. Then, to take care of the independent refinement performed by the cutting process on each trapezoid $\sigma \in T_{i-1}$, we perform *two levels of clean-up*; both levels are similar to Step 3 in the algorithm of the previous section. The first clean-up obtains T_σ and $T_\sigma[S]$ from $\hat{T}_\sigma$ and $\hat{T}_\sigma[S]$, and the second one obtains the final decomposition $T_i = \mathcal{T}(S_i)$ and its conflict lists $T_i[S]$ from T_i^I and $T_i^I[S]$. Similarly, for the last round. In the analysis, the essential fact needed is that t_σ behaves as a constant in the expectations in which it appears. See [11] for detailed description and analysis.

THEOREM 2. *The trapezoidal decomposition problem on a set of N line segments with K pairwise intersections can be solved using $O(n \log_m n + k)$ expected I/Os, which is optimal.*

6. Concluding Remarks

We have described a technique that leads to an optimal algorithm for the trapezoidal decomposition problem. Making use of the deviation bound in Eqn. (2), we can avoid the complication introduced in Sec. 5 (the cuttings), and yet obtain an algorithm that uses an optimal expected number of I/Os, under some practical conditions on N, M and B. The technique is very general and can be applied to other geometric problems, like 3-d half space intersections, 2-d and 3-d convex hulls, 2-d abstract Voronoi diagrams and batched planar point location [11]. Furthermore, the approach can be easily generalized to the D-disk model and it seems even possible that it is also efficient for a model with D disks and p processors.

References

[1] A. Aggarwal and J. S. Vitter. The Input/Output complexity of sorting and related problems. *Communications of the ACM*, 31(9):1116–1127, 1988.

[2] N. M. Amato, M. T. Goodrich, and E. A. Ramos. Computing faces in segment and simplex arrangements. In *Proc. 27th Annu. ACM Sympos. Theory Comput.*, 672–682, 1995.

[3] L. Arge, D. E. Vengroff, and J. S. Vitter. External-memory algorithms for processing line segments in geographic information systems. In *Proc. Annual European Symposium on Algorithms, LNCS 979*, 295–310, 1995.

[4] I.J. Balaban. An optimal algorithm for finding segment intersections. In *Proc. 11th Annu. ACM Sympos. Comput. Geom.*, 1995, 211–219.

[5] H. Brönnimann, B. Chazelle, and J. Matoušek. Product range spaces, sensitive sampling, and derandomization. In *Proc. 34th Annu. IEEE Sympos. Found. Comput. Sci.*, 1993, 400–409.

[6] B. Chazelle. An optimal convex hull algorithm in any fixed dimension. *Discrete Comput. Geom.* **10** (1993) 377–409.

[7] B. Chazelle and H. Edelsbrunner. An optimal algorithm for intersecting line segments in the plane. *J. ACM* **39** (1992) 1–54.

[8] B. Chazelle and J. Friedman. A deterministic view of random sampling and its use in geometry. *Combinatorica* **10** (1990) 229–249.

[9] K.L. Clarkson. Randomized geometric algorithms. In D.-Z. Du and F.K.Hwang, eds., *Computing in Euclidean Geometry*, Vol.1 of *Lecture Notes Series on Computing*, pages 117–162. World Scientific, Singapore, 1992.

[10] K. L. Clarkson and P. W. Shor. Applications of random sampling in computational geometry, II. *Discrete Comput. Geom.*, **4** (1989) 387–421.

[11] A. Crauser, P. Ferragina, K. Mehlhorn, U. Meyer, E.Ramos. Randomized External-Memory Algorithms for Some Geometric Problems. http://www.mpi-sb.mpg.de/~ramos

[12] R. F. Cromp. An intelligent information fusion system for handling the archiving and querying of terabyte-sized spatial databases. In *S. R. Tate ed., Report on the Workshop on Data and Image Compression Needs and Uses in Scientific Community, CESDIS TR-93-99*, 75–84, 1993.

[13] M. T. Goodrich, J.-J. Tsay, D. E. Vengroff, and J. S. Vitter. External-memory computational geometry. In *Proc. IEEE Symp. on Foundations of Comp. Sci.*, 714–723, 1993.

[14] R. Laurini and A. D. Thompson. *Fundamentals of Spatial Information Systems*. A.P.I.C. Series, Academic Press, NY 1992.

[15] J. Matoušek. Cutting hyperplane arrangements. *Discrete Comput. Geom.* **6** (1991) 385–406.

[16] K. Mulmuley. A fast planar partition algorithm, I. In *Proc. 29th Annu. IEEE Sympos. Found. Comput. Sci.*, 1988, 580–589.

[17] K. Mulmuley. *Computational Geometry: An Introduction Through Randomized Algorithms*. Prentice Hall, Englewood Cliffs, NJ, 1993.

[18] F. P. Preparata and M. I. Shamos. *Computational Geometry: An Introduction*. Springer-Verlag, New York, 1985.

[19] H. Samet. *Applications of Spatial Data Structures: Computer Graphics, Image Processing, and GIS*. Addison Wesley, MA 1989.

[20] J. S. Vitter and E. A. M. Shriver. Algorithms for parallel memory, I: Two-level memories. *Algorithmica*, 12(2–3):110–147, 1994.

DIMACS Series in Discrete Mathematics
and Theoretical Computer Science
Volume **50**, 1999

On Showing Lower Bounds for External-Memory Computational Geometry Problems

Lars Arge and Peter Bro Miltersen

ABSTRACT. In this paper we consider lower bounds for external-memory computational geometry problems. We find that it is not quite clear which model of computation to use when considering such problems. As an attempt of providing a model, we define the external memory Turing machine model, and we derive lower bounds for a number of problems, including the element distinctness problem. For these lower bounds we make the standard assumption that records are indivisible. Waiving the indivisibility assumption we show how to beat the lower bound for element distinctness. As an alternative model, we briefly discuss an external-memory version of the algebraic computation tree.

1. Introduction

The Input/Output (or just *I/O*) communication between fast internal memory and slower external storage is the bottleneck in many large-scale computations. The significance of this bottleneck is increasing as internal computation gets faster, and as parallel computation gains popularity. Currently, technological advances are increasing CPU speeds at an annual rate of 40–60% while disk transfer rates are only increasing by 7–10% annually [**RW94**]. Internal memory sizes are also increasing, but not nearly fast enough to meet the needs of important large-scale applications. In recent years a lot of research has therefore been done in the area of external-memory algorithms. In this paper we study lower bounds for external-memory computational geometry problems.

1.1. The I/O-model and lower bounds for fundamental problems. We will be working in variations of the external-memory model of computation introduced by Aggarwal and Vitter [**AV88**]. The model has the following parameters:

$$N = \text{\# of records in the problem instance;}$$
$$M = \text{\# of records that can fit into main memory;}$$
$$B = \text{\# of records per block,}$$

1991 *Mathematics Subject Classification.* Primary 68Q05; Secondary 65Y25.

The first author was supported in part by U.S. Army Research Office MURI grant DAAH04–96–1–0013 and by National Science Foundation ESS grant EIA–9870734.

The second author was supported in part by the ESPRIT Long Term Research Programme of the EU under project number 20244 (ALCOM-IT).

where $1 < B \leq M < N$. The model captures the essential parameters of many of the I/O-systems in use today, and depending on the size of the data records, typical values for workstations and file servers are on the order of $M = 10^6$ or 10^7 and $B = 10^3$. Large scale problem instances can be in the range $N = 10^{10}$ to $N = 10^{12}$. Aggarwal and Vitter [AV88] defined an I/O operation in the model to be a swap of B records from internal memory with B consecutive records from external memory. The measure of performance is then the number of such I/Os needed to solve a given problem. Internal computation is free.

Aggarwal and Vitter proved lower bounds on the I/O-complexity of fundamental problems such as PERMUTING and SORTING. The number of I/Os needed to rearrange N records according to a given permutation is $\Omega(\min\{N, \frac{N}{B} \log_{\frac{M}{B}} \frac{N}{B}\})$. To prove this bound Aggarwal and Vitter made the *indivisibility of records* assumption (or indivisibility assumption for short), stating that records are moved in their entirety and that except for copying of records one is not allowed to create new records. The basic idea of the proof is to count how many permutations can be generated with a given number of I/Os and compare this number to $N!$. In other words the proof is based on the fact that we are *building* an object (a permutation) and counting arguments are used to estimate the number of I/Os required to do so. The indivisibility of records assumption ensures that a unique permutation of the records in internal and external memory can be identified throughout the life of an algorithm, and that it is well defined which permutations can be obtained from a given permutation by a single I/O. The lower bound can be shown without making any assumptions about the way the algorithm works and how it gathers information about the input (that is, without stating precisely which operations are allowed on the records in internal memory) because of the *physical* nature of the task at hand; we are moving uniquely identifiable "pebbles" around and have to arrange them in a specific configuration. We refer to this general method of proving lower bounds as the *permutation technique*.

As PERMUTING is a special case of SORTING, the permutation lower bound also applies to SORTING. Note that for realistic values in practice, N will be larger than $\frac{N}{B} \log_{\frac{M}{B}} \frac{N}{B}$. However, for extremely small values of M and B the permutation bound becomes N, and the optimal algorithm for PERMUTING is to move the records one at a time using one I/O on each record. Assuming that comparison of two records is the only allowed operation on the elements in internal memory (the *comparison I/O-model*) and using an adversary argument, Aggarwal and Vitter proved that the $\Omega(\frac{N}{B} \log_{M/B} \frac{N}{B})$ lower bound still holds for SORTING in this case. Aggarwal and Vitter conjectured that their lower bounds for SORTING and PERMUTATION hold even if the indivisibility assumption is waived, but were not able to prove so.

The model where comparisons are the only allowed computation on the elements in internal memory was later formalized by Arge et al. [AKL93]. They proved that any problem with an $\Omega(N \log_2 N)$ comparison model lower bound has a lower bound in the comparison I/O-model of $\Omega(\frac{N}{B} \log_{\frac{M}{B}} \frac{N}{B})$ I/Os. As information theoretic arguments and not the permutation technique was used in their proofs, their model was also capable of handling *decision problems*. Indivisibility of records is still assumed in their model, but the Aggarwal and Vitter conjecture may not seem so obvious for decision problems.

Recently, Arge et al. [**AFGV97**] studied the complexity of sorting strings of records in the comparison I/O-model (assuming indivisibility of records), and obtained different lower bounds depending on if one is allowed to break the strings into their characters (records) or not. Some lower bounds for external-memory graph problems are discussed in [**CGG$^+$95, Arg95**].

1.2. Computational geometry lower bounds. Our interest in this paper is lower bounds for external-memory computational geometry problems. Let us first recall how internal-memory computational geometry lower bounds are usually proved. Some lower bounds are proved in the comparison model. However, often the solution of computational geometry problems require computations like e.g. deciding if a point lies to the right or the left of a line given by two other points. For such problems, comparison model lower bounds are not interesting (because a lower bound of ∞ trivially holds in that model!). Therefore the majority of the lower bounds are proved in the *algebraic computation tree* model [**Rei72, Rab72, BO83, PS85**]. In [**PS85**] three fundamental "prototype" problems are identified for this model; SORTING, ELEMENT DISTINCTNESS and EXTREME POINTS (the problem of computing if a set of points in the plane all are vertices of their convex hull). These problems all have complexity $\Omega(N \log_2 N)$ and almost all lower bound proofs use reductions from one of them (or variations of them).

Let us see what external memory lower bounds we can derive for the three prototype problems. As discussed, the permuting bound $\Omega(\min\{N, \frac{N}{B} \log_{\frac{M}{B}} \frac{N}{B}\})$ I/Os is a lower bound on SORTING as well, and if the comparison I/O-model is assumed an $\Omega(\frac{N}{B} \log_{\frac{M}{B}} \frac{N}{B})$ I/O bound can be obtained. The result in [**AKL93**] implies an $\Omega(\frac{N}{B} \log_{\frac{M}{B}} \frac{N}{B})$ lower bound on the ELEMENT DISTINCTNESS problem in the comparison I/O-model as well. However, no lower bound is known for the problem in the general I/O-model. Finally, it is not clear how to show a lower bound for the EXTREME POINTS problem, because it is a decision problem (so the permutation technique does not apply, as no object is constructed), for which the comparison model is not strong enough.

In [**GTVV93, AVV98, APR$^+$98**] external-memory algorithms for a large number of computational geometry problems are developed. The problems include PAIRWISE ORTHOGONAL LINE SEGMENT INTERSECTION, PAIRWISE RECTANGLE INTERSECTION, answering BATCHED RANGE QUERIES, computing 3-D MAXIMA, SEGMENT SORTING, RED-BLUE LINE SEGMENT INTERSECTION, ALL NEAREST NEIGHBORS, answering BATCHED PLANAR POINT LOCATION queries, the construction of the CONVEX HULL of a set of 2 or 3-d points, and computing the MEASURE OF UNION OF RECTANGLES. A typical I/O bound on the algorithms developed for these problems is the sorting bound $O(\frac{N}{B} \log_{\frac{M}{B}} \frac{N}{B})$.

For the first four of the above problems the comparison model makes perfect sense. For example one can easily solve the pairwise ORTHOGONAL LINE SEGMENT INTERSECTION problem using comparisons only. Thus comparison I/O-model lower bounds matching the upper bounds can be obtained from their comparison model counterparts using the result in [**AKL93**]. However, the rest of the problems cannot be solved using comparisons only. Thus, reductions from e.g. ELEMENT DISTINCTNESS will not yield interesting lower bounds. However, most of the problems (including SEGMENT SORTING, RED-BLUE LINE SEGMENT INTERSECTION,

ALL NEAREST NEIGHBORS, CONVEX HULL, and BATCHED PLANAR POINT LOCA-TION) require us to build an object which can be regarded as a permutation of the input records, making the permutation technique applicable. For instance, when solving the ALL NEAREST NEIGHBORS problem, one "builds" a solution consisting of all the pairs of nearest points. This allows us to prove the permutation lower bound $\Omega(\min\{N, \frac{N}{B} \log_{\frac{M}{B}} \frac{N}{B}\})$ on these problems.

For the MEASURE OF UNION OF RECTANGLES problem it is difficult to see how to show a lower bound. In this problem no large object is constructed and the comparison model does not apply. In the algebraic computation tree model, an $\Omega(N \log_2 N)$ lower bound can be shown for this problem by reduction to a problem closely related to element distinctness, but we know of no external memory version of this model in which interesting lower bounds are known. Actually, it seems that one cannot even *solve* the problem in the Aggarwal and Vitter model, as *reporting* the result requires the creation of a new record. This is not allowed when the indivisibility of records assumption is made.

1.3. Our results. The above discussion leads us to the following question: Is it possible to define a sensible model of computation formalizing the indivisibility of records assumption (and only that) so that we can prove lower bounds, also for those computational geometry decision problems that cannot be solved by comparisons? We do come up with such a model called the *external memory Turing machine* which does make it possible to formally prove the sorting lower bound $\Omega(\frac{N}{B} \log_{\frac{M}{B}} \frac{N}{B})$ for a number of problems, including decision problems for which neither the permutation based bounds or the comparison based bounds are applicable. We believe the model is an adequate formalization of the Aggarwal and Vitter model for many computational geometry problems. In particular we show that the sorting lower bound holds for *all* the prototype problems discussed previously. Thus we remove the comparison I/O-model assumption for small B and M in the bound for SORTING proved by Aggarwal and Vitter [**AV88**], we extend the ELEMENT DISTINCTNESS lower bound proved in [**AKL93**] to the external memory Turing machine model, and we prove a new lower bound on the EXTREME POINTS problem.

Using these lower bounds we obtain alternative and improved lower bounds for the computational geometry problems discussed previously. We improve the permutation lower bound on the SEGMENT SORTING, ALL NEAREST NEIGHBORS, CONVEX HULL, and the various LINE SEGMENT INTERSECTION problems to the sorting bound. We extend the result in [**AKL93**] to the general I/O model, that is, we prove that the sorting lower bound holds for any problem requiring $\Omega(N \log_2 N)$ comparisons. This in particular means that we extend the comparison I/O-model lower bounds on ORTHOGONAL LINE SEGMENT INTERSECTION, BATCHED RANGE QUERIES, RECTANGLE INTERSECTION and 3-D MAXIMA to the external memory Turing machine model. Furthermore, we obtain a number of new lower bounds on decision as well as construction problems; CLOSEST PAIR, EUCLIDEAN MINI-MUM SPANNING TREE, INTERVAL OVERLAP, LINE SEGMENT INTERSECTION TEST, RECTANGLE CONTOUR, TRIANGULATION, DELAUNAY TRIANGULATION, SET DIS-JOINTNESS and SET DIAMETER.

We extend the model in order also to consider randomized algorithms of Las Vegas variety. We prove that the lower bounds mentioned above hold in this model as well. However, we show that if we remove the indivisibility of records assump-tion, the ELEMENT DISTINCTNESS problem (among other problems) can be solved in $O(N/B)$ I/Os by a Las Vegas algorithm! For this, we use a modification of the internal *signature sort* of Andersson et al. [**AHNR95**]. The algorithm is not unrealistic though it does take advantage of a certain amount of non-uniformity allowed in our model. This indicates that the indivisibility of records assumption *may* be a harmful restriction of the class of algorithms we want to consider. (A similar observation is made by Adler for a problem related to matrix transposition [**Adl96**]).

Finally, as an alternative model for external memory computational geometry, we define the *external memory algebraic computation tree*. This model does not include any indivisibility assumption and is concerned with computation over the reals, rather than integers as the external memory Turing Machine. We show that in this model, any continuous function can be approximated arbitrarily well using $O(N/B)$ I/Os. The algorithm is unrealistic, but suggests that showing lower bounds in the external memory algebraic computation tree model may be difficult.

2. The external memory Turing machine

In this section we define the *external memory Turing machine model*. Let us first formalize the kind of problem we want to look at. By a *decision problem* we mean a decision problem parameterized by N, denoting as usual the number of records in the input, and w, denoting the number of bits in each record. For convenience we assume that our records are bit strings (other assumptions are possible). The set of bit strings $\{0,1\}^w$ is equipped with the natural total ordering (interpreting each string as a number in binary notation). An example of a decision problem is the ELEMENT DISTINCTNESS problem, deciding if N records, each containing w bits, are distinct.

The external memory Turing machine is a finite control equipped with four tapes. Each tape consists of a string of cells, each holding a letter in a finite alphabet. The Turing machine has 4 associated fixed parameters, N, L, M and B with $B \leq M < N < L$. The machine solves a decision problem with fixed parameter N but for varying value of w. A description of each tape follows.

- The *external* tape. Intuitively, this tape contains the contents of the external memory. The tape has length exactly $(w+1)L$, where L can be any value $\geq N$ (but it must be independent of w). When the computation starts, the N input records are written in sequence at the start of the tape with #-symbols separating them. The rest of the tape contains only #-symbols. By the i'th *record* of the tape we mean cells $(w+1)(i-1)+1 \ldots (w+1)i$. A record containing only #-symbols is called *blank*. By the i'th *track* on the tape we mean records $B(i-1)+1 \ldots Bi$. There is *no* head on this tape, so the finite control cannot access it directly.

- The *internal* tape. Intuitively, this tape contains the contents of the internal memory. The tape has length exactly $(w+1)M$. Records and tracks are defined in the same way as for the external tape. There is a read/write head on this tape. The finite control can move the head and thereby read and overwrite any desired segment of the tape. When the computation starts, all records on the internal tape are blank.

- The *work* tape. This tape has unbounded length and a read/write head. The Turing machine can perform arbitrary computations using this tape.

- The *instruction* tape. There is a write only head on this tape. Some of the states of the finite control are marked as EXECUTE states. When such a state is entered, the instruction tape is examined. If the tape has contents of the form "READ *i*", where *i* is a number in binary notation, the contents of the first track of the internal tape is erased and replaced with the contents of the *i*'th track of the external tape. If the tape has contents of the form "WRITE *i*", the *i*'th track of the external tape is erased and replaced with the first track of the internal tape. If the tape has contents of the form "ASSIGN *i* TO *j*", the *j*'th *record* of the internal tape is erased and replaced with a copy of the *i*'th record of the internal tape. After an EXECUTE state has been visited, the instruction tape is erased, and its head is moved to the first cell on the tape. Furthermore, whenever an EXECUTE state is entered with a READ or WRITE instruction on the instruction tape (I/O instruction for short), the work tape is erased and the heads of the work tape and the internal tape are moved to the first cells on the tapes (this is *not* the case for ASSIGN instructions).

The machine has two special states, ACCEPT and REJECT, one of which is entered when the machine has made up its mind about the input. The complexity of a computation is the number of I/O instructions executed before this happens, and we say that an external memory Turing machine with parameters N, M, B and L, *solves* a decision problem if the machine for *all* values of w correctly accepts or rejects each input $(x_1, \ldots, x_N) \in (\{0,1\}^w)^N$.

We say that a Turing machine satisfies the *indivisibility* assumption if the head on the internal tape never writes. Note that if the Turing machine does *not* satisfy the indivisibility assumption, then the ASSIGN instruction is superfluous and can be replaced by an explicit copying routine in the finite control. Note also that if the Turing machine *does* satisfy the indivisibility assumption, each record of the external and internal tapes will at all time either contain a copy of an identifiable input record or be blank. This leads to the following definition: By a *memory configuration* π, we mean an assignment of items $x_1, x_2, \ldots, x_N$ to the records of the internal and external memories. Each x_i may appear several times or may not appear at all. For two memory configurations π_1 and π_2, we say that π_2 is *obtainable from π_1 in one step* if there is a permutation of the records in internal memory, followed by an I/O operation, followed by another permutation of the records in internal memory, turning π_1 into π_2. The *standard* memory configuration π_0 is the initial one with the N input records appearing only as the first N records of the external tape.

Let us discuss and motivate the different features of the model. First, note the parameter L, denoting the total number of records in the external memory. In previous work on external memory algorithms and lower bounds, such a fixed parameter has not been necessary, but unfortunately, we have to assume a bound on L in our lower bound proofs. Note however, that since *any* bound is sufficient (like $L < 2^{2^N}$) the restriction is not too serious. We believe that a lower bound for unbounded L can probably be obtained and leave this as a small open problem. Second, note the role of the work tape: We allow arbitrary computations on the tape but erase it after each I/O. This prevent the machine from "cheating" by storing useful information on this tape. Third, note that we allow non-uniformity in the parameters N, M and B, but not in w. If we allowed non-uniformity in w as well, any problem considered would be finite, and thus solvable in N/B I/Os using table-lookup. This would trivialize everything. It also seems quite reasonable and consistent with other assumptions to require uniformity in w. For instance, the intention of the indivisibility assumption is to prohibit that the algorithm does something very clever to each individual record. Taking non-uniform advantage of the record length would certainly violate this intention. However, why do we allow non-uniformity in N, M and B? When we assume indivisibility of records, we tie the hands of the machine behind its back: After an I/O the only information left on the tapes is some permutation of the original input records, the machine cannot even make a note on a tape about what to do next. It seems that no interesting algorithm can be implemented with such a severe restriction. However, by allowing non-uniformity in N, M, and B, we can encode some information in the *state* of the machine—as long as the amount of information is independent of w. This is quite a powerful feature, for instance, the machine can remember the current memory configuration completely without using any of the tapes. It could be argued that this is too strong, but we can still show lower bounds in the model. The non-uniformity makes it essential that w can be varied independently of N. Unfortunately, this is *not* the case for most graph problems (where $w = O(\log N)$) [**CGG**$^+$**95**], so the external memory Turing machine is not an appropriate model for these problems. An approach to modifying it would be to require uniformity in N, M and B as well, but to add an additional tape of (very) restricted length which would *not* be erased during I/Os. We shall not pursue this approach in this paper, but instead concentrate on computational geometry problems. Since the model is discrete it could be argued that it is not particularly suited for computational geometry problems either. However, most computational geometry problems still make perfect sense in a finite precision setting, and we believe that such a setting is just as interesting (and more realistic) than a "Real RAM" setting.

As mentioned, we shall in addition to decision problems also consider "construction problems" like the CLOSEST PAIR problem. Therefore, we augment the model above with a FINISHED state, and say that the machine solves a construction problem if for any input the answer to the problem resides first on the external tape (with blank records following) when the machine enters this state. In order to consider randomized algorithms in the Las Vegas variety as well, we define an *external memory Las Vegas Turing Machine* as follows: We augment the model with a fifth tape, the *random* tape, with a read-only, move-right-only head. We also include another special final state, the DON'T KNOW state. When the computation starts, the random tape is filled with an infinite sequence of random, independent, unbiased coin tosses. An external memory Turing machine solves a problem (decision or construction) if for any input, the machine answers DON'T KNOW with probability $\leq 1/2$ and gives the correct answer otherwise. In order to obey a complexity bound t, *all* possible computations on a given input x must execute at most t I/O instructions.[1]

3. Lower bounds with the indivisibility assumption

In this section we show lower bounds on the number of I/Os required to solve various problems by external memory Las Vegas Turing machines satisfying the indivisibility assumption. In Section 3.1 we first prove lower bounds on fundamental problems like SORTING and ELEMENT DISTINCTNESS. In Section 3.2 we then define

---

[1] An often seen equivalent definition of Las Vegas algorithms demands a correct answer by any terminating execution but requires only an upper bound on the *expected* number of I/Os. The worst case version with the DON'T KNOW-option is more convenient for us.

reductions in our model and derive several lower bounds from the lower bounds on the fundamental problems.

3.1. Lower bounds for fundamental problems. Our discussion of the lower bounds for fundamental problems is divided in two; in Section 3.1.1 we first prove a connection between the number of comparisons required to solve a given problem in the comparison model and the number of I/Os required by a deterministic external Turing machine. This results allows us to transform $\Omega(N \log_2 N)$ comparison lower bounds to $\Omega(\frac{N}{B} \log_{\frac{M}{B}} \frac{N}{B})$ I/O lower bounds. In Section 3.1.2 we extend the result to external memory Las Vegas Turing machines. This is done primarily to make a direct comparison with our upper bound in Section 4 meaningful.

3.1.1. *Deterministic Turing machine model.* The proof technique we will use to prove the lower bound in this section is an adaption of the Ramsey theoretic technique of Moran et al. [MSM85]. Let the parameters N, M and B be fixed. Furthermore, let a domain D be given. An *I/O decision* tree is a rooted tree. Each node has two associated labels π_v and f_v. The label π_v is a memory configuration. The label of the root must be the standard memory configuration π_0. Furthermore, if v is a son of u, π_v must be obtainable from π_u in one step. The label f_v is a function $f_v : D^M \to \{1, 2, \ldots, k\}$, where k is the out-degree of v. The out-edges of v are labelled with the integers from 1 to k. Each leaf in the tree is labelled either "yes" or "no". The I/O decision tree computes a function $D^N \to \{\text{yes}, \text{no}\}$ in the following way: The computation takes as input $(x_1, \ldots, x_N)$, starts in the root of the tree and proceeds towards one of the leaves. When a node v is encountered the function f_v is applied to the tuple $(x_{i_1}, x_{i_2}, \ldots, x_{i_M})$, where $i_1, i_2, \ldots, i_M$ are the indices of items in the internal memory according to π_v. The value of the function determines which son of v to go to next. The result of the computation is found in the leaf finally encountered.

LEMMA 3.1. *If there is an external memory Turing machine satisfying the indivisibility assumption solving a decision problem using t I/Os in the worst case, then there is a constant d, so that for any w, there is an I/O decision tree of height t and out-degree d, solving the problem for domain $\{0, 1\}^w$.*

PROOF. The proof is a relatively straightforward simulation, we simply "fold out" all possible computations of the Turing machine. For instance, the root node in the tree represents the initial configuration of the machine. For fixed N, M, B and L the first I/O instruction is fixed (because the internal memory is empty). Thus, we make a single son v of the root note, representing the configuration after this first I/O instruction. The function f_0 of the root is just the trivial function mapping everything to 1.

The next I/O instruction we perform, the memory configuration when it is performed, and the state of the Turing machine when it is performed, depends on the elements in the internal memory. For each such (instruction,configuration,state)-triple, we make a son of v and encode the relationship between the records of internal memory and the relevant (instruction,configuration,state)-triple in the function f_v. This construction is now performed recursively on each son of v. Whenever the state part of a new son is the ACCEPT or REJECT state, we replace the son with a leaf. Clearly, the tree solves the same decision problem as the machine and the height of the tree is equal to the worst case number of I/Os performed, while the outdegree of the tree is bounded by the number of states of the machine multiplied by the number of possible I/O-instructions, and by the number of possible memory configurations, all of which are independent of w. □

Remark: Our lower bounds for external memory Turing machines are shown by proving lower bounds for I/O decision trees and applying Lemma 3.1. Some people might find it more natural to have I/O decision trees as the basic model. However, the lower bound we shall show only holds if the trees are restricted to have constant out-degree and we find it more natural to motivate this restriction through the external memory Turing machine model.

The following lemma is essentially Theorem 3.5 of [**MSM85**]. A function is *order invariant* if its value can be determined by performing a number of pairwise comparisons of its arguments, each comparison yielding $<$, $>$ or $=$.

LEMMA 3.2. *For each N, d, and t, there exists a number $M(N, d, t)$ such that the following holds. Let T be an I/O decision tree of out-degree d and height t that solves an order invariant decision problem defined on D^N where $|D| > M(N, d, t)$. Then the functions f_v labelling the nodes of T can be modified so that they are all order invariant, and so that the resulting tree still solves the problem.*

By an order invariant I/O decision tree we mean an I/O decision tree where all functions f_v are order invariant. The following Lemma is a version of a theorem found in [**AKL93**].

LEMMA 3.3. *If an order invariant I/O decision tree of height t solves a problem, then there is a comparison based algorithm solving the problem using*

$$N \log B + t \cdot T_{\mathrm{merge}}(M - B, B)$$

comparisons in the worst case. Here, $T_{\mathrm{merge}}(n, m)$ denotes the number of comparisons needed for merging a sorted list of n elements with one of m elements.

PROOF. First, use $N \log B$ comparisons to determine the total order of each individual track of the external memory. The algorithm we design will maintain the invariant that the total order of the records of each individual track of the external memory, as well as the total order of the records in the internal memory, are known. Now simulate the I/O decision tree as follows, starting in the root. Each node contains a new memory configuration, obtainable from the previous one in one step. To maintain the invariant for the new memory configuration, at most $T_{\mathrm{merge}}(M - B, B)$ comparisons are needed. Given the invariant, the order invariant function f_v in the node can now be evaluated without using any further comparisons, and we can proceed to the next node. □

Combining Lemma 3.1, 3.2, and 3.3, we obtain the following Theorem.

THEOREM 3.4. *Let T be an external memory Turing machine with parameters (N, M, B, L) satisfying the indivisibility assumption. If T solves an order invariant decision problem using t I/Os in the worst case, then there is a comparison based algorithm solving the problem using $N \log B + t \cdot T_{\mathrm{merge}}(M - B, B)$ comparisons in the worst case.*

According to Knuth [**Knu73**], if $n \geq m$, $T_{\mathrm{merge}}(n, m) \leq m + \lfloor n/2^s \rfloor - 1 + sm$, where $s = \lfloor \log \frac{n}{m} \rfloor$. This means that $T_{\mathrm{merge}}(M - B, B) \leq B \log \left(\frac{M-B}{B}\right) + 3B$. Combining this with Theorem 3.4 we get the following.

THEOREM 3.5. *Let T be an external memory Turing Machine with parameters (N, M, B, L) satisfying the indivisibility assumptions. If T solves an order invariant problem with complexity $\Omega(N \log_2 N)$ in the comparison model then it has complexity at least $\Omega(\frac{N}{B} \log_{\frac{M}{B}} \frac{N}{B})$.*

COROLLARY 3.6. *Any external Turing Machine with parameters (N, M, B, L) satisfying the indivisibility assumptions and solving the* SORTING *or the* ELEMENT DISTINCTNESS *problem has complexity at least $\Omega(\frac{N}{B} \log_{\frac{M}{B}} \frac{N}{B})$.*

3.1.2. *Las Vegas Turing machine model.* We now describe how to extend the lower bound to Las Vegas algorithms. First, we define a Las Vegas I/O decision tree. This is defined as an I/O decision tree except that the label f_v does not denote a function $f_v : D^M \to \{1, \dots, k\}$, but a function from D^M into the space of probability distributions on $\{1, \dots, k\}$, i.e. the value of $f_v(x)$ is a vector $(p_1, p_2, \dots, p_k)$ with $\sum_{i=1}^{k} p_i = 1$. Furthermore, we allow leaves labelled DON'T KNOW. When executing the tree, we choose a random son according to the probability distribution. The output of the tree now becomes a random variable. We say that the tree solves a problem, if for all inputs, the output DON'T KNOW is given with probability $\leq 1/2$ and the correct answer is given otherwise.

The following can be proved in a way similar to the proof of Lemma 3.1.

LEMMA 3.7. *If there is an external memory Las Vegas Turing machine satisfying the indivisibility assumption solving a decision problem using t I/Os in the worst case, then there is a d, so that for all w, there is a Las Vegas I/O decision tree of height t and out-degree d, solving the problem for domain $\{0, 1\}^w$.*

Since Lemma 3.2 is also shown for probabilistic decision trees in [MSM85], one might now think that we just proceed as before. Unfortunately, the model of probabilistic trees given in [MSM85] is not compatible with our model. In [MSM85] the trees contain explicit coin-tossing nodes and these nodes contribute to the depth of the tree. With our definition of Las Vegas I/O trees, it becomes difficult to apply the Ramsey theoretic arguments of [MSM85] directly, the problem being that there is an infinite number of possibilities of probability distributions in a given node, even if we fix the domain and the out-degree of the node. We get around this using the following Lemma.

LEMMA 3.8. *Suppose a problem is decided by a Las Vegas I/O decision tree of height t and out-degree d. Then it is also decided by a Las Vegas I/O decision tree of height $3t$ and out-degree $d + 1$ with the constraint that for all v, x, all entries in the vector $f_v(x)$ are integer multiples of $\frac{1}{10td}$.*

PROOF. Let $\epsilon = \frac{1}{10td}$. We modify the old tree as follows. Suppose $f_v(x) = (p_1, p_2, \dots, p_k)$. We replace it with $f'_v(x) = (p'_1, p'_2, \dots p'_k, 1 - \sum_{i=1}^{k} p'_i)$, where $p'_i = \lfloor p_i/\epsilon \rfloor \epsilon$. Doing so we introduce a new son of v (son number $k + 1$) which we make a leaf marked DON'T KNOW.

Clearly, whenever the new tree gives a non-DON'T KNOW answer, it is correct. Let us estimate the error probability of the new tree. Note that all paths in the new tree which also existed in the old tree are now followed with a smaller probability. Therefore, their contribution to the DON'T KNOW probability is at most $\frac{1}{2}$. Also, note that standing in a node v, the probability of ending in the new leaf, is

$$1 - \sum_{i=1}^{k} p_i' \leq 1 - \sum_{i=1}^{k} (p_i - \epsilon) = 1 - \sum_{i=1}^{k} p_i + k\epsilon = k\epsilon \leq \frac{1}{10t}.$$

Since the tree has depth t, the probability of the entire execution ending in one of the new leaves is at most $\frac{1}{10}$. Therefore, the DON'T KNOW probability of the new tree is at most $\frac{1}{2} + \frac{1}{10}$.

We now make a tree consisting of executing the new tree independently two times and return a non-DON'T KNOW answer whenever one of the executions do. In order to do so, between the two executions, we have to reset the memory configuration to the original π_0. Thus the total depth of the tree becomes $3t$. The DON'T KNOW probability is at most $(\frac{1}{2} + \frac{1}{10})^2 < \frac{1}{2}$, as desired. □

Using Lemma 3.8 we can ensure that for fixed d and t each node in our Las Vegas I/O decision tree only has a finite (though large) number of different possible appearances. This makes it possible to use the Ramsey theoretic technique of [**MSM85**]. Emulating the proof of their Theorem 3.5 now gives us the following Lemma.

LEMMA 3.9. *For each N, d, and t, there exists a number $M(N, d, t)$ such that the following holds. Let T be a Las Vegas I/O decision tree of out degree d and height t solving an order invariant decision problem defined on D^N where $|D| > M(N, d, t)$. Then there is an order invariant Las Vegas I/O decision tree of out-degree $d + 1$ and height $3t$ solving the problem.*

We can now in analogy with Theorem 3.4 prove the following.

THEOREM 3.10. *If an external memory Las Vegas Turing machine with parameters (N, M, B, L) solves an order invariant problem using t I/Os in the worst case, then there is a comparison based Las Vegas algorithm solving the problem using $N \log B + 3t \cdot T_{\mathrm{merge}}(M - B, B)$ comparisons in the worst case.*

In order to obtain the Las Vegas variant of Corollary 3.6 we need the following.

LEMMA 3.11. *Any comparison based (internal) Las Vegas algorithm solving the* ELEMENT DISTINCTNESS *problem uses at least $\log(N!) - 1$ comparisons in the worst case.*

PROOF. Suppose a Las Vegas algorithm using less than $\log(N!) - 1$ comparisons is given. Furthermore, let a random permutation of $\{1, 2, \ldots, N\}$ be given as input. We will first argue that with probability strictly greater than $1/2$, the partial order defined by the results of the performed comparisons contains two incomparable elements. Suppose this is not the case. Then the total ordering of the elements are determined with probability at least $\geq \frac{1}{2}$. We can fix the probabilistic choices made by the algorithm so that the total ordering is still determined with probability $\geq \frac{1}{2}$, for a random permutation. Thus, we have a decision tree with at least $N!/2$ leaves of depth $< \log(N!/2)$, a contradiction.

So with probability strictly greater than $\frac{1}{2}$, two elements are incomparable, and it is thus still possible that these elements are equal. Thus, given as input a random permutation which should be accepted, the algorithm must answer DON'T KNOW with probability strictly greater than $1/2$, contradicting the definition of a Las Vegas algorithm. □

Now combining Theorem 3.10, Lemma 3.11 and the fact that an external memory Las Vegas Turing Machine solving the SORTING problem can easily be modified to solve ELEMENT DISTINCTNESS we obtain the following.

COROLLARY 3.12. *Any external memory Las Vegas Turing Machine with parameters (N, M, B, L) satisfying the indivisibility assumptions and solving the* ELEMENT DISTINCTNESS *or the* SORTING *problem has complexity at least* $\Omega(\frac{N}{B} \log_{\frac{M}{B}} \frac{N}{B})$.

3.2. Lower bounds by reduction. In this section we will show several lower bounds on computational geometry problems by reduction. Note that we need to be careful when proving these bounds, as we are not allowed to create new records.

Consider first the simple decision problem INTERVAL OVERLAP. In this problem we are given a number of intervals and should decide if any of them overlap. The standard reduction from ELEMENT DISTINCTNESS consists of converting each of the N elements x_i to the interval $[x_i, x_i]$ [PS85]. This set of intervals overlap if and only if the elements are not distinct. If we represent an interval by a pair of elements this reduction will also work in the external memory Turing machine model, because the intervals can be produced using only copying of the original elements. A machine for the ELEMENT DISTINCTNESS problem can thus be constructed from a machine for the INTERVAL OVERLAP problem simply by first scanning through the input, copying each of the records, and then simulating the INTERVAL OVERLAP machine.

The above reduction is very simple and in general standard reductions use creation of new records and not just copying of the original ones. Thus it is not clear that the class of problems solvable by external memory Turing machines satisfying the indivisibility assumption is closed under any sufficiently strong kind of reduction. However, it turns out that under some restrictions we can actually allow reductions that create new records. As a simple example of the idea behind what we will call *syntactic reductions* consider the INTERVAL OVERLAP problem again, but where intervals are now given by *one* record containing the two endpoints. It seems that the indivisibility assumption prevents us from reducing ELEMENT DISTINCTNESS to this problem. However, we can still obtain an external memory Turing machine for ELEMENT DISTINCTNESS from a machine for the INTERVAL OVERLAP problem by making the first machine simulate the latter, such that the two machines make the same I/Os but on different record types. The key point is that there is a simple connection between these types. Thus even though the new machine does I/O on records with single elements, it can after each I/O transform the records in *internal memory* to the corresponding two element records.

In general the connection between the record types can of course be much more complex than in the above example and some reductions may also need a number of extra records which are not functions of the input. We also need to formalize how a reduction from a decision problem to a construction problem is done. We combine all these considerations in the following definition.

DEFINITION 3.13. *Consider a map* $l : \mathbf{N} \to \mathbf{N}$, *maps* $h_1, h_2, \ldots h_N : \{0,1\}^* \to \{0,1\}^*$, *and maps* $c_{N+1}, \ldots, c_{N'} : \mathbf{N} \to \{0,1\}^*$, *so that for all* w *and* x *where* $|x| = w$, $|c_i(w)| = |h_i(x)| = l(w)$. *Furthermore, consider a map* r *mapping instances of a problem* P *on* N *records to instances of a problem* P' *on* $N' \geq N$ *records, such that for all instances* $(x_1, \ldots, x_N) \in (\{0,1\}^w)^N$ *of* P, $r(x_1, \ldots, x_N) = (h_1(x_1), h_2(x_2), \ldots, h_N(x_N), c_{N+1}(w), \ldots, c_{N'}(w))$.

The mapping r *is called a* **syntactic** *reduction from the decision problem* P *to the decision problem* P' *if for all instances* x, $x \in P$ *if and only if* $r(x) \in P'$.

The mapping r *is called a* **syntactic** *reduction from the decision problem* P *to the construction problem* P' *if for all instances* x, *an external memory Turing Machine satisfying the indivisibility assumption can decide if* $x \in P$ *using* $O(N/B)$ *I/Os given the answer to* $r(x)$ *as input.*

In the definition of the mapping r, the h_i's define the connection between the records in the two problems, and the c_i's define the extra fixed records needed in the reduction. The map l gives a connection between the size of the records in the two problems.

LEMMA 3.14. *There is a syntactic reduction from* ELEMENT DISTINCTNESS *on* N *records to* INTERVAL OVERLAP *on* N *records.*

PROOF. An instance of the ELEMENT DISTINCTNESS problem consists of bit-vectors of length w which can be read as numbers between 0 and $2^w - 1$. As discussed, we reduce an instance of ELEMENT DISTINCTNESS to an instance of INTERVAL OVERLAP, consisting of elements (e_1, e_2) represented as bit-vectors of length $2w$, that is, we let $l(w) = 2w$ and $h_i(x) = (x, x)$ for all $i = 1 \ldots N$. The elements are distinct if and only if the intervals do not overlap. □

The definition of syntactic reductions now allows us to prove the following.

LEMMA 3.15. *If a problem* P' *can be solved using* t *I/Os by an external memory (Las Vegas) Turing Machine satisfying the indivisibility assumption, and there is a syntactic reduction* r *from* P *to* P', *then* P *can be solved using* $t + O(N/B)$ *I/Os by such a machine.*

PROOF. We convert the machine M' solving P' to a machine M solving P as follows. M pretends that it is really M' operating on $r(x)$. This is done by maintaining, on the work-tape, a "virtual" internal memory, representing the internal memory of M'. We maintain the invariant than when the i'th record of the virtual internal memory is $h_j(x_j)$, the i'th record in the *real* internal memory is x_j, and when the i'th record of the virtual memory is $c_j(w)$, the i'th record of the real internal memory is blank. This invariant is explicitly maintained after each assignment or I/O-instruction (which clears the work-tape). In order to do so, we have to know which h_i or c_i to apply in order to recompute the records of the virtual internal memory. Therefore, we let the state of M encode full information about the current memory configuration.

If p' is a construction problem, we after the simulation use $O(N/B)$ I/Os to compute the final result by simulating the $O(N/B)$ algorithm in a similar way. □

This immediately leads to the following.

COROLLARY 3.16. *Any external memory (Las Vegas) Turing Machine with parameters* (N, M, B, L) *satisfying the indivisibility assumptions and solving the* INTERVAL OVERLAP *problem has complexity at least* $\Omega(\frac{N}{B} \log_{\frac{M}{B}} \frac{N}{B})$.

As an example of a syntactic reduction to a construction problem consider
the CLOSEST PAIR problem. An external memory Turing machine solving this
problem contains the two closest points first on the external tape when entering
the FINISHED state.

LEMMA 3.17. *There is a syntactic reduction from* ELEMENT DISTINCTNESS *on*
N *records to* CLOSEST PAIR *on* N *records.*

PROOF. We map each of the elements to points (x, y) with $0 \leq x, y \leq 2^w - 1$
given as bit-vectors of length $2w$. An element x_i is mapped to the point $(x_i, 0)$.
Thus we use $l(w) = 2w$ and $h_i(x) = (x, 0)$ for all $i = 1 \ldots N$. The elements are
distinct if and only if the two closest points are distinct.   □

THEOREM 3.18. *Any external memory (Las Vegas) Turing Machine with pa-
rameters* (N, M, B, L) *satisfying the indivisibility assumptions and solving the* CLOS-
EST PAIR, EUCLIDEAN MINIMUM SPANNING TREE, ALL NEAREST NEIGHBORS,
LINE SEGMENT INTERSECTION TEST, LINE SEGMENT INTERSECTION, RECTAN-
GLE INTERSECTION *or the* SEGMENT SORTING *problem has complexity at least*
$\Omega(\frac{N}{B} \log_{\frac{M}{B}} \frac{N}{B})$.

PROOF. For the CLOSEST PAIR problem the theorem follows immediately from
Lemma 3.15 and Lemma 3.17. Using the well know reductions [**PS85**] from this
problem to the EUCLIDEAN MINIMUM SPANNING TREE problem or the ALL NEAR-
EST NEIGHBORS problem we then obtain the lower bound on the latter two (for both
problems we assume that the result is given as a sequence of pairs of points first on
the external tape). The lower bound on the LINE SEGMENT INTERSECTION TEST
problem follows using a slightly modified version of the reduction used to prove the
bound on the INTERVAL OVERLAP problem. The bounds on the LINE SEGMENT
INTERSECTION, RECTANGLE INTERSECTION and the SEGMENT SORTING problems
then follow using simple reductions.   □

After these simple syntactic reductions we now turn to reductions where we use
extra points as well as the knowledge about the number of bits in the representation
of the input elements. First we consider the problem of computing the contour of
a set of rectangles. We assume that a rectangle is given by two points, namely its
lower left and upper right corner. Furthermore, for convenience we assume that a
set of N rectangles is given by the N upper corner points followed by the N bottom
corner points. In order to satisfy the indivisibility assumption the contour should
then be given as a sequence of line segments each given by four of the original
points; thus the points p_1, p_2, p_3, p_4 encode the segment with endpoints (x_{p_1}, y_{p_2})
and (x_{p_3}, y_{p_4}).

LEMMA 3.19. *There is a syntactic reduction from* ELEMENT DISTINCTNESS *on*
N *records to* RECTANGLE CONTOUR *on* $2N$ *records.*

PROOF. For each elements x_i in the ELEMENT DISTINCTNESS problem we create a rectangle with bottom left corner $(0, 0)$ and upper right corner $(x_i, 2^w - x_i)$.
More precisely we use $l(w) = 2w+2$, $h_i(x) = (x, 2^w - x)$ and $c_{N+1}, c_{N+2}, \ldots, c_{2N} =
(0, 0)$—refer to Figure 1. The elements are distinct if and only if the contour of the
rectangles consists of $2N+2$ segments. Thus we can solve ELEMENT DISTINCTNESS
in $O(N/B)$ I/Os with a scan through the representation of the contour.   □

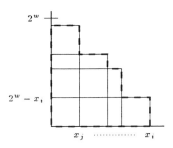

FIGURE 1. Proof of Lemma 3.19.

Next we consider the EXTREME POINTS problem. Usually the lower bound on this problem is proved using a (complicated) connected component proof [**SY82, BO83**]. Here we take advantage of the bounded domain to prove a lower bound using a syntactic reduction from ELEMENT DISTINCTNESS.

LEMMA 3.20. *There is a syntactic reduction from* ELEMENT DISTINCTNESS *on* N *records to* EXTREME POINTS *on* $N + 2$ *records.*

PROOF. We let $l(w) = 2(4w + \lceil \log N \rceil)$. We reduce an ELEMENT DISTINCT-NESS instance to an instance of EXTREME POINTS consisting of points (x, y) with $0 \le x, y \le 2^{4w + \lceil \log N \rceil} - 1$, represented as bit-vectors of length $2(4w + \lceil \log N \rceil)$. The reduction is as follows: For $i = 1, \ldots, N$, we let

$$h_i(x) = ((x + 1)2^{3w + \lceil \log N \rceil} + i, (x + 1)^2),$$

$$c_{N+1} = (0, 0)$$

and

$$c_{N+2} = ((2^w + 1)2^{3w + \lceil \log N \rceil}, (2^w + 1)^2).$$

No-instances of ELEMENT DISTINCTNESS reduce to instances where not all points are extreme, in particular if $i < j$ and $x_i = x_j$ then $h_i(x_i)$ will be in the convex hull of c_{N+1}, c_{N+2} and $h_j(x_j)$. Refer to Figure 2. For the converse, consider a yes-instance of ELEMENT DISTINCTNESS and consider three consecutive elements $x < y < z$ in the sorted version of the instance. We have to ensure that no matter what indices i_x, i_y and i_z that x, y and z have, then $h_{i_x}(x), h_{i_y}(y), h_{i_z}(z)$ form a left turn, i.e. the point $h_{i_y}(y)$ is below the line between $h_{i_x}(x)$ and $h_{i_z}(z)$. Tedious arithmetic ensures this. □

Next we consider the TRIANGULATION problem. In this problem the input is given as a number of points and the output triangulation is represented as a sequence of pairs of the original points (the edges in the triangulation). To prove a lower bound on this problem we use ideas similar to the ones used in the previous lemma.

LEMMA 3.21. *There is a syntactic reduction from* ELEMENT DISTINCTNESS *on* N *records to* TRIANGULATION *on* $N + 1$ *records.*

PROOF. We let $l(w) = 2(w + \lceil \log N \rceil)$ and reduce an instance of ELEMENT DISTINCTNESS to an instance of TRIANGULATION, consisting of points (x, y) with $0 \le x, y \le 2^{w + \lceil \log N \rceil} - 1$. We choose $h_i(x) = (x2^{\lceil \log N \rceil} + i, 0)$, that is, we map the N

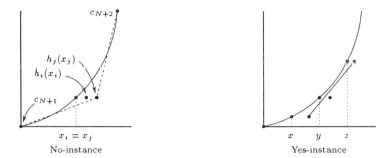

FIGURE 2. Proof of Lemma 3.20

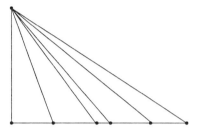

FIGURE 3. Proof of Lemma 3.21.

elements to N collinear points. Furthermore, we use an extra point $c_{N+1} = (1, 1)$. The produced set of points possesses only one triangulation which is pictured in Figure 3. As the mapping ensures that equal elements end up next to each other in the sorted sequence of the collinear points, we can solve the ELEMENT DISTINCT-NESS problem by scanning through the pairs of points defining the triangulation and check if any of these are equal. As the triangulation is of linear size this can be done in $O(N/B)$ I/Os.   □

The TRIANGULATION problem is a good example of a problem where the usual internal memory reduction from sorting [**PS85**] does not work in the external model. The reason is that in internal memory one can easily and in linear time obtain the sorted sequence of a set of points from a list of neighbor pairs. In external memory it is in general not possible to obtain the sequence in $O(N/B)$ I/O [**CGG$^+$95**].

For our final reduction we require a lower bound for the SET DISJOINTNESS problem. A Las Vegas lower bound for this problem can be shown in essentially the same way as for ELEMENT DISTINCTNESS. We omit the proof here. If a worst case lower bound is sufficient, an $\Omega(N \log N)$ comparison lower bound can be found in the literature [**BO83**], yielding an $\Omega(\frac{N}{B} \log_{\frac{M}{B}} \frac{N}{B})$ bound on the number of I/Os by Theorem 3.5.

Consider the SET DIAMETER problem: Given a set of points return the pair of points P_{i_1}, P_{i_2} maximizing $\text{dist}(P_{i_1}, P_{i_2})$.

LEMMA 3.22. *There is a syntactic reduction from* SET DISJOINTNESS *on N records to* SET DIAMETER *on N records.*

PROOF. We are given an instance $u_1, u_2, \ldots, u_{N/2}, v_1, v_2, \ldots, v_{N/2}$ of SET DIS-JOINTNESS with $u_i, v_i \in \{0, \ldots, 2^w - 1\}$. For convenience, when describing the reduction below, we shall assume that all the values have been divided by 2^w, so that they are really numbers in $[0, 1)$. We reduce the instance to an instance of SET DIAMETER as follows: Each element is mapped to a point $(x, y) \in (2^{2w+10})^2$. When describing the reduction, we shall actually map to points in $[-1, 1]^2$ with coordinates which are integer multiples of 2^{-2w-9}, this transformation is immediate. Let $\epsilon = 2^{-2w-9}$. We map u_i to (x, y) such that $|x - \cos(\pi u_i)| \leq \epsilon$ and $|x - \sin(\pi u_i)| < \epsilon$, and we map v_i to (x, y) such that $|x - \cos(\pi(1 + v_i))| \leq \epsilon$ and $|y - \sin(\pi(1 + v_i))| \leq \epsilon$. Let $\phi = \pi 2^{-w}$. Suppose ϵ had been 0. Then, if the original instance represents disjoint sets, the point pair with maximal distance has distance at most $\sqrt{(\cos(\phi) + 1)^2 + \sin^2(\phi)} = \sqrt{2 + 2\cos(\phi)}$. If the original instance represents non-disjoint sets, the point pair with maximal distance has distance exactly 2. Now, actually $\epsilon > 0$, but the the first value can be at most $2\sqrt{2}\epsilon$ bigger and the second value can be at most $2\sqrt{2}\epsilon$ smaller. Freshman calculus shows that we can still separate the two cases. $\qquad \square$

Finally, note that as a Delaunay triangulation is just a special triangulation, and as the CONVEX HULL problem is a special case of the EXTREME POINTS problem, we have obtained the following.

COROLLARY 3.23. *Any external memory (Las Vegas) Turing Machine with parameters* (N, M, B, L) *satisfying the indivisibility assumptions and solving the* RECTANGLE CONTOUR, TRIANGULATION, EXTREME POINTS, DELAUNAY TRIANGULATION, CONVEX HULL, SET DISJOINTNESS *or the* SET DIAMETER *problem has complexity at least* $\Omega(\frac{N}{B} \log_{\frac{M}{B}} \frac{N}{B})$.

4. Upper bound without the indivisibility assumption

In this section we will show that by waiving the indivisibility assumption we can beat some of the lower bounds proved in the last section.

Consider the following problem REPRESENT PERMUTATION: Given records $x_1, \ldots, x_N$ in the external memory, construct a set of numbers $y_1, y_2, \ldots, y_N$ with $y_i \in \{1, \ldots, N\}$ in the internal memory, so that for all i, j, we have $y_i < y_j$ if and only if $x_i < x_j$. The y_i's will be stored in the internal memory in a packed fashion, with $w/\log N$ values per record. With a little abuse of the notation of Section 2 we will consider an external memory Turing machine solving the REPRESENT PERMUTATION problem. Note that it is only meaningful to consider solving the problem for sufficiently large records, since we must have $N/(w/\log N) < M$ in order to store the representation in internal memory.

LEMMA 4.1. *For sufficiently large records, the* REPRESENT PERMUTATION *problem can be solved by an external memory Las Vegas Turing machine with parameters* (N, M, B, L) *using at most* $2\frac{N}{B}$ *I/Os in the worst case.*

PROOF. The algorithm is a simple variant of *signature sort* by Andersson et al. [**AHNR95**].

Let a universal [**CW79**] class of hash functions $h_k : \{0, 1\}^r \to \{0, 1\}^s$ be given, where $r = \sqrt{w \log N}$ and $s = \log(N^2 w)$. Pick a random h_k. First, we make a single pass through all the records. For each record x, we store, in the internal memory, a value $\tilde{h}(x)$ defined as follows. Divide x into w/r fields $\alpha_1, \alpha_2, \ldots, \alpha_{w/r}$, each containing r bits. For $i = 1 \ldots w/r$, compute $z_i = h_k(\alpha_i)$. Let $\tilde{h}(x) = (z_1, z_2, \ldots, z_{w/r})$. The number of bits in $\tilde{h}(x)$ is sw/r. We can pack the $\tilde{h}$-values of $w/(sw/r) = r/s$ records into a single record in the internal memory, so the total number of records used in the internal memory is Ns/r. For sufficiently large w, this becomes smaller than $M/6$.

Since the total number of fields in all the records is $Nw/r = N\sqrt{w/\log N} < \sqrt{N^2 w}$, the hash function is collision free on the set of all fields with high probability [CW79]. Assume that it is. We now construct a *path compressed trie* (see, e.g., [Knu73], page 490) with alphabet $\{0,1\}^s$ in internal memory, containing all the $\tilde{h}(x)$-vectors. Each node in the trie corresponds to some fixed prefix $(z_1, \ldots, z_i)$. The edges from that tree corresponds to all possible values of $z_{i+1} = h_k(\alpha_{i+1})$. The leaves of the tree corresponds to the individual records of the external memory. For a node u in the tree, let S_u be the set of possible values of α_{i+1} for the set of records x using that node. Since we assumed that the hash function is collision free, this number is equal to the number of edges leaving the node. The number of additional records needed for representing the trie is smaller than $M/6$ for sufficiently large w.

In order to find the total order of all the elements in the external memory, it is sufficient, for all nodes u in the tree, to determine the set S_u, thereby making it possible to sort the edges u. Since there are only $O(N)$ edges in the trie, we only need to get from the external memory $O(N)$ fields taken from a variety of records. Which fields we need is determined by the trie. We obtain the fields we need during a second pass through the external memory and store them packed in the internal memory. The total number of internal records required to store all the fields is Nr/w which is less than $M/6$ for sufficiently large records. In total we use less than $M/2$ records to store the compressed records, the trie, and the extra fields.

Finally, we derive a representation of the permutation in the external memory from the trie, by simply letting y_i be the *rank* of x_i in $x_1, x_2, \ldots, x_N$.

Above we assumed that the hash function was collision free on the set of fields. In order to get a Las Vegas algorithm, we must be able to detect when it is not. Note that the above algorithm only requires that the hash function is collision free on each particular S_u. Such collisions are easily detected in the second pass through the external memory. If a collision is detected we answer DON'T KNOW.   □

THEOREM 4.2. *Any order invariant decision problem can be solved by an external memory Las Vegas Turing machine with parameters (N, M, B, L) using $2\frac{N}{B}$ I/Os in the worst case.*

PROOF. We simply combine the solution to REPRESENT PERMUTATION with an internal algorithm for the problem. If the record size is too small for the REPRESENT PERMUTATION algorithm to be applicable we solve the problem by table lookup (the table being encoded in the state space of the Turing Machine).   □

Combining our lower and upper bounds, we have provided some evidence that the indivisibility assumption may be harmful: The ELEMENT DISTINCTNESS problem can be solved by an external memory Las Vegas Turing machine using $2N/B$ I/Os. On the other hand, no external memory Las Vegas Turing machine satisfying the indivisibility assumption solves the ELEMENT DISTINCTNESS problem using less than $\Omega(\frac{N}{B} \log_{\frac{M}{B}} \frac{N}{B})$ I/Os.

5. External memory algebraic computation trees

In this section we discuss an alternative model for formalizing external memory geometric algorithms, an external memory variant of the algebraic computation tree [**PS85**]. This model does not make the indivisibility assumption.

Let parameters N, B, M and L be given. An *external memory algebraic computation tree* is a rooted tree, containing four kinds of nodes; *decision nodes, computation nodes, read nodes* and *write nodes*. The tree operates on *internal* real variables $z_1, z_2, \ldots, z_M$, and *external* real variables $x_1, x_2, \ldots, x_L$. The tree is given as input $x_1, x_2, \ldots, x_N$ with $N \leq L$. All other variables are initially 0. It returns as output a real or Boolean value.

The computation starts in the root of the tree and proceeds towards one of the leafs where the output of the computation is given. A decision node has the form $z_i \leq z_j$?, where z_i and z_j are internal variables. A decision node has two children. When such a node is encountered, the comparison is made, and if true, the computation proceeds to the left child, otherwise it proceeds to the right child. Computation nodes, read nodes, and write nodes are unary, and after the execution of one of these nodes the computation proceeds to the child of the node. A computation node has the form $z_i \leftarrow z_j \circ z_k$ where z_i, z_j and z_k are internal variables and $\circ \in \{+, *, -, /\}$. When executed, the variable z_i is assigned the value $z_j \circ z_k$. A read node has the form "Read i" where i is a positive integer. When executed, the variables $z_1, z_2, \ldots, z_B$ are assigned the values $x_{B(i-1)+1} \ldots x_{Bi}$. A write node has the form "Write i" where i is a positive integer. When executed, the variables $x_{B(i-1)+1} \ldots x_{Bi}$ are assigned the values $z_1, z_2, \ldots, z_B$. The leaves of the tree are marked "Return z_i" where z_i is an internal variable or "Return true" or "Return false". The real or Boolean value indicated is returned when the leaf is executed.

The I/O complexity of an external memory computation tree is the maximum number of read and write nodes on a path from the root to the leaves. Note that it is possible for an infinite tree to have finite I/O-complexity. We do allow external memory computation trees to be infinite, i.e., we allow the amount of internal computation performed to depend upon the input, without some global upper bound. This seems to be a reasonable decision, given that internal computation is usually considered as "free" in external memory algorithms. See, however, the remarks following Theorem 5.1.

A non-trivial lower bound for, e.g, ELEMENT DISTINCTNESS in the external memory algebraic computation tree model would be very interesting indeed. We conjecture that the problem requires more than $O(N/B)$ I/Os in the model, but have not been able to prove so. In the following we illustrate some of the difficulties that need to be overcome in order to prove such a bound. We provide a counterintuitive upper bound for finding approximations to solutions to function problems such as the MEASURE OF UNION OF RECTANGLES problem. The upper bound is not realistic but exposes the power of the model. It works by computing an approximation to each input value and "packing" all the approximations into a single variable in internal memory, after which all further computation is internal.

THEOREM 5.1. *Given a continuous function* $f : \Re^N \to \Re$. *For all* $10 \leq B \leq M \leq N$, *there is an algebraic computation tree with parameters* L, N, M *and* B *and I/O complexity* $O(N/B)$ *which on input* $(x_1, x_2, \ldots, x_N)$ *and* $\epsilon > 0$, *computes an approximation to* $f(x_1, x_2, \ldots, x_N)$ *correct within an additive error of* ϵ.

PROOF. We shall use the following fact extensively in the proof. Given an *arbitrary* (not necessarily computable) function $g : \mathbf{N}^k \to \Re$ with $k < M/2$, we can compute g by an (infinite) external memory computation tree with no read or write nodes: We simply go through any enumeration $\mathbf{x}_1, \mathbf{x}_2, \ldots$ of $\mathbf{N}^k$ and check, for each $\mathbf{x}_i$, if it matches the actual input. If it does, we give the correct output.

Now we present the approximation-algorithm for f: On a first pass through the external memory, compute $m = \max\{|x_i|\}$ and store it in the internal memory. Also store ϵ in the internal memory. Now, compute $m' = \lceil m \rceil$ using an infinite external memory computation tree with no read or write nodes (the tree first checks if $m \leq 1$, then if $m \leq 2$, etc.) Now find the numerator p and denominator q of a positive rational number smaller than ϵ (with $r_1, r_2, \ldots$ being an enumeration of the positive rational numbers, the tree first check if $r_1 < \epsilon$, then if $r_2 < \epsilon$, etc.) Since f is continuous, it is absolutely continuous on any compact set, in particular on the box $[-m', m']^N$. Thus, given ϵ, there is a δ, so that for any input vector $\mathbf{x}$ in the box, computing f on a δ-approximation to the x_i-coordinates gives an ϵ-approximation to the correct output, as desired. Let $h : \mathbf{N}^3 \to \Re$ be a function, which, given m' and numerator and denominator of an ϵ returns an appropriate δ. Compute $\delta = h(m', p, q)$.

Let $\langle \cdot, \cdot \rangle : \mathbf{N} \times \mathbf{N} \to \mathbf{N}$ be a pairing function with projections π_1 and π_2. That is, for all natural numbers x and y, $\pi_1(\langle x, y \rangle) = x$ and $\pi_2(\langle x, y \rangle) = y$. In a second pass through the external memory, compute, for each x_i, the numerator p_i and denominator q_i of a rational δ-approximation to x_i, the approximation being found by a search as above. Also, define $r_0 = 0$, and compute $r_i = \langle \langle r_{i-1}, p_i \rangle, q_i \rangle$.

The value r_n contains all the δ-approximations of the input variables. Let $g : \mathbf{N} \to \Re$ be a function which, given such a packed vector, outputs an ϵ-approximation of $f(x_1, x_2, \ldots, x_N)$. Compute $g(r_n)$ and output the result. $\square$

Note that the proof takes advantage of the fact that we allow external memory algebraic computation trees to be infinite, i.e., we do not have a global bound on the internal computation performed, though for any fixed input, the algorithm only performs a finite amount of internal work. If we require the trees to be finite, it is still possible to obtain a somewhat weaker approximation result, where ϵ and some upper bound on the numbers of the input are fixed parameters of the problem, instead of being part of the input.

6. Open problems

As already pointed out by Aggarwal and Vitter [**AV88**] the most important open problem in the area of I/O lower bounds is to show lower bounds *without* assuming indivisibility of records.

With the models introduced in this paper, two concrete questions are:

- Can PERMUTING be done using $O(N/B)$ I/Os by an external memory Turing Machine that does not satisfy the indivisibility assumption?
- Can ELEMENT DISTINCTNESS be solved by an external memory algebraic decision tree using $O(N/B)$ I/Os?

We believe that the answers to both questions are no.

Acknowledgment. The first author would like to thank Michael Ben-Or for directing his attention to the result in [**MSM85**].

References

[Adl96]   M. Adler, *New coding techniques for improved bandwidth utilization*, Proc. IEEE
            Symp. on Foundations of Comp. Sci., 1996, pp. 173–182.

[AFGV97]   L. Arge, P. Ferragina, R. Grossi, and J. Vitter, *On sorting strings in external memory*,
            Proc. ACM Symp. on Theory of Computation, 1997, pp. 540–548.

[AHNR95]   A. Andersson, T. Hagerup, S. Nilsson, and R. Raman, *Sorting in linear time?*, Proc.
            ACM Symp. on Theory of Computation, 1995, pp. 427–436.

[AKL93]   L. Arge, M. Knudsen, and K. Larsen, *A general lower bound on the I/O-complexity
            of comparison-based algorithms*, Proc. Workshop on Algorithms and Data Structures,
            LNCS 709, 1993, pp. 83–94.

[APR+98]   L. Arge, O. Procopiuc, S. Ramaswamy, T. Suel, and J. S. Vitter, *Theory and practice
            of I/O-efficient algorithms for multidimensional batched searching problems*, Proc.
            ACM-SIAM Symp. on Discrete Algorithms, 1998, pp. 685–694.

[Arg95]   L. Arge, *The I/O-complexity of ordered binary-decision diagram manipulation*, Proc.
            Int. Symp. on Algorithms and Computation, LNCS 1004, 1995, pp. 82–91. A complete
            version appears as BRICS technical report RS-96-29, University of Aarhus.

[AV88]   A. Aggarwal and J. S. Vitter, *The Input/Output complexity of sorting and related
            problems*, Communications of the ACM **31** (1988), no. 9, 1116–1127.

[AVV98]   L. Arge, D. E. Vengroff, and J. S. Vitter, *External-memory algorithms for processing
            line segments in geographic information systems*, Algorithmica (to appear in special
            issues on Geographical Information Systems) 1998. Extended abstract appears in Proc.
            of Third European Symposium on Algorithms, 1995.

[BO83]   M. Ben-Or, *Lower bounds for algebraic computation trees*, Proc. ACM Symp. on The-
            ory of Computation, 1983, pp. 80–86.

[CGG+95]   Y.-J. Chiang, M. T. Goodrich, E. F. Grove, R. Tamassia, D. E. Vengroff, and J. S.
            Vitter, *External-memory graph algorithms*, Proc. ACM-SIAM Symp. on Discrete Al-
            gorithms, 1995, pp. 139–149.

[CW79]   J.L. Carter and M.N. Wegman, *Universal classes of hash functions*, J. Comput. Syst.
            Sci. **18** (1979), 143–154.

[GTVV93]   M. T. Goodrich, J.-J. Tsay, D. E. Vengroff, and J. S. Vitter, *External-memory compu-
            tational geometry*, Proc. IEEE Symp. on Foundations of Comp. Sci., 1993, pp. 714–723.

[Knu73]   D. Knuth, *The art of computer programming, vol. 3: sorting and searching*, Addison-
            Wesley, 1973.

[Min61]   M. L. Minsky, *Recursive unsolvability of Post's problem of 'tag' and other topics in
            the theory of Turing machines*, Ann. Math. **74** (1961), no. 3, 437–454.

[MSM85]   S. Moran, M. Snir, and U. Manber, *Applications of Ramsey's theorem to decision tree
            complexity*, Journal of the ACM **32** (1985), 938–949.

[PS85]   F. P. Preparata and M. I. Shamos, *Computational geometry: An introduction*,
            Springer-Verlag, 1985.

[Rab72]   M. O. Rabin, *Proving simultaneous positivity of linear forms*, J. Comput. Syst. Sci. **6**
            (1972), 639–650.

[Rei72]   E. M. Reingold, *On the optimality of some set algorithms*, J. ACM **19** (1972), 649–659.

[RW94]   Chris Ruemmler and John Wilkes, *An introduction to disk drive modeling*, IEEE
            Computer **27** (1994), no. 3, 17–28.

[SY82]   J. M. Steele and A. C. Yao, *Lower bounds for algebraic decision trees.*, Journal of
            Algorithms **3** (1982), 1–8.

DEPARTMENT OF COMPUTER SCIENCE, DUKE UNIVERSITY, DURHAM 27708, USA (PART OF
THIS WORK WAS DONE WHILE AT BRICS, DEPARTMENT OF COMPUTER SCIENCE, UNIVERSITY OF
AARHUS, DENMARK.)
     E-mail address: `large@cs.duke.edu`
     URL: `http://www.cs.duke.edu/~large`

BRICS, DEPARTMENT OF COMPUTER SCIENCE, UNIVERSITY OF AARHUS, 8000 AARHUS C,
DENMARK. (BRICS IS AN ACRONYM FOR BASIC RESEARCH IN COMPUTER SCIENCE, A CENTER OF
THE DANISH NATIONAL RESEARCH FOUNDATION.)
     E-mail address: `bromille@brics.dk`
     URL: `http://www.brics.dk/~bromille`

DIMACS Series in Discrete Mathematics
and Theoretical Computer Science
Volume **50**, 1999

A Survey of Out-of-Core Algorithms in Numerical Linear Algebra

Sivan Toledo

ABSTRACT. This paper surveys algorithms that efficiently solve linear equations or compute eigenvalues even when the matrices involved are too large to fit in the main memory of the computer and must be stored on disks. The paper focuses on scheduling techniques that result in mostly sequential data accesses and in data reuse, and on techniques for transforming algorithms that cannot be effectively scheduled. The survey covers out-of-core algorithms for solving dense systems of linear equations, for the direct and iterative solution of sparse systems, for computing eigenvalues, for fast Fourier transforms, and for N-body computations. The paper also discusses reasonable assumptions on memory size, approaches for the analysis of out-of-core algorithms, and relationships between out-of-core, cache-aware, and parallel algorithms.

1. Introduction

Algorithms in numerical linear algebra solve systems of linear equations and eigenvalue problems. When the data structures that these algorithms use are too large to fit in the main memory of a computer, the data structures must be stored on disks. Accessing data that is stored on disks is slow. To achieve acceptable performance, an algorithm must access data stored on disks in large contiguous blocks and reuse data that is stored in main memory many times. Algorithms that are designed to achieve high performance when their data structures are stored on disks are called *out-of-core* algorithms.

This paper surveys out-of-core algorithms in numerical linear algebra. The first part of the paper, Section 2, describes algorithms that are essentially clever schedules of conventional algorithms. The second part of the paper, Section 3, describes algorithms whose dependences must be changed before they can can be scheduled for data reuse. The final part of the paper, Section 4, explains several issues that pertain to all the algorithms that are described. In particular, we

1991 *Mathematics Subject Classification.* Primary 65Y10, 65Y20, 65F05, 65F10, 65F15, 65F25, 65F50, 65T20, 65–02, 68–02.

Key words and phrases. out-of-core algorithms, external-memory algorithms, numerical linear algebra.

This research was performed at the Xerox Palo Alto Research Center and supported in part by DARPA contract number DABT63-95-C-0087 and by NSF contract number ASC-96-26298. The author is on leave from Tel Aviv University.

discuss typical memory sizes, approaches to the analysis of out-of-core algorithms, differences between out-of-core algorithms and cache-aware and parallel algorithms, and abstractions that are useful for designing schedules. The rest of the introduction explains why out-of-core algorithms are useful and why they are not as common as one might expect.

Out-of-core algorithms are useful primarily because they enable users to efficiently solve large problems on relatively cheap computers. Disk storage is significantly cheaper than main-memory storage (DRAM).[1] Since many out-of-core algorithms deliver performance similar to that of in-core algorithms, an out-of-core algorithm on a machine with a small main memory delivers a better price/performance ratio than an in-core algorithm on a machine with enough main memory to solve large problems in core. Furthermore, ubiquitous parallel computing in the form of scalable parallel computers, symmetric multiprocessors, and parallel I/O means that buying enough memory to solve a problem in-core is rarely the best way to speed up an out-of-core computation. If the computation spends most of its time in in-core computations, the best way to speed it up is to add more processors to the computer. If the computation spends most of its time waiting for I/O, the best way to speed it up is often to attach more disks to the computer to increase I/O bandwidth.

Still, one must acknowledge two trends that cause out-of-core algorithms to become less attractive to users. First, the size of computer memories, both DRAM and disks, grows rapidly. Hence, problems that could not be solved in-core in a cost-effective manner a few years ago can be solved in-core today on machines with standard memory sizes. Problems that could not be solved in core on the world's largest supercomputer 20 years ago, the Cray-1, can now be solved in core on a laptop computer. (And not because the Cray-1 had a small memory for its period; it was designed with a memory large enough to make virtual memory unnecessary.) While users' desire to solve ever-larger problems does not seem to wane, including problems too large to fit in memory, growing main memory sizes imply that out-of-core algorithms have become a specialty.

Second, improvements in numerical methods sometimes render out-of-core algorithms obsolete. Some methods are inherently more difficult to schedule out-of-core than others. For example, the conjugate gradient method for iteratively solving sparse linear systems is inherently more difficult to schedule out of core than the iterative methods that were in use prior to the its discovery. When such a method is invented, it renders out-of-core algorithms obsolete. The algorithms described in Sections 2.4 and 3.2 exemplify this phenomenon.

2. Scheduling and Data Layout for Out-of-Core Algorithms

When an algorithm is to to be executed out of core, the ordering of independent operations must be chosen so as to minimize I/O. The schedule must also satisfy the data-flow and control-flow constraints of the algorithm. In addition, the layout of data structures on disks must be chosen so that I/O is performed in large blocks, and so that all or most of the data that is read in one block-I/O operation is used before it is evicted from main memory. When there are multiple processors, the schedule must also expose enough parallelism, and when the data is laid out on

---

[1]Today (early 1998), DRAM costs about \$5000 per GByte, whereas disks cost about \$100 per Gbyte.

multiple disks, a good schedule reads and writes from multiple disks simultaneously. This section describes techniques for scheduling and data layout that are used in out-of-core numerical linear algebra and the algorithms that use these techniques.

Some algorithms cannot be scheduled for efficient out-of-core execution. These algorithms must be modified, in the sense that their data- or control-flow constraints must be altered so as to admit an efficient schedule. Techniques for modifying algorithms so that their data-flow graphs admit efficient schedules are presented in Section 3.

2.1. An Example: Dense Matrix Multiplication. Several important computations on dense matrices can be scheduled out of core using a simple technique that was discovered very early. We illustrate this technique in some detail using matrix multiplication as an example. Specifically, we show that a naive schedule is inefficient, and that there is a simple schedule that is significantly more efficient than the naive schedule and asymptotically optimal.

The product C of two n-by-n matrices A and B is defined as $c_{ij} = \sum_{k=1}^{n} a_{ik} b_{kj}$. It is easy to see that each term $a_{ik} b_{kj}$ appears in just one summation, namely, the computation of c_{ij}. In other words, the output values share no common subexpressions, so the algorithm must compute n^3 products. The challenge is to compute them using a schedule that minimizes I/O.

A naive way to implement this algorithm uses a set of three nested loops:

```
naive-matrix-multiply(n,C,A,B)
for i = 1 to n
  for j = 1 to n
    C[i,j] = 0
    for k = 1 to n
      C[i,j] = C[i,j] + A[i,k] * B[k,j]
    end for
  end for
end for
```

Let us analyze the amount of I/O that this schedule performs. We denote the size of main memory by M, and we assume that main memory cannot store more than one half of one matrix. That is, we assume that $M < n^2/2$. (We use this assumption throughout the paper.) All the elements of B are accessed in every iteration of the outer loop. Since at most half of them can remain in memory from one iteration to the next, at least half of them must be read from disk in each iteration. Hence, the total number of words read from disk is at least $n \cdot n^2/2 = n^3/2$. In other words, the number of words transferred[2] is proportional to the work in the algorithm. It is clear that no schedule needs to transfer more than $4n^3$ words, or three reads and one write in every inner iteration. Therefore, our analysis is asymptotically tight: this schedule transfers $\Theta(n^3)$ words.

Can we do better? We can. There are schedules that transfer only $\Theta(n^3/\sqrt{M})$ words. The key to these schedules is considering the three matrices as block matrices with block size $b = \sqrt{M/3}$. We denote the b-by-b submatrix of A whose upper left corner is A_{ij} by `A[i:i+b-1,j:j+b-1]`. The following pseudo code implements this

---

[2]When we refer to "transfers" from here on we mean data transfers between main memory and disk.

improved schedule; it is not hard to prove that the code implements the same data flow graph as `naive-matrix-multiply`, if n is a multiple of b.

```
blocked-matrix-multiply(n,C,A,B)
b = square root of (memory size/3)
for i = 1 to n step b
  for j = 1 to n step b
    fill C[i:i+b-1,j:j+b-1] with zeros
    for k = 1 to n step b
      naive-matrix-multiply(b,
                            C[i:i+b-1,j:j+b-1],
                            A[i:i+b-1,k:k+b-1],
                            B[k:k+b-1,j:j+b-1])
    end for
  end for
end for
```

This schedule is more efficient because each iteration of the inner loop of `blocked-matrix-multiply` accesses only $3b^2 = M$ words but performs b^3 multiplications. The words that are accessed in the inner iteration need to be read from disk only once, because all of them fit within main memory, and written back once.(Actually, a block of C needs to be read and written only every n/b iterations.) The total number of words transferred is bounded by the number of inner iterations times the number of transfers per iteration, or

$$(n/b)^3 \times 2M = (n/\sqrt{M/3})^3 \times 2M = O(n^3/\sqrt{M}) \ .$$

This schedule performs a factor of $\Theta(\sqrt{M})$ less I/O than the naive schedule.

Can we do better yet? We cannot. The following theorem states that when main memory cannot store more than about one sixth of one input matrix, the blocked schedule described above is asymptotically optimal. This result was originally proved by Hong and Kung [**33**]; we give here a novel proof which is both simple and intuitive. Note that the theorem does not bound the number of transfers required to multiply two matrices, only the number of transfers that is required if we multiply the matrices using the conventional algorithm. Other matrix multiplication algorithms, such as Strassen's algorithm, may require fewer transfers.

THEOREM 1. *Any schedule of the conventional matrix multiplication algorithm must transfer* $\Omega(n^3/\sqrt{M})$ *words between main memory and disk, where the input matrices are n-by-n, M is the size of main memory, and* $M < n^2/5.2415$.

Proof: We decompose a given schedule into phases that transfer exactly M words each, except perhaps the last phase.

We say that c_{ij} is *alive* during phase p if during phase p the schedule computes products $a_{ik}b_{kj}$ for some k, and if some partial sum of these products is either in main memory at the end of the phase or is written to disk during the phase. By the definition of a live c_{ij}, there can be at most $2M$ live c_{ij}'s during a phase. Note that for each triplet (i,j,k), the multiplication $a_{ik}b_{kj}$ must be performed in some phase in which c_{ij} is alive. A multiplication $a_{ik}b_{kj}$ that is performed in a phase in which c_{ij} is not alive does not represent progress toward the scheduling of the data-flow graph, so without loss of generality we assume that there are no such multiplications in the schedule.

There can also be at most $2M$ distinct elements of A and at most $2M$ distinct elements of B in main memory during phase p, because every such element must be either in memory at the beginning of the phase or it must be read from disk during the phase. We denote the set of elements of A that are in main memory during phase p by A_p.

How many multiplications can the schedule perform in phase p?

We denote by S_p^1 the set of rows of A with $\sqrt{M}$ or more elements in A_p, and by S_p^2 the set of rows with fewer elements in A_p. Clearly, the size of S_p^1 is at most $2\sqrt{M}$.

We begin by bounding the number of scalar multiplications during phase p that involve elements of A from rows in S_p^1. Each element of B is used exactly once when a row of A is multiplied by B. Therefore, the number of scalar multiplications in phase p that involve elements of A from rows in S_p^1 is bounded by the size of S_p^1 times the number of elements of B in main memory during the phase, or $2\sqrt{M} \times 2M = 4M^{3/2}$.

We now bound the number of scalar multiplications during phase p that involve elements of A from rows in S_p^2. Each c_{ij} is a product of a row of A and a column of B. The number of multiplications that involve elements of A from rows in S_p^2 is bounded by the number of live c_{ij}'s times the maximal number of elements in a row of S_p^2 that are in A_p, or $2M \times \sqrt{M} = 2M^{3/2}$.

We conclude that the number of multiplications per phase is at most $6M^{3/2}$. Since the total number of multiplications is n^3, the number of phases is at least $\lceil n^3/6M^{3/2} \rceil$, so the total number of words transferred is at least $M(n^3/6M^{3/2} - 1)$ (since the last phase may transfer fewer than M words). For $M < n^2/5.2415$, the number of transfers is $\Omega(n^3/\sqrt{M})$. $\qquad\square$

2.2. Dense Matrix Computations.

Considering matrices to be block matrices leads to high levels of data reuse in other dense matrix algorithms. In out-of-core implementations the matrices are stored on disks such that b-by-b blocks are contiguous, so I/O operations are done in large blocks.

A Cholesky factorization algorithm, for example, factors a square symmetric positive definite matrix A into a product of triangular factors LL^T. Here the basic operations in the conventional algorithm are multiplication, square roots, and division by square roots. If we interpret the square root operation on a block as computing $A_{II} = L_{II}L_{II}^T$ (the operation is only applied to diagonal elements), and division by a square root as a triangular solve, we again obtain an efficient out-of-core schedule for the conventional algorithm. Interpreting these basic operations in other ways, however, can change the data flow in the algorithm and hence the output that the algorithm computes. For example, if we implement the square root operation by computing a block B_{II} such that $A_{II} = B_{II}^2$, the new algorithm will follow a different data flow path and will compute a different result. In fact, the factor that the new algorithm computes will be block triangular and not triangular.

Algorithms that merely schedule the data flow in the original algorithm are called *partitioned*, and those that follow a different data flow graph and potentially produce different results are referred to as *blocked*. When a partitioned algorithm exists it is generally preferable to a blocked one, because blocked algorithms can be numerically less stable and because they sometimes do more work than the original algorithms.

Both partitioned and blocked algorithms were discovered very early. Rutledge and Rubinstein [**45, 46**] describe the linear algebra library of the Univac. The library was operational since 1952, and Rutledge and Rubinstein attribute its basic design to Herbert F. Mitchell. The library used a partitioned algorithm for matrix multiplication. This algorithm is still widely used today. The library used a blocked algorithm for matrix inversion which is now obsolete. The Univac had a 1,000-word main memory and magnetic tapes served as secondary memory. The algorithms used 10-by-10 blocks, and the data layout and schedule were organized so that both in-core and out-of-core accesses are sequential, since both levels of memory were sequential (the Univac used acoustic delay lines for its main memory).

Other important dense algorithms that can be scheduled for out-of-core execution include the LU factorization with partial pivoting and the QR factorization. In these algorithms, the schedule cannot be generated using a partitioning of the matrix into square blocks, because these algorithms have steps that must access full columns (or full rows). The partitioning scheme that was developed for these algorithms is based on a partitioning of the matrix into vertical slabs of groups of columns.

Barron and Swinnerton-Dyerm [**4**] described in 1960 a partitioned algorithm for LU factorization with partial pivoting. They implemented a factorization algorithm for the EDSAC 2 using magnetic tapes to store the matrix. Their algorithm, which is now called the block-column right-looking algorithm, is still used today to improve data reuse in caches. The most widely used out-of-core LU factorization algorithm today, the block-column left-looking algorithm, was proposed in 1981 by Du Cruz, Nugent, Reid and Taylor [**17**]. Their algorithm was developed for the NAG library as a factorization algorithm for virtual-memory machines. The I/O behavior of the right- and left-looking algorithms is similar in terms of reads but the right-looking algorithm performs more writes. A somewhat different algorithm was proposed by the author recently [**53**]. This algorithm partitions the matrix recursively, rather than into fixed block columns, and performs asymptotically fewer reads and writes than either the right- or left-looking algorithms. It is always faster than the right-looking algorithm and it is faster than the left-looking algorithm when factoring large matrices on machines with a small main memory.

A steady stream of recent implementations of dense out-of-core factorization codes [**7, 16, 22, 26, 30, 34, 49, 50, 52, 54, 61**] serves as a testimony to the need for out-of-core dense solvers; most of these papers describe parallel out-of-core algorithms and implementations.

The first systematic comparison of conventional and partitioned schedules was published by McKellar and Coffman in 1969 [**39**], in the context of virtual memory. They assumed that the matrices are stored by blocks for the partitioned algorithms and showed that the partitioned algorithms generate asymptotically fewer page faults (I/O transfers) than conventional algorithms. Fischer and Probert [**25**] performed a similar analysis for Strassen's matrix multiplication algorithm.

Hong and Kung proved in 1981 [**33**] that the partitioned schedule for matrix multiplication is asymptotically optimal in a model of I/O that allows redundant computation and does not assume that I/O is performed in blocks. They also proved lower bounds for other problems that are discussed in this survey, such as FFTs. Their bounds were later extended to more complex models that assume that I/O is performed in blocks on multiple independent disks by Aggarwal and Vitter [**1**] and by Vitter and Shriver [**58**].

The misperception that partitioned schedules require a data layout by blocks caused partitioned schedules to go out of fashion in the 1970s. Virtual memory was becoming a standard way to solve problems larger than memory. The convenience of laying out matrices in two-dimensional arrays that are laid out by a compiler in column- or row-major order favors storing matrices by column or row, not by block. The trend to abandon partitioned schedules was apparently also fueled by the fact that on machines of that period, simply using all the words that are fetched by a page fault once or twice was enough to achieve high processor utilization. The high level of data reuse offered by partitioned algorithms was not deemed necessary. An influential 1972 paper by Moler [**40**] is typical of that period. In it, he describes a schedule for LU factorization that access entire columns of the matrix, so that all the words fetched in a page fault are used by the algorithm. The schedule is not efficient in the sense that when a typical page is fetched, its content is used once, and the page is evicted. Moler dismisses the analysis of McKellar and Coffman on the basis that it requires layout by blocks, which is inconvenient in Fortran. He ignores the 1960 paper by Barron and Swinnerton-Dyerm [**4**] that showed a partitioned schedule that accesses entire columns and does not require a block layout. Papers describing similar schedules for other problems were published in the following years [**18, 19, 59**].

Schedules that guarantee sequential access or access to large contiguous blocks were always important. In the early days, out-of-core storage used tapes, a sequential medium, so sequential access was of paramount importance. When disks became the main medium for out-of-core storage, completely sequential access lost its importance, but access to large contiguous blocks remains important. But sequential or mostly sequential access can be achieved even with partitioned algorithms, and even when the matrix is laid out in columns.

2.3. Sparse Matrix Factorizations. Sparse factorization algorithms can be classified into categories based on two criteria: what assumptions they make on sparsity patterns and whether they are row/column methods or frontal methods. Some algorithms assume that the nonzeros are clustered in relatively large dense blocks (block methods), some assume that the nonzeros are clustered around the main diagonal (band and envelope methods), while others do not assume any particular sparsity pattern (general sparse methods). We explain the difference between row/column and frontal methods later.

Band solvers were popular until well into the 1970s, and were later refined into envelope solvers. These methods have the property that at any given point in the factorization, the rows and columns of the matrix can be decomposed into three classes: done, active, and inactive. The rows/columns that are done have been factored and they need not be read again. The active part consists of rows/columns that either have been factored but are needed for updates to later rows/columns (in a left-looking algorithm), or of rows/columns that have been updated but not factored (in a right-looking algorithm). The inactive part of the matrix has not been yet read at all. Some band methods assume that the active part of the matrix can be stored in core. These methods require $\Theta(m^2)$ words of in core storage, where m is the bandwidth of the matrix, and compute the factors in one pass over the matrix. Riesel [**43**] describes a 1955 implementation of a sparse elimination algorithm on the BESK computer at the SAAB Aircraft Company in Sweden. It seems that he used such a banded algorithm, although he does not describe the algorithm in detail. His solver was used to solve linear systems with 214 variables, which did not fit within the BESK's 8192-word magnetic drum. Cantin [8] describes another such algorithm from the mid 1960s, which he attributed to E. Wilson.

When the bandwidth of the matrix is large, the active part of the matrix cannot be stored in core. The matrix is then partitioned into dense blocks and a partitioned algorithm is used. These algorithms are essentially identical to dense partitioned algorithms, except that they do not operate on zero blocks. Cantin [8], Mondkar and Powell [41] and Wilson, Bathe and Doherty [60] all describe such algorithms in the early 1970s. The same idea can be used when the nonzeros are clustered in large dense blocks even when the blocks themselves are not clustered near the main diagonal. Crotty [14] described such a solver in 1982 for problems arising from the boundary-elements method in linear elasticity.

General sparse algorithms that are row/column based are usually left-looking. These algorithms can, at least in principle, be implemented out of core using a block-column partitioned approach, similar to that used for dense matrices. The performance of such a method can be significantly worse than the performance of similar methods for dense and band matrices, however. In dense and band matrices, the columns that are needed to update a given column or block of columns are contiguous. In a general sparse algorithm, they are not. Therefore, a simple out-of-core left-looking algorithm may need to read many small noncontiguous blocks of data. To the best of my knowledge, implementations of such algorithms have not been described in the literature, perhaps for this reason.

The general sparse algorithms that have been described are frontal or multi-frontal, rather than row/column based algorithms. Unlike row/column algorithms that only store original rows and columns of A, partially updated rows/columns of A, and rows/columns of the factors, multifrontal algorithms (including simple frontal ones) store original rows/columns, rows/columns of the factors, and a stack of sparse matrices called frontal matrices. The frontal matrices are sparse, but all their rows and columns have the same structure, so they can be stored and manipulated as dense matrices. If the stack of frontal matrices fits in core, the algorithm can be implemented out-of-core as a single-pass algorithm. Interestingly, the maximum size of the stack depends on the ordering of independent operations in the algorithm. Liu [36] describes techniques to schedule the algorithm so as to minimize the maximum size of the stack so that the need for an out-of-core algorithm is eliminated in many cases. When the stack is stored out of core, the same algorithm can reduce the amount of I/O required for out-of-core stack operations. In another paper [37], Liu describes another technique that can reduce the amount of memory required for in-core factorization. This technique switches between column Cholesky and multifrontal Cholesky to reduce running time and/or memory requirements.

Another technique, similar in motivation to Liu's techniques, is described by Eisenstat, Schultz and Sherman [20]. They propose a technique that discards elements of the factor and recomputes them as necessary, instead of writing them to disk and reading them later. The method performs more work than the conventional factorizations, it solve a linear system but does not produce the factors, and it may fail for very large matrices. But like Liu's techniques, it does allow one to solve in-core linear systems that would otherwise have to be solved out of core.

Out-of-core implementations of multifrontal and related factorization methods are described by George and Rashwan [28], by George, Heath, and Plemmons [29], by Bjørstad [**5**], and by Rothberg and Schreiber [**44**]. Some of these implementations can handle frontal matrices larger than main memory by partitioning them, and recent ones (e.g. [**44**]) use Liu's technique for reducing the size of the stack and hence the amount of I/O performed. Rothberg and Schreiber use another technique to minimize I/O. Their algorithm can switch from a multifrontal method to a left-looking method based on a criterion for estimating I/O costs. The general idea is similar to that of Liu [**37**], except that Liu switches from a column factorization to a multifrontal one to save space and Rothberg and Schreiber switch in the other direction to minimize I/O.

2.4. Sparse Iterative Methods. Iterative algorithms maintain a state vector x and repeatedly update it, $x^{(t)} = F x^{(t-1)}$, where F represents the update rule. When the update in each iteration is a multiplication by a sparse matrix A, such as in Jacobi relaxation, a simple technique called *covering* can be used to schedule the computations out-of-core. The main idea is to load into main memory a subset x_I of the state vector x, perform more than one iteration on at least some elements of x_I, then to write back the updated elements of x to disk. For example, when the underlying graph of A is a $\sqrt{n}$-by-$\sqrt{n}$ mesh, this method can perform $\Theta(\sqrt{M})$ iterations while transferring only $\Theta(n)$ words. In contrast, the naive schedule that applies each iteration to the entire state vector before starting the next iteration transfers $\Theta(n)$ words per iteration. The same technique can be used when the state-vector update represents a sparse triangular or block triangular linear system.

A program called PDQ-5 used this technique to implement an out-of-core solver for two-dimensional problems arising in nuclear reactor design [**42**]. The program was written in the early 1960's by W. R. Cadwell, L. A. Hageman, and C. J. Pfeifer. It ran on a Philco-2000 using magnetic tapes to store the state vector. The program used a line successive overrelaxation (line-SOR) method, in which the update rule represents the solution of a block triangular linear system. Once the first two lines in the domain (first two blocks) are relaxed, the first line is relaxed again.

Hong and Kung [**33**] showed upper and lower bounds on the use of this technique on regular meshes.

Unfortunately, the majority of modern iterative methods cannot be scheduled for data reuse. Methods that cannot be scheduled for significant in-core data reuse include implicit time-stepping schemes, symmetric successive overrelaxation, conjugate gradient, alternating directions, and multigrid algorithms (see [**55**] for lower bounds for several of these). Indeed, in a 1964 survey paper on computers in nuclear reactor design [**15**], Cuthill remarked that alternating directions iterative methods converge faster than line-SOR methods, but are more difficult to implement efficiently out-of-core.

Consequently, since the mid 1960s most iterative solvers have been implemented in core. Schedules that access data mostly sequentially have been developed for virtual memory systems, however. Wang, for example, shows such a schedule for alternating directions [**59**]. Another idea that has emerged recently is to modify the data flow of iterative algorithms that cannot be scheduled for data reuse so that they can be scheduled. This idea is explored in Section 3.2.

2.5. Fast Fourier Transforms (FFTs). The FFT can be scheduled for out-of-core execution fairly efficiently. Theoretically, both lower and upper bounds for the number of words transferred is $\Theta(n \log n / \log M)$ [**33**, **1**]. In practice, the

$\log n / \log M$ factor is a small constant, rarely larger than 2, so asymptotic bounds are not particularly useful.

Soon after the introduction of the FFT algorithm by Cooley and Tukey, Gentleman and Sande [27] presented an FFT algorithm which is suitable for out-of-core applications. The algorithm works by arranging the input n vector in a two-dimensional array, performing an FFT on every row, scaling the array, transposing the array and performing an FFT on every row again (every column of the original array). Assuming that primary memory can hold at least one row and one column of the array, the algorithm requires only two passes over the data and an out-of-core matrix transposition. The algorithm did not gain much acceptance, but it was rediscovered and implemented on several machines. See [3] for a survey and a description of an implementation on the Cray X-MP.

Although the total amount of I/O that this algorithm performs is not large relative to the size of the data set, it often suffers from low processor utilization. The total amount of work per datum in the FFT, $\Theta(\log n)$, does not allow for extensive data reuse. With only a moderate level of data reuse, I/O often dominates the running time.

Other, less efficient, out-of-core FFT algorithms were proposed by Singleton [51] and Brenner [6]. Eklundh [21] noticed that an efficient out-of-core matrix transposition algorithm is a key to out-of-core two-dimensional FFT algorithms, and suggested one. Others have also proposed algorithms for matrix transposition, for example [2, 48, 56].

More recently, a series of papers by Cormen, Wegmann, and Nicols describe parallel out-of-core FFT algorithms that use multiple disks [11, 12, 13].

The monograph by Van Loan [57] describes numerous FFT algorithms, including out-of-core algorithms, together with extensive references. Bailey's paper [3] also presents a good survey of out-of-core FFT algorithms.

2.6. N-Body Computations. N-body or fast multipole algorithms are tree based algorithms for evaluating integrals. In a recent paper, Salmon and Warren [47] describe a parallel out-of-core N-body code. The code uses several scheduling techniques to achieve high out-of-core performance. The most important technique is a special ordering of points in space, and hence data accesses to them, which enhances data reuse. The ordering is based on a self-similar space-filling curve. This ordering was developed to enhance locality in caches and in distributed memory parallel computers, but it proved suitable for an out-of-core implementation. The other important algorithmic technique that they use is essentially loop fusion. They fuse several passes over the data that were originally separate into as few traversals as possible, in order to minimize I/O. They comment that this scheduling technique enhances performance but reduces the modularity of the code, because tasks that are conceptually unrelated are now coded together. They report good performance using a cluster of Pentium Pro computers.

Most implementations of out-of-core algorithms in numerical linear algebra either perform I/O data transfers explicitly or rely on the virtual memory system to perform data transfers. In contrast, Salmon and Warren use a hybrid scheme that can best be described as a user-level paging mechanism.

3. Data-Flow Transformation Techniques

When the data flow of an algorithm does not admit an efficient out-of-core schedule, the designer must develop an algorithm with a more suitable data-flow graph. The new algorithm may be a "transformation" of the original algorithm, or it may be completely new. By transformation I mean that the two algorithms are equivalent in some sense, although there is no clear distinction between transformed algorithms and completely new algorithms.

Numerous parallel algorithms can be thought of as transformations of conventional sequential algorithms. Parallel summation is probably the simplest example. An algorithm that adds n numbers to a running sum has a deep data-flow graph that does not admit a good parallel schedule. An algorithm that sums two $n/2$ groups of numbers recursively and then adds up the two partial sums has a data flow graph with depth $\log n$ that admits an efficient parallel schedule. If addition is associative, the two algorithms are algebraically equivalent. Floating-point addition is not associative, but the two algorithms are essentially equivalent for most practical purposes. This is not always the case. Some sequential algorithms are significantly more stable than their (transformed) parallel versions. For example, the data-flow graph of Modified Gram-Schmidt orthogonalization contains less parallelism than the data-flow graph of Classical Gram-Schmidt, but it is significantly more accurate. Transformed algorithms sometimes perform significantly more work (arithmetic) than their sequential counterparts. For example, parallel algorithms for solving tridiagonal systems of linear equations usually perform at least twice the amount of arithmetic that sequential algorithms perform. Another example is Gauss-Seidel relaxation, which often converges more slowly when the unknowns are reordered to maximize parallelism (e.g., red-black ordering of a grid instead of a natural ordering).

Algorithms can sometimes be transformed so that their data-flow admits an efficient out-of-core schedule, but they are not as common as algorithms that are transformed to introduce parallelism. One obvious reason for this is that parallel algorithms have been investigated much more intensively than out-of-core algorithms. Another possible reason is that work inefficiency, and to some extent instability, are often tolerated in parallel algorithms. The common argument goes something like this: "this parallel algorithm performs 4 times more work than the sequential algorithm, but if you have 100 processors, it would still 25 be times faster than a sequential algorithm that uses only one processor." Although this argument fails if you have only 4 or 8 processors, it is valid if you have a 100, and some users do have applications that they must parallelize on large numbers of processors. The same reasoning cannot be applied to out-of-core algorithms, because the in-core/out-of-core performance ratio is fixed, not unbounded like the number of processors. For example, if running an algorithm which cannot reuse in-core data out of core is 10 times slower than running it in core, and if a "transformed" algorithm with improved data reuse performs 4 times more arithmetic, it will solve problems at most 2.5 times faster than the original algorithm. For many users the difference is not significant enough to switch to an out-of-core algorithm, especially if the out-of-core algorithm is numerically less stable.

Still, there are interesting examples of algorithms that have been transformed to improve data reuse, which this section describes.

3.1. Dense Eigensolvers. The most widely-used dense eigensolvers are based on the QR algorithm and its variants. In these algorithms, the matrix is first reduced to either a tridiagonal matrix (in the symmetric case) or to an upper Hessenberg matrix (in the nonsymmetric case) using symmetric orthogonal transformations. The algorithm then computes the eigenvalues of the reduced matrix iteratively.

It appears that there is no way to schedule the symmetric reduction with a significant level of data reuse, although I am not aware of any formal lower bound. But the algorithm can be replaced by a block algorithm that offers a high level of data reuse. The transformed algorithm performs more work than the original. It is not considered to be less stable, since both apply only orthogonal transformations to the matrix. In the symmetric case, the rest of the algorithm can be performed in-core, since it operates on a tridiagonal matrix (with $\Theta(n)$ nonzeros, as opposed to the original dense matrix with has $\Theta(n^2)$ nonzeros). In the nonsymmetric case, computing the eigenvalues of the Hessenberg matrix involves an iterative algorithm in which every iteration accesses $\Theta(n^2)$ words and performs only $\Theta(n^2)$ work. I am not aware of an efficient way to implement this algorithm out-of-core.

In general terms, The blocked algorithm for the reduction of a symmetric matrix to a tridiagonal one works as follows. It first reduces the matrix to a block tridiagonal matrix, rather than to a true tridiagonal matrix. The reduced matrix can then be reduced to a new block tridiagonal matrix with a smaller block size, until we get a true tridiagonal matrix. The blocked algorithm performs more work than the original scalar algorithm.

Grimes and Simon [**31**] describe such an algorithm for dense symmetric generalized eigenvalue problems, motivated by quantum mechanical bandstructure computations. Their algorithm works by reducing the matrix into a band matrix using block Householder transformation. They assume that the band matrix fits within primary memory and reduce it to a tridiagonal matrix using an in-core algorithm. This variant is more efficient than the general scheme we described above because their algorithm reduces the original matrix not to a general block tridiagonal matrix, but to a band matrix. They report on the performance of the algorithm on a Cray X-MP with a solid state storage device (SSD) as secondary storage.

Dubrulle [**18**] describes an implementation of the symmetric reduction to tridiagonal algorithm with a sequential access pattern, but without introducing any data reuse. This paper is similar in motivation to Moler's paper [**40**].

3.2. Iterative Linear Solvers. Early algorithms were all implemented out-of-core, but advances in iterative linear solvers made out-of-core implementations increasingly difficult. Early algorithms were relatively easy to schedule out of core. More advanced techniques that were introduced later, such as conjugate gradient and multigrid, however, are difficult to schedule out-of-core. The same techniques can sometimes be applied to conjugate-gradient-like algorithms that compute eigenvalues of sparse matrices.

Many modern iterative methods cannot be scheduled for significant data reuse because each element of their state vectors in iteration t depends on all the elements of the state vector in iteration $t-1$. In conjugate gradient and similar algorithms, these dependences occur through a global summation; in multigrid and other multilevel algorithms, the dependences occur through a traversal of a hierarchical data structure.

Three different techniques have been proposed to overcome these problems. The first technique retains the basic structure of the algorithm, but tries to reduce the number of iterations by making each iteration more expensive. When each iteration requires one pass over the out-of-core data structure, this approach reduces I/O. Mandel [**38**], for example, describe an out-of-core Krylov-subspace algorithm (i.e., conjugate-gradient-like) with a domain-decomposition preconditioner for solving finite-elements problems. The preconditioner accelerates convergence by making every iteration computationally more expensive.

The second approach is to algebraically transform algorithms so that they can be scheduled for data reuse. The effect of these transformations is to delay information propagation through the state vector. All of the algorithms in this category perform more work than the original algorithms, and they are generally less numerically stable. On the other hand, they are equivalent to the original algorithms in exact arithmetic (so convergence is not a concern, only accuracy) and transformed algorithms can usually be derived almost mechanically from existing ones.

Leiserson, Rao, and the author [**35, 55**] introduced the notion of blocking covers, which they used to design out-of-core multigrid algorithms. These algorithms have not been implemented. The author also developed transformation techniques for conjugate gradient and related algorithms [**55**]. These algorithms have been successfully implemented out-of-core. These conjugate-gradient algorithms are closely related to multistep conjugate-gradient algorithms [**9**] whose goal is to eliminate synchronization points (i.e., global summations) in parallel algorithms.

The third technique for removing dependencies from iterative algorithms relies on the structure of the conjugate gradient algorithm. In the conjugate gradient and related algorithms, global summations are used to incrementally construct a polynomial that minimizes some error norm, and hence, ensures optimal convergence. It is also possible to choose the polynomial without computing global sums. The polynomial that we choose can be either fixed for all linear systems (as in so-called stationary iterations and in certain polynomial preconditioners), or it can depend on certain properties of the coefficient matrix (as in Chebychev iteration, which depends on the extreme eigenvalues of the matrix).

Fischer and Freund proposed such out-of-core methods based on polynomial preconditioning [**24**] and on an inner-product free Krylov-subspace method [**23**]. Both methods perform a small number of conjugate-gradient iterations to approximate the spectrum of the matrix. This approximation is used to construct a family of polynomials that is used in a polynomial preconditioner in one method, and in an inner-product free Krylov-subspace algorithm in another method. Both methods compute far fewer inner products than popular Krylov-subspace algorithms and should therefore be easier to implement out-of-core. Generally speaking, these methods are not as numerically robust as the (hard to schedule) algorithms that they are supposed to replace.

4. Concluding Remarks

This section discusses a few general issues that pertain to all of the algorithms described in this paper.

4.1. How Much Main Memory Should Computers Have? The costs of main memory and disks dictate a certain reasonable range of ratios of disk to main memory sizes. Disk storage is about 50 times cheaper today than DRAM storage.[3] Therefore, computers are likely to have main memory sizes that are, say, 10 to 1000 times smaller than their disks. Computers with less memory than this are poorly designed, since their main memories, and hence their performance, can be significantly increased at a modest increase in their total cost.

In terms of out-of-core algorithms, and especially in terms of asymptotic data-reuse bounds, these disk/memory size rations imply that main memories are very large. For example, using the conservative assumptions that computers have main memories that are at least $1/1000$ the size of their disks, and that the data set size n is at least $1,000,000$, we conclude that computers have enough main memory to hold $\Theta(\sqrt{n})$ data items in core. Smaller data sizes can usually be handled in core today, and smaller memories imply poor design, as explained above.

This analysis implies that in some cases we cannot evaluate algorithms by comparing asymptotic data-reuse bounds. The asymptotic I/O bound for the FFT, for example, $\Theta(n \log n / \log M)$, essentially reduces to $\Theta(n)$ since $\log n / \log M \leq 2$ on any reasonable computer. We cannot use the asymptotic bound to compare the algorithm to another algorithm that performs asymptotically only $\Theta(n)$ I/O's.

4.2. Analyzing I/O and Data Reuse. Four approaches for analyzing I/O counts and levels of data reuse have been used in the literature.

The most obvious approach is to simply count read and write operations exactly. This approach is the most tedious, and it rarely predicts the impact of I/O on the total execution time accurately. The time it takes to service I/O requests depends on many factors, including the size of the requests, whether data is prefetched or fetched on demand, whether the requests access data sequentially, etc. Therefore, exactly analyzing the number of I/O requests or the total amount of data that is read and written usually does not produce accurate running-time estimates.

A more relaxed approach is to asymptotically bound the number of I/O's in the algorithm. This is easier, of course, than counting I/O's exactly, especially when one wishes to prove lower bounds. The usual warnings that apply to any asymptotic bound are applicable here, with the added caveat that one must assume that main memories are large, as explained above. This reduces some asymptotic bounds, such as the ones for FFT and for sorting, to $\Theta(n)$.

Another approach, which is applicable only when the algorithm performs repeated sequential accesses to the data, is to count the number of times the algorithm accesses the data sets, commonly referred to as the number of passes. When applicable, this approach can predict running times very accurately. The approach can be extended somewhat by considering additional access patterns, such as out-of-core matrix transposition, as "primitives". We can then state, for example, that a certain FFT algorithm performs two passes over the data and one matrix transposition. In my opinion, such statements are more useful in practice than asymptotic bounds.

The last approach, which is widely used in the literature on dense-matrix algorithms, analyzes the so-called *level-3 fraction* in the algorithm. The level-3 fraction is the fraction of arithmetic operations that can be represented by matrix-matrix operations, mainly multiplication, factorization, and triangular solves. The rationale behind this approach is that matrix-matrix operations can be scheduled for a

---

[3]The DRAM/disk price ratio fluctuates (it was 100 two years ago, for example), but these fluctuations are too small to invalidate the argument that we present.

high level of data reuse, and that thus, an algorithm with a high level-3 fraction is likely to run well on a computer with caches or as an out-of-core algorithm.

The trouble with this approach is that it predicts what fraction of the work can be expected to enjoy *some* level of data reuse, but it does not predict how much data reuse can be achieved. One can thus find in the literature algorithms with a high level-3 fraction, but in which the matrix-matrix operations have a constant level of data reuse, say 2 or 4. These algorithms will clearly be inefficient when executed out-of-core. On the other hand, the level-3 fraction is a good indicator of cache performance, since even a modest level of data reuse is sufficient to reduce cache miss rates significantly on many computers (this is changing as cache miss penalties are growing).

4.3. Out-of-Core and Other Architecture-Aware Algorithms. Main memory and disks are just two levels of what is often an elaborate memory hierarchy. The memory systems of today's workstations and personal computers consist of a register file, two levels of cache, main memory, and disks. High-end machines often have three levels of cache, and their main memories and file systems are distributed. Algorithms that are designed to exploit one level of the memory system often exploit other levels well. For example, an out-of-core algorithm that is designed to exploit the speed of main memory when the data is stored on disks can sometimes be used to exploit caches and even the register file. Similarly, an algorithm that is designed to run well on a distributed-memory parallel computer can sometimes be efficient as an out-of-core algorithm. In particular, many partitioned dense-matrix algorithms can effectively exploit all of these architectural features. But this is not always true. The main memory/disk relationship is different enough from the cache/main memory relationship that some algorithms that are efficient out-of-core do not exploit caches efficiently, or vice versa. Algorithms that run well on a distributed-memory parallel computer do not always translate into efficient out-of-core algorithms, and vice versa.

At least until recently, the cache/main-memory bandwidth ratio has been much lower than the main-memory/disk bandwidth ratio. Therefore, algorithms with a modest level of data reuse exploit caches well, but are inefficient when implemented out-of-core. Algorithms which trade more work for a high level of data reuse are often useful out of core, but are inefficient in core, because the gain in cache effectiveness is more than offset by the loss in work efficiency. The cache/main-memory bandwidth ratio is increasing, so it is possible that in the future there will not be any difference between cache-aware and out-of-core algorithms. I believe, however, that it is more likely that new main memory technologies will improve main memory bandwidth.

Some algorithms have schedules that are efficient on distributed-memory parallel computers but have no efficient out-of-core schedules. The most important group of algorithms in this class are iterative linear solvers. Many of these algorithms can exploit a large number of processors and can be scheduled so that the amount of interprocessor communication is asymptotically smaller than the amount of work. For example, a conjugate gradient solver for a 2-dimensional problem of size n can theoretically exploit $n/\log n$ processors in a work-efficient manner, with a computation/communication ratio of $\sqrt{n/p}$. The high computation/communication ratio ensures that the algorithm runs efficiently even when interprocessor communication is significantly slower than the computation speed, which is common.

As explained in Section 3.2, these algorithms cannot be scheduled for any significant level of data reuse. Even though in both computational environments a processor can compute much faster than it can communicate (with other processors or with the disk), there is a crucial difference between the two environments. On the parallel computer, data that is not local is still being updated, but in the out-of-core environment, data that is not in core is not being updated.

There are also algorithms that can be scheduled for a high level of data reuse but cannot be scheduled in parallel, but they are not common in practice.

4.4. Abstractions for Scheduling. Discovering and implementing good schedules is difficult. Abstractions help us discover, understand, and implement complex schedules that allow algorithms to run efficiently out-of-core and in parallel. Many of the schedules and algorithms that this paper describes can be expressed in terms of simple abstractions.

Many efficient schedules for dense matrix algorithms can be expressed in terms of block-matrix algorithms. That is, the schedule is expressed as an algorithm applied to matrices consisting of submatrices rather than matrices of scalars.

Efficient out-of-core FFT algorithms are expressed in terms of multiple FFTs on smaller data sets and a structured permutation, namely, matrix transposition.

Schedules for general sparse matrix factorizations are usually expressed (and generated) using *elimination trees*, which are essentially compact approximate representations of the data-flow graphs of the algorithms. They are compact mostly because they represent dependences between rows and/or columns, not between individual nonzeros of the factors. Some elimination trees represent actual dependences, some represent a superset of the dependences.

While these are the most common abstractions one finds in the literature to describe schedules, there are a few others that deserve mention.

Recursion is a powerful abstraction that has not been used extensively by designers of numerical software, perhaps because of the influence of Fortran (which did not support procedural recursion until recently). The author designed a recursive schedule for dense LU factorization with partial pivoting that is more efficient than schedules based on block algorithms [**53**]. Gustavson applied the same technique to several other dense matrix algorithms [**32**].

Cormen developed an abstraction for describing structured permutations in terms of matrix operations on indices [**10**].

Van Loan uses an abstraction based on Kronecker products to describe FFT algorithms and their schedules [**57**].

References

[1] Alok Aggarwal and Jeffrey S. Vitter. The input/output complexity of sorting and related problems. *Communications of the ACM*, 31(9):1116–1127, August 1988.

[2] Mordechai Ben Ari. On transposing large $2^n \times 2^n$ matrices. *IEEE Transactions on Computers*, C-28(1):72–75, 1979. Due to a typo, the pages in that issue are marked as Vol. C-27.

[3] David H. Bailey. FFTs in external or hierarchical memory. *The Journal of Supercomputing*, 4(1):23–35, 1990.

[4] D. W. Barron and H. P. F. Swinnerton-Dyerm. Solution of simultaneous linear equations using a magnetic tape store. *Computer Journal*, 3:28–33, 1960.

[5] Petter E. Bjørstad. A large scale, sparse, secondary storage, direct linear equation solver for structural analysis and its implementation on vector and parallel architectures. *Parallel Computing*, 5(1):3–12, 1987.

[6] N. M. Brenner. Fast Fourier transform of externally stored data. *IEEE Transactions on Audio and Electroacoustics*, AU-17:128–132, 1969.

[7] Jean-Philippe Brunet, Palle Pedersen, and S. Lennart Johnsson. Load-balanced LU and QR factor and solve routines for scalable processors with scalable I/O. In *Proceedings of the 17th IMACS World Congress*, Atlanta, Georgia, July 1994. Also available as Harvard University Computer Science Technical Report TR-20-94.

[8] Gilles Cantin. An equation solver of very large capacity. *International Journal for Numerical Methods in Engineering*, 3:379–388, 1971.

[9] A. T. Chronopoulos and C. W. Gear. s-step iterative methods for symmetric linear systems. *Journal of Computational and Applied Mathematics*, 25:153–168, 1989.

[10] Thomas H. Cormen. *Virtual Memory for Data-Parallel Computing*. PhD thesis, Massachusetts Institute of Technology, 1992.

[11] Thomas H. Cormen. Determining an out-of-core FFT decomposition strategy for parallel disks by dynamic programming. In Michael T. Heath, Abhiram Ranade, and Robert S. Schreiber, editors, *Algorithms for Parallel Processing, volume 105 of IMA Volumes in Mathematics and its Applications*, pages 307–320. Springer-Verlag, 1998.

[12] Thomas H. Cormen and David M. Nicol. Out-of-core FFTs with parallel disks. *ACM SIGMETRICS Performance Evaluation Review*, 25(3):3–12, December 1997.

[13] Thomas H. Cormen, Jake Wegmann, , and David M. Nicol. Multiprocessor out-of-core FFTs with distributed memory and parallel disks. In *Proceedings of IOPADS '97*, pages 68–78, San Jose, California, November 1997.

[14] J. M. Crotty. A block equation solver for large unsymmetric matrices arising in the boundary integral equation method. *International Journal for Numerical Methods in Engineering*, 18:997–1017, 1982.

[15] Elizabeth Cuthill. Digital computers in nuclear reactor design. In Franz L. Alt and Morris Rubinoff, editors, *Advances in Computers*, volume 5, pages 289–348. Academic Press, 1964.

[16] J. J. Dongarra, S. Hammarling, and D. W. Walker. Key concepts for parallel out-of-core LU factorization. Technical Report CS-96-324, University of Tennessee, April 1996. LAPACK Working Note 110.

[17] J. J. Du Cruz, S. M. Nugent, J. K. Reid, and D. B. Taylor. Solving large full sets of linear equations in a paged virtual store. *ACM Transactions on Mathematical Software*, 7(4):527–536, 1981.

[18] A. A. Dubrulle. Solution of the complete symmetric eigenproblem in a virtual memory environment. *IBM Journal of Research and Development*, pages 612–616, November 1972.

[19] A. A. Dubrulle. The design of matrix algorithms for Fortran and virtual storage. Technical Report G320-3396, IBM Palo Alto Scientific Center, November 1979.

[20] S. C. Eisenstat, M. H. Schultz, and A. H. Sherman. Software for sparse Gaussian elimination with limited core memory. In Ian. S. Duff and C. W. Stewart, editors, *Sparse Matrix Proceedings*, pages 135–153. SIAM, Philadelphia, 1978.

[21] J. O. Eklundh. A fast computer method for matrix transposing. *IEEE Transactions on Computers*, C-21(7):801–803, 1972.

[22] Charbel Farhat. Large out-of-core calculation runs on the IBM SP2. *NAS News*, 2(11), August 1995.

[23] Bernd Fischer and Roland W. Freund. An inner product-free conjugate gradient-like algorithm for hermitian positive definite systems. In *Proceedings of the Cornelius Lanczos 1993 International Centenary Conference*, pages 288–290. SIAM, December 1993.

[24] Bernd Fischer and Roland W. Freund. On adaptive weighted polynomial preconditioning for hermitian positive definite matrices. *SIAM Journal on Scientific Computing*, 15(2):408–426, 1994.

[25] Patrick C.. Fischer and Robert L. Probert. A not one matrix multiplication in a paging environment. In *ACM '76: Proceedings of the Annual Conference*, pages 17–21, 1976.

[26] Nikolaus Geers and Roland Klee. Out-of-core solver for large dense nonsymmetric linear systems. *Manuscripta Geodetica*, 18(6):331–342, 1993.

[27] W. M. Gentleman and G. Sande. Fast Fourier transforms for fun and profit. In *Proceedings of the AFIPS*, volume 29, pages 563–578, 1966.

[28] Alan George and Hamza Rashwan. Auxiliary storage methods for solving finite element systems. *SIAM Journal on Scientific and Statistical Computing*, 6(4):882–910, 1985.

[29] J. A. George, M. T. Heath, and R. J. Plemmons. Solution of large-scale sparse least squares problems using auxiliary storage. *SIAM Journal on Scientific and Statistical Computing*, 2(4):416–429, 1981.

[30] Roger G. Grimes. Solving systems of large dense linear equations. *The Journal of Supercomputing*, 1(3):291–299, 1988.

[31] Roger G. Grimes and Horst D. Simon. Solution of large, dense symmetric generalized eigenvalue problems using secondary storage. *ACM Transactions on Mathematical Software*, 14(3):241–256, 1988.

[32] F. G. Gustavson. Recursion leads to automatic variable blocking for dense linear-algebra algorithms. *IBM Journal of Research and Development*, pages 737–755, November 1997.

[33] J.-W. Hong and H. T. Kung. I/O complexity: the red-blue pebble game. In *Proceedings of the 13th Annual ACM Symposium on Theory of Computing*, pages 326–333, 1981.

[34] Kenneth Klimkowski and Robert van de Geijn. Anatomy of an out-of-core dense linear solver. In *Proceedings of the 1995 International Conference on Parallel Processing*, pages III:29–33, 1995.

[35] Charles E. Leiserson, Satish Rao, and Sivan Toledo. Efficient out-of-core algorithms for linear relaxation using blocking covers. *Journal of Computer and System Sciences*, 54(2):332–344, 1997.

[36] Joseph W. H. Liu. On the storage requirement in the out-of-core multifrontal method for sparse factorization. *ACM Transactions on Mathematical Software*, 12(3):249–264, 1986.

[37] Joseph W. H. Liu. The multifrontal method and paging in sparse Cholesky factorization. *ACM Transactions on Mathematical Software*, 15(4):310–325, 1989.

[38] Jan Mandel. An iterative solver for p-version finite elements in three dimensions. *Computer Methods in Applied Mechanics and Engineering*, 116:175–183, 1994.

[39] A. C. McKeller and E. G. Coffman, Jr. Organizing matrices and matrix operations for paged memory systems. *Communications of the ACM*, 12(3):153–165, 1969.

[40] Cleve B. Moler. Matrix computations with Fortran and paging. *Communications of the ACM*, 15(4):268–270, 1972.

[41] Digambar P. Mondkar and Graham H. Powell. Large capacity equation solver for structural analysis. *Computers and Structures*, 4:699–728, 1974.

[42] C. J. Pfeifer. Data flow and storage allocation for the PDQ-5 program on the Philco-2000. *Communications of the ACM*, 6(7):365–366, 1963.

[43] Hans Riesel. A note on large linear systems. *Mathematical Tables and other Aids to Computation*, 10:226–227, 1956.

[44] Edward Rothberg and Robert Schreiber. Efficient, limited memory sparse Cholesky factorization. Unpublished manuscript, May 1997.

[45] J. Rutledge and H. Rubinstein. High order matrix computation on the UNIVAC. Presented at the meeting of the Association for Computing Machinery, May 1952.

[46] Joseph Rutledge and Harvey Rubinstein. Matrix algebra programs for the UNIVAC. Presented at the Wayne Conference on Automatic Computing Machinery and Applications, March 1951.

[47] John Salmon and Michael S. Warren. Parallel out-of-core methods for N-body simulation. In *Proceedings of the 8th SIAM Conference on Parallel Processing for Scientific Computing*, *CD-ROM*, Minneapolis, March 1997.

[48] Ulrich Schumann. Comments on "A fast computer method for matrix transposing" and application to the solution of Poisson's equation. *IEEE Transactions on Computers*, C-22(5):542–544, 1973.

[49] David S. Scott. Out of core dense solvers on Intel parallel supercomputers. In *Proceedings of the Fourth Symposium on the Frontiers of Massively Parallel Computation*, pages 484–487, 1992.

[50] David S. Scott. Parallel I/O and solving out of core systems of linear equations. In *Proceedings of the 1993 DAGS/PC Symposium*, pages 123–130, Hanover, NH, June 1993. Dartmouth Institute for Advanced Graduate Studies.

[51] R. C. Singleton. A method for computing the fast Fourier transform with auxiliary memory and limited high-speed storage. *IEEE Transactions on Audio and Electroacoustics*, AU-15:91–98, 1967.

[52] M. M. Stabrowski. A block equation solver for large unsymmetric linear equation systems with dense coefficient matrices. *International Journal for Numerical Methods in Engineering*, 24:289–300, 1982.

[53] Sivan Toledo. Locality of reference in LU decomposition with partial pivoting. *SIAM Journal on Matrix Analysis and Applications*, 18(4):1065–1081, 1997.

[54] Sivan Toledo and Fred G. Gustavson. The design and implementation of SOLAR, a portable library for scalable out-of-core linear algebra computations. In *Proceedings of the 4th Annual Workshop on I/O in Parallel and Distributed Systems*, pages 28–40, Philadelphia, May 1996.

[55] Sivan A. Toledo. *Quantitative Performance Modeling of Scientific Computations and Creating Locality in Numerical Algorithms*. PhD thesis, Massachusetts Institute of Technology, 1995.

[56] R. E. Twogood and M. P. Ekstrom. An extension of Eklundh's matrix transposition algorithm and its application in digital image processing. *IEEE Transactions on Computers*, C-25(9):950–952, 1976.

[57] Charles Van Loan. *Computational Frameworks for the Fast Fourier Transform*. SIAM, Philadelphia, 1992.

[58] Jeffry Scott Vitter and Elizabeth A. M. Shriver. Algorithms for parallel memory (in two parts) I: Two-level memories, and II: Hierarchical multilevel memories. *Algorithmica*, 12(2–3):110–169, 1994.

[59] H. H. Wang. An ADI procedure suitable for virtual storage systems. Technical Report G320-3322, IBM Palo Alto Scientific Center, January 1974.

[60] Edward L. Wilson, Klaus-Jürgen Bathe, and William P. Doherty. Direct solution of large systems of linear equations. *Computers and Structures*, 4:363–372, 1974.

[61] David Womble, David Greenberg, Stephen Wheat, and Rolf Riesen. Beyond core: Making parallel computer I/O practical. In *Proceedings of the 1993 DAGS/PC Symposium*, pages 56–63, Hanover, NH, June 1993. Dartmouth Institute for Advanced Graduate Studies. Also available online from `http://www.cs.sandia.gov/~dewombl`.

XEROX PALO ALTO RESEARCH CENTER, 3333 COYOTE HILL ROAD, PALO ALTO, CA 94304.
E-mail address: `toledo@parc.xerox.com, sivan@math.tau.ac.il`

DIMACS Series in Discrete Mathematics
and Theoretical Computer Science
Volume **50**, 1999

Concrete Software Libraries

Kiem-Phong Vo

ABSTRACT. The availability of high quality reusable components is a key for
fast and accurate construction of large software systems. My colleagues and
I have written a number of software libraries dealing with fundamental programming aspects such as buffered I/O, memory allocation, and container
data types. These libraries are examples of "Concrete Software", i.e., software
components that are efficient, robust, easy to use and widely portable. Efficiency and robustness mean careful selection and implementation of scalable
algorithms and data structures while ease of use and portability mean captur-
ing such algorithms and data structures in code and interfaces that are highly
reusable and extensible. Much of our software is widely used around the world.
This paper gives an overview of a few of the major libraries and discusses the
design principles followed in building them.

1. Introduction

Modern information systems must deal with very large quantities of data. A
number of systems have been built at AT&T Laboratories to process data from
the operational streams of events generated in the AT&T network including both
telephone and web usages. These systems produce information used for a variety
of purposes including marketing, fraud detection, network maintenance, etc. The
daily volumes of data to be processed exceed several tens of gigabytes. The analyses often demand sophisticated statistical techniques, graph algorithms and display
technologies. The combination of large-scale analyses and stressed computing resources often means that algorithms and data structures must be invented and
customized for particular situations. However, not everything must be invented
anew each time. System construction can be greatly aided by having basic building blocks that deal with fundamental programming issues such as I/O, memory
allocation, sorting, container data types, and others. Further, even with ad-hoc
algorithms, there is significant advantage in trying to encapsulate them in reusable
code with extensible interfaces. This both increases the chance that they would

1991 *Mathematics Subject Classification.* Primary 68N05; Secondary 68N99.

be used and improved and reduces maintenance work when the code is upgraded with new advances. The key problems to be solved are to design clean and flexible abstract interfaces, and to implement code in ways that make it widely usable and interoperable.

Over the years, my colleagues and I have gained much experience in building scalable and reusable software libraries. Many of our libraries including Sfio [1] for buffered I/O, Vmalloc [2] for region-based memory allocation, and Cdt [3] for container data types are widely used around the world. These libraries are among the best of their types and typically reusable on any platform without any needs for adaptation by users. For these reasons, I call them *Concrete Software*. The characteristics that we have come to expect from Concrete Software Libraries are:

- Each serves a niche with some unique functionality,
- Each implements scalable algorithms and data structures,
- Each possesses a general and extensible interface,
- Each is easily integratable with other components,
- Each is single source and single binary image per platform, and
- Each is widely portable.

A number of considerations in library design and implementation enable the above characteristics. The interfaces of many of the libraries follow a *discipline and method architecture*.[4] This architecture both increases interface generality and allows applications to customize library usage based on specific needs. Other design principles and coding and naming conventions [5] were developed to ensure robustness and interoperability among the libraries.

The rest of this paper discusses our way of building software libraries and is organized as follows. Section 2 gives an overview of three main libraries, Sfio, Vmalloc and Cdt, and compares their performance to that of their competitors. Section 3 gives an overview of the discipline and method library architecture and other design and implementation considerations. Section 4 concludes the paper.

2. Three concrete software libraries

2.1. Sfio: a buffered I/O library. C and C++ programs routinely perform I/O via Stdio, the Standard I/O package defined by the ANSI-C language [6]. However, Stdio has numerous shortcomings both at the interface level and in its various implementations. [1] The Sfio library was first introduced in 1991 [7] as a better way to perform buffered I/O. Aside from providing a clean interface and an efficient implementation, Sfio also includes a number of new features that give applications a much wider array of tools to customize buffered I/O to their needs.

2.1.1. *Sfio architecture and features.* Figure 1 shows the architecture of the Sfio library. Sfio provides applications with streams to perform I/O. I/O operations

Stream	I/O operations

Buffered data
Stream states
	`Sfio_t* sfopen(Sfio_t* f, char* file, char* mode);`
	`sfclose(Sfio_t* f);`
	`sfread(Sfio_t* f, void* buf, size_t n);`
	`sfwrite(Sfio_t* f, const void* buf, size_t n);`
	`sfseek(Sfio_t* f, Sfoff_t pos, int type);`
	`sfprintf(Sfio_t* f, const char* format, ...);`
	`sfscanf(Sfio_t* f, const char* format, ...);`

Discipline	`void* sfreserve(Sfio_t* f, size_t size, int type);`
Raw I/O functions Event handler	`void* sfgetr(Sfio_t* f, int separ, int type);` `...`

FIGURE 1. The Sfio library architecture

include the familiar functions for reading, writing, seeking and formatting and new functions for buffer reservation, stream pools and stream stacks. Each stream can be equipped with a stack of disciplines, each of which defines functions to process raw data and to process exceptions. Below is a summary of the main Sfio features:

- *Unlimited streams:* Unlike Stdio, which limits open streams to a small number (typically 20 on older Unix systems or a few hundred on newer ones), Sfio allows an application to open as many streams as it needs. The number of streams is bounded only by resource limitations such as available memory and open file descriptors as imposed by the underlying platforms.
- *String streams:* Aside from the familiar file streams, Sfio provides string streams, i.e., streams based entirely on main memory. This enables the use of all I/O facilities including the formatting functions on main memory.
- *Zero-copy I/O:* Unlike Stdio which always requires an application buffer to transfer data into and out of streams, Sfio allows direct accesses to stream buffers. For example, the call `sfgetr()` to get a record from a stream constructs a record in the stream buffer and returns a pointer to it. Coupling with low-level system facilities such as memory mapping, [8] this means that applications can avoid most, if not all, memory copying between buffers.
- *Management of groups of streams:* Sfio introduces the notions of stream pooling and stream stacking. A stream pool guarantees automatic buffer synchronization among streams as I/O operations are performed on specific streams. On the other hand, a stream stack allows some streams to be processed as if they are logically parts of another.
- *Extended formatting:* Applications can extend the families of scanning and formatting functions to include new formatting patterns as well as to redefine the meaning of existing ones.

- *Discipline stack:* Applications can associate each stream with a discipline stack to redefine the semantics of methods to read and write raw data. This allows data filtering without altering the high-level application logic. Disciplines are a good source of reusable code. Sfio provides a number of standard disciplines for tasks ranging from duplicating stream outputs to automatic decompression of certain common types of compressed data.
- *Data portability:* The library provides functions to encode and decode integer and floating point values in formats that are portable among platforms with different scalar formats. This has proved to be useful in networking applications where data must be transported among heterogeneous hardwares.
- *Stdio compatibility:* The Sfio interface is much more effective than that of Stdio. However, there is a legion of existing code based on Stdio. Thus, Sfio provides two Stdio compatibility packages at both source and binary code levels. The source compatibility package provides a `stdio.h` header that translates Stdio calls to Sfio calls while the binary compatibility package allows code already compiled with Stdio to transparently link with Sfio.

```
1. #include  <sfio.h>

2. ssize_t lower(Sfio_t* f, Void_t* buf, size_t n, Sfdisc_t* disc)
3. {    int i;
4.       n = sfrd(f,buf,n,disc);
5.       for(i = 0; i < n; ++i)
6.            buf[i] = tolower(buf[i]);
7.       return n;
8. }

9. Sfdisc_t Disc = { lower, 0, 0, 0};

10. main()
11. {
12.       sfdisc(sfstdin,&Disc);
13.       ...data processing...
14. }
```

FIGURE 2. Translation from upper-case to lower-case

2.1.2. *Examples of using Sfio.* Figure 2 gives an example of using Sfio to process data from the standard input stream. Here, a discipline is used to translate upper case letters to lower case before the application performs any data processing. Lines 2-8 define a discipline reading function `lower()` that reads raw data and performs translation. Line 9 creates a discipline Disc with lower() as it reading function. Line 12 installs the discipline on the standard input stream. As the application processes data from the standard input stream, the function lower() is called to fill the stream buffer. It calls the Sfio function sfrd() on line 4 to read raw data, then performs the translation task on lines 5-6.

The discipline type Sfdisc_t provides the below four fields to define functions for reading, writing, seeking and processing exceptions:

```
ssize_t (*readf)(Sfio_t* f, void* buf, size_t n, Sfdisc_t* disc);
ssize_t (*writef)(Sfio_t* f, const void* buf, size_t n, Sfdisc_t* disc);
Sfoff_t (*seekf)(Sfio_t* f, Sfoff_t offset, int type, Sfdisc_t* disc);
int     (*exceptf)(Sfio_t* f, int type, void* data, Sfdisc_t* disc);
```

Disciplines allow applications to extend stream processing and keep their data processing logic separate from other routine data filtering tasks. Note also that the lower() function and the discipline Disc in the above example are in a form that is easily reusable elsewhere.

```
1. #include <sfio.h>
2. main()
3. {
4.       sfdclzw(sfstdin);
5.       sfmove(sfstdin,sfstdout,SF_UNBOUND,-1);
6. }
```

FIGURE 3. An example of automatic decompression

Figure 3 shows an example of using a standardly provided discipline, sfdclzw, to decompress a data source compressed by the Unix *compress* program. Line 4 calls sfdclzw to install the decompression discipline on the standard input stream. Line 5 calls sfmove() to move uncompressed data from the standard input stream to the standard output stream.

Beside being simple, the above program is also efficient. The decompression logic is carried out in main memory so that no secondary storage and I/O is required for this purpose. Further, the call sfmove() moves data between streams without any extra buffer copying.

2.1.3. *Sfio performance.* A study was performed to compare Sfio against various implementations of Stdio. The below fifteen tests were used. They were designed to exercise different implementation aspects of an I/O library.

- write: Write 10 megabytes in chunks of 8K bytes.
- read: Read 10 megabytes in chunks of 8K bytes.
- revread: Read 10 megabytes in reverse order in chunks of 8K bytes.

- w757: Write 10 megabytes in chunks of 757 bytes. 757 is an arbitrarily chosen value that satisfies two conditions. First, it is small enough to be smaller than the buffer sizes used in typical Stdio implementations. Second, it does not divide any power of two so that most buffering schemes are forced to deal with I/O requests that overlap buffer boundary.
- r757: Read 10 megabytes in chunks of 757 bytes.
- rev757: Read 10 megabytes in reverse order in chunks of 757 bytes.
- copy: Copy 10 megabytes from one file to another in chunks of 8K bytes.
- seek: Seek randomly 2,500 times in a 10 megabytes file and copy 8K blocks to location 0.
- putc: Write 5 megabytes using fputc().
- getc: Read 5 megabytes using fgetc().
- puts: Write 100,000 text lines, each 100 bytes long.
- gets: Read 100,000 text lines, each 100 bytes long.
- revgets: Read in reverse order 100,000 text lines, each 100 bytes long.
- printf: Format 25,000 text lines of integers, strings, and floating point values.
- scanf: Scan 25,000 text lines of integers, strings, and floating point values.

A single program was written based on the Stdio interface to exercise all fifteen performance tests on a variety of platforms. For each platform, the test program was compiled and run in three different ways, based on the native Stdio and the source and binary compatibility packages provided by Sfio. The source compatibility version can be thought of as Sfio since Stdio calls were redefined to Sfio calls via simple macros.

Platform	Total CPU+Sys			No Reverse Reads		
	Sfio	Binary	Stdio	Sfio	Binary	Stdio
INTEL586/166, SCO SYSV3.2	91.76	93.88	157.13	68.18	69.54	84.70
INTEL486/50, BSDI2.1	65.34	73.29	68.43	51.01	57.14	55.11
SPARC4/670, SUN OS4.1.2	69.50	73.32	180.34	56.04	58.46	69.41
HP9000/877, HP-UXA.09.04	35.33	38.42	82.40	29.25	31.14	34.39
SGI-MIPS3, IRIX5.3	19.78	22.99	55.00	15.10	17.35	32.17
INTEL686/200, LINUX2.0.18	11.10	11.66	12.65	9.34	9.59	11.21
SPARC-ULTRA, SUN OS5.5.1	10.27	11.15	25.96	8.72	9.32	10.93
DEC-ALPHA, Digital UNIX V3.2G	3.62	4.14	7.59	2.99	3.40	3.80

TABLE 1. Sfio vs. Stdio: CPU+System times (in seconds)

Table 1 presents total CPU and system times. Most Stdio implementations do not handle reading data in reverse well. Since such operations are relatively infrequent, two different times are shown, one with and one without reverse reading. Sfio is the overall winner on all platforms, sometimes by large margins. The performance gains for Sfio can be attributed to the following factors:

- When appropriate, Sfio uses memory mapping for raw I/O while Stdio always uses reading and writing system calls.
- Sfio always aligns buffered data by page boundaries. This both speeds up raw I/O and reduces system calls when applications perform many operations of random seeking or reverse reading.
- The functions for data scanning and formatting employ new and faster algorithms for dealing with integer and floating point value conversions.

Note that Table 1 also shows relative performance among the platforms. For example, the DEC Alpha processor running Digital Unix performed best. In particular, the Intel686/200 processor running Linux2.0.18 performed extremely well against much more expensive platforms such as the HP9000 and SGI-MIPS3.

2.2. Vmalloc: a region-based memory allocator. Dynamic memory allocation is an integral part of programming. Most C and C++ programs use the ANSI-C Malloc interface to manage *heap memory*. Although the Malloc interface is simple to use, it does not address the full range of memory allocation. For example, Malloc cannot be used to manage any types of memory beyond the heap. It also cannot be customized for frequent allocation modes such as allocating objects of the same sizes. The Vmalloc library solves these problems by providing a general framework for memory allocation.

Region	Allocation operations

Memory segments
Free lists
	`Vmalloc_t* vmopen(Vmdisc_t* disc,`
	`Vmethod_t* meth, int type);`
	`int vmclose(Vmalloc_t* vm);`
	`int vmclear(Vmalloc_t* vm);`
	`void* vmalloc(Vmalloc_t* vm, size_t size);`
	`int vmfree(Vmalloc_t* vm, void* mem);`
	`void* vmresize(Vmalloc_t* vm, void* mem,`
	`size_t size, int type);`
	`...`

Discipline	**Methods**
Raw memory manager Event handler	`Vmbest, Vmpool, Vmlast` `Vmdebug, Vmprofile`

FIGURE 4. The Vmalloc library architecture

2.2.1. *Vmalloc architecture and features.* Figure 4 shows the architecture of the Vmalloc library. Allocation operations are carried out on regions. Each region has a memory discipline that specify functions to obtain raw memory and to handle exceptions. Allocation functions are parameterized by methods that define allocation strategies. The library provides five methods:

- Vmbest: This is a general purpose allocator based on the best-fit strategy and a few heuristics to speed up common memory allocation scenarios.
- Vmpool: This method allocates blocks of fixed sizes.
- Vmlast: This method only allows freeing and resizing of the most recently allocated block.
- Vmdebug: This is like Vmbest but with aids for detecting memory misuses.
- Vmprofile: This is like Vmbest but with aids for profiling memory usage.

Below is a brief summary of the main Vmalloc features:

- *Flexible memory organization:* Unlike Malloc which manages only heap memory, Vmalloc allows creation of multiple regions, each of which parameterized with a memory discipline and an allocation method. In this way, applications can tailor memory usage to their needs. For example, by writing a discipline that obtains shared memory, an application can reuse all of Vmalloc allocation methods for allocating shared memory.
- *On-line debugging and profiling of memory usage:* The methods Vmdebug and Vmprofile allow selectively monitoring of memory usage per region to detect memory errors or wastage.
- *Efficiency:* The general purpose allocation method Vmbest performs well against current implementations of Malloc (see below). Applications can also create regions with the special purpose allocators Vmpool and Vmlast to further tune for efficiency.
- *Malloc compatibility:* To make Vmalloc functionality available to programs written based on Malloc, the library also provides a Malloc interface. This interface is instrumented with Vmalloc methods which can be set by environment variables. In this way, an application can run fast in normal mode yet it can also debug and/or profile memory usage by simply setting an environment variable before execution.

2.2.2. *Examples of using Vmalloc.* Figure 5 shows an example of using Vmalloc to manage memory. Here, the outer while() loop is a part of a language interpreter that construct a parse tree for some expression, interpret the expression, then free up the tree. Since a tree is constructed from many pieces but always freed as a whole, Line 1 creates a region vm using the discipline Vmdcheap and the method Vmlast. This means that memory for vm will be recursively obtained from the

```
1. Vmalloc_t* vm = vmopen(Vmdcheap,Vmlast,0);
2. while(...have expression...)
3. {   while(...constructing...)
4.     {   b = vmalloc(vm, size);
5.         ...
6.     }
7.     ...interpret tree...
8.     vmclear(vm);
9. }
10. vmclose(vm);
```

FIGURE 5. An example of memory allocation with Vmalloc

heap. Lines 3-6 shows the inner loop that calls `vmalloc()` to allocate memory and constructs the parse tree. Line 8 frees the entire parse tree.

In the above example, performance is gained in two different ways. First, since `Vmlast` is used, no extra structure is constructed to manage busy and free blocks. This saves both space and time. Second, the parse tree is composed from many parts. If Malloc was used to allocate these parts, the tree must be freed in postorder with a `free()` call per part. By contrast, Vmalloc allows the entire tree to be freed in a single `vmclear()` call. This saves time.

The parameterization of allocation functions by an allocation method also affords flexibility. For example, should some memory usage errors are suspected, it is a simple matter to change `Vmlast` to `Vmdebug` to sniff them out.

Disciplines allow applications to customize memory acquisition for regions. The full Vmalloc discipline structure, `Vmdisc_t`, defines the below fields for a memory acquisition function, an exception handler, and a value indicating the natural amount to round memory requests to:

```
void* (*memoryf)(Vmalloc_t* vm, void* addr, size_t csz, size_t* nsz,
                Vmdisc_t* disc);
int   (*exceptf)(Vmalloc_t* vm, int type, void* data, Vmdisc_t* disc);
ssize_t round;
```

Figure 6 shows two example memory obtaining functions, one for UNIX systems and one for WINDOWS systems. The simplicity of such code shows that it is easy to port Vmalloc among different platforms. It is equally easy to write disciplines to deal with shared and/or mapped memory.

2.2.3. *Vmalloc performance.* A study was performed to compare the general purpose allocation method `Vmbest` against the below Malloc implementations:

```
1. void* unixmem(Vmalloc_t* vm, void* addr, size_t csz, size_t nsz,
               Vmdisc_t* disc)
2. {   unsigned char* seg;
3.     if(csz > 0 && sbrk(0) != (unsigned char*)addr+csz)
4.         return (void*)0;
5.     if((seg = sbrk((ssize_t)nsz-(ssize_t)csz)) == (unsigned char*)-1)
6.         return (void*)0;
7.     else return csz == 0 ? addr : seg;
8. }


9. void* windowsmem(Vmalloc_t* vm, void* addr, size_t csz, size_t nsz,
               Vmdisc_t* disc)
10. {   if(csz == 0)
11.         return (void*)VirtualAlloc(0,nsz,MEM_COMMIT,PAGE_READWRITE);
12.     else if(nsz == 0)
13.         return VirtualFree(addr,0,MEM_RELEASE) ? addr : (void*)0;
14.     else return (void*)0;
15. }
```

FIGURE 6. Vmalloc discipline examples

- *V*: by Phong Vo, distributed with various derivatives of Unix System V such as Solaris, Irix and SCO Unix systems. This *malloc* is based on a best-fit strategy using a bottom-up splay tree for free blocks.
- *S*: by Chris Aoki and C. Adams, distributed with SUN OS. This *malloc* is Stephenson's better-fit allocator [9].
- *P*: by Chris Kingsley, modified and distributed with the Perl language interpreter. This *malloc* uses a power-of-two buddy system.
- *X*: by Doug McIlroy, used in the 10th Edition Bell Labs Research UNIX system. This *malloc* is based on a first-fit strategy with a roving pointer. Small blocks are cached on freeing to speed up subsequent allocations.
- *H*: by Mike Haertel, distributed with the GNU C library, dated Mar 1 1994. This allocator segregates blocks of same size in same pages.
- *L*: by Doug Lea, distributed with the GNU C++ library, [10] version 2.5.3b.
- *B*: by Hans Boehm and Mark Weiser, a conservative garbage collector, [11] version 4.5. Here, GC_malloc_uncollectable() and GC_free() are used so that garbage collection is bypassed and only allocation performance is measured. Allocated space is not cleared because that is already done by the allocator.

- *C*: the same Boehm-Weiser garbage collector. Here, `GC_malloc()` is used and objects are freed by removing them from the linked list discussed above. Thus, in this case, the garbage collection performance is measured.

Dataset	Allocate	Free	Resize	MaxAllocate	MaxBusy
gawk	723,470	722,922	150,888	47,684K	38K
db.2X	880,688	879,648	0	10,953K	20K
db.ioQ	66,626	11,912	0	1,777K	1,411K
mt.ioQ	69,387	10,677	0	1,867K	1,575K
C++parser	44,730	5,381	0	1,024K	848K
graph	111,782	14,882	0	1,706K	1,590K
S	102,146	83,124	56	800,369K	5,887K
ciao	163,044	145,113	3,246	912,507K	6,839K
fragment	10,001	0	10,000	1,563,203K	547K

TABLE 2. Summary of datasets in the simulation study

Traces of allocation requests from a variety of application programs are used in simulations that exercised the various allocators. The applications include parsers, database queries, data analyses and interactive graphics. Table 2 summarizes information about the traces. The first three numerical columns display total numbers of different types of operations. The fourth column shows the total of space requested via `malloc()` or `realloc()`. The last column shows the maximum busy space at any time.

Figure 7 shows time and space performance comparisons normalized to `Vmbest`. Measurements were done by averaging several runs on an idle Sparc-5 running SUNOS4.1. The bottoms of the graphs were labeled with the time values (CPU plus system times) and arena sizes for `Vmbest`. Each data point was constructed by dividing the time or space value of the respective allocator by that of `Vmbest` if the former was larger; otherwise, the reverse was done. Thus, the horizontal lines at 1 in both graphs represented `Vmbest`. An allocator was slower or faster (less or more space efficient) than `Vmbest` if the corresponding data point was above or below this line. A data point between *inf0* and *inf1* meant that the respective allocator was able to service all allocation requests but its time or space value was at least 4 times that of `Vmbest`. A datapoint beyond *inf1* meant that the respective allocator failed – typically because it ran out of memory. `Vmbest` was competitive against the best of the other allocators thanks to its allocation strategy:

- The basic allocation mechanism is best-fit, i.e., allocating from a smallest free area that fits a requested size. Free blocks are kept in a top-down splay tree [12] for fast search.

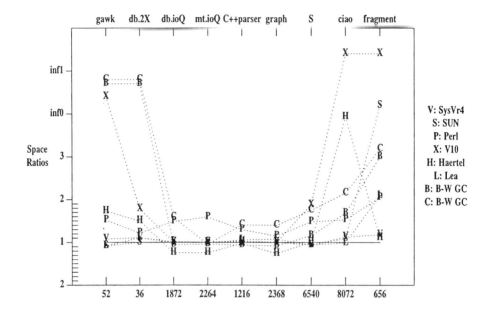

Vmbest arena size in Kbytes

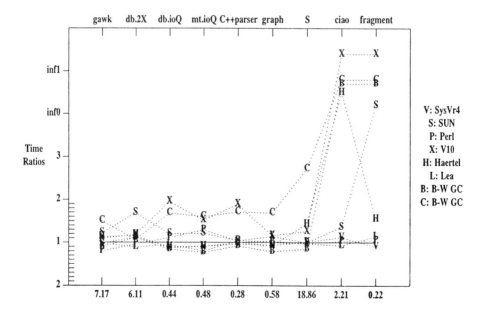

Vmbest user+sys time in seconds

FIGURE 7. Time and space allocation performances normalized to Vmbest

- Free blocks are cached and processed only at some future allocation. This heuristic saves potentially expensive coalescing operations in programs that free everything before exiting. It also benefits programs that continually free and allocate the same block types by avoiding searches if requested sizes fit recently freed blocks.

- Free blocks of frequent and small allocation sizes are adaptively constructed in lists. This heuristic benefits programs that allocate small data types.

- When a block has to be moved during resizing because it cannot be extended in place, a small amount is added to the size before searching for new space. This heuristic reduces data movement and fragmentation because a resized block is often resized multiple times.

- The "wilderness preservation heuristic" [13] is observed. This means that the free area with highest address, or the wilderness, is only allocated as a last resource. This heuristic prevents unnecessary growth of the arena.

	C++parser		graph	
	Vmlast	Vmbest	Vmlast	Vmbest
Arena size	1,076K	1,216K	1,724K	2,368K
Cpu+Sys	0.22s	0.28s	0.50s	0.58s

TABLE 3. Performance comparison of Vmlast and Vmbest

Tailoring an allocation strategy to actual usage can have significant benefit. For example, it is a folklore that parsers and compilers do well with an allocator that ignore free calls. The datasets *C++parser* and *graph* are examples of allocation traces from such parsers. Table 3 shows that, indeed, using Vmlast improves both time and space performance over Vmbest. The space improvement is somewhat surprising since Vmlast loses all space freed but not reclaimable. However, since Vmlast does not need to manage allocated blocks, the administrative space saved is more than enough to cover that lost through leakage.

2.3. Cdt: a container data type library. Container data types are ubiquitous in programming. They are used to manage object collections. An instance of a container data type is often called a *dictionary*. The basic operations for a dictionary are insertion, deletion, search, and iteration. However, these operations can be realized via different data structures and algorithms with different sematics and performance levels. Not accidentally, such diversity has led to a plethora of library packages. Unix/C environments provide functions such as tsearch, hsearch and lsearch to manipulate respectively objects stored in binary trees, hash tables, and lists. C++ provides classes such as Map [14] and Set [15] to deal with ordered maps and unordered sets. C++ has also standardized on STL, the Standard

Template Libraries,[16, 17] a set of templates dealing with ordered and unordered maps and sets. However, these packages do not provide a full range of flexibility. For example, STL statically binds container and object types, therefore, reducing flexibility in object interpretation and making it difficult to match container types to usage contexts. To address such issues, the Cdt library provides a comprehensive set of container data types with a uniform and flexible interface implemented on top of efficient data structures.

Dictionary	Dictionary operations

Objects
Data structures
	`Dt_* dtopen(Dtdisc_t* disc, Dtmethod_t* meth);`
	`int dtclose(Dt_t* dt);`
	`void* dtsearch(Dt_t* dt, void* obj);`
	`void* dtinsert(Dt_t* dt, void* obj);`
	`void* dtdelete(Dt_t* dt, void* obj);`
	`void* dtfirst(Dt_t* dt);`
	`void* dtnext(Dt_t* dt, void* obj);`
	...

Discipline	**Methods**

Key, comparator,
Object managers,
Event handler
	`Dtset, Dtbag,`
	`Dtoset, Dtobag,`
	`Dtlist, Dtstack, Dtqueue`

FIGURE 8. The Cdt library architecture

2.3.1. *Cdt architecture and features.* Figure 8 shows the architecture of the Cdt library. Objects are stored in dictionaries. Discipline structures are used to specify characteristics of objects such as their keys and functions to create, delete, compare, and hash them. A few representative dictionary operations are shown in the top right part of the figure. For example, object insertion is done with `dtinsert()` while object iteration is done with `dtfirst()` and `dtnext()`. Note that dictionary operations are abstract and do not carry any information about the underlying data structures. In actual usage, they must be made concrete per dictionary by selecting a method from the below set:

- `Dtset` and `Dtbag`: These methods implement unordered sets and multisets respectively. The underlying data structure is a hash table [18] with move-to-front collision chains.
- `Dtoset` and `Dtobag`: These methods implement ordered sets and multisets respectively. The underlying data structure is a splay tree. [12]
- `Dtlist`, `Dtstack` and `Dtqueue`: These methods implement lists, stacks and queues on top of doubly linked lists. [19]

Below is a brief summary of Cdt features:

- *Comprehensiveness:* Cdt provides a comprehensive set of container data types under a uniform abstract interface. In this way, programs can be written without too much awareness of the underlying data structures and it is easy to experiment with different container types to find the right matches.
- *Dynamically changeable container data types:* The container data type of a dictionary can be changed any time during execution, for example, from an unordered set to an ordered set. This means that applications can tailor dictionary usage to processing needs based on contexts instead of having the dictionary and object types statically bound.
- *Set-like vs. map-like dictionaries:* Object attributes are described in discipline structures that support both set-like dictionaries, i.e., dictionaries that identify objects by matching, and map-like dictionaries, i.e., dictionaries that identify objects by keys.
- *Dynamically changable disciplines:* A dictionary can change its discipline any time to alter object descriptions. Therefore, the same objects can be treated in multiple ways depending the processing contexts.
- *Direct object iteration:* Unlike packages such as STL that require a separate iterator type to iterate over objects, Cdt allows applications to directly iterate over the objects themselves. The underlying data structure of a dictionary maintains a pointer to the most recently accessed object so that computing its next or last neighbor can be done quickly.

2.3.2. *Examples of using Cdt.* Figure 9 shows an example application that reads text from the standard input stream, partitions it into tokens, and counts their frequencies. Here, a token is any sequence of characters not containing spaces, tabs and new-lines. Lines 3-6 define a Token_t data structure that associates a token name and its frequency count freq. Line 7 defines a discipline Tkdisc to describe Token_t objects. Here, the key of a Token_t object is its name whose offset is computed via the ANSI-C macro offsetof(). The next two fields of Tkdisc are set to -1 to indicate that the key is a null-terminated string and Cdt should allocate its own structures to hold objects. Line 10 open a dictionary dt based on the discipline Tkdisc and the container method Dtset. Lines 11-20 form the main loop that obtains tokens and counts their frequencies. The function readtoken() is straightforward so its implementation has been omitted. Finally, lines 21-22 iterate over all objects in dt to output data.

Since Dtset is used, the code in Figure 9 will output tokens in some random order. To output tokens in a lexicographic order, the ordered set method Dtoset should be used. This can be simply done by changing line 10 to:

```
Dt_t* dt = dtopen(&Tkdisc, Dtoset);
```

```
1. #include    <sfio.h>
2. #include    <cdt.h>

3. typedef struct
4. {   char*    name;
5.      int      freq;
6. } Token_t;

7. Dtdisc_t      Tkdisc = { offsetof(Token_t,name), -1, -1 };

8. main()
9. {   char*    s;
10.      Dt_t*    dt = dtopen(&Tkdisc,Dtset);

11.      while((s = readtoken(sfstdin)) )
12.      {   Token_t* tk = dtmatch(dt,s);
13.          if(!tk)
14.          {   tk = malloc(sizeof(Token_t));
15.              tk->name = strdup(s);
16.              tk->freq = 1;
17.              dtinsert(dt,tk);
18.          }
19.          else tk->freq += 1;
20.      }

21.      for(tk = dtfirst(dt); tk; tk = dtnext(dt,tk) )
22.          sfprintf(sfstdout,"%s:\t%d\n", tk->name, tk->freq);
23. }
```

FIGURE 9. A program to count token frequencies

However, ordered sets are more expensive to maintain than unordered sets so they are not suitable for processing input sources with many duplicated tokens. A better strategy for such cases is to continue using Dtset during the dictionary construction phase and switch to Dtoset just before output. This can be simply implemented by inserting the call dtmethod(dt, Dtoset) before the output loop on lines 21-22.

Finally, suppose that we wish to output tokens so that the most frequent ones come first while the least frequent ones come last. Figure 10 shows the necessary code. Lines 1-3 show a function to compare integers in reverse order. Lines 4-8 should be inserted before the output loops on lines 21-22 of Figure 9. Line 4 changes the storage method to Dtobag because tokens will be stored as an ordered collection

```
1. int intcompare(Dt_t* dt, void* i1, void*i2, Dtdisc_t* disc)
2. {    return *((int*)i2) - *((int*)i1);
3. }


4. dtmethod(dt, Dtobag);
5. Tkdisc.key = offsetof(Token_t,freq);
6. Tkdisc.size = 0;
7. Tkdisc.comparef = intcompare;
8. dtdisc(dt,&Tkdisc,0);
```

FIGURE 10. Output tokens by frequency

in which elements may repeat (i.e., if they have the same frequency). Lines 5–8 redefine the discipline Tkdisc to compare objects by their frequencies, then call dtdisc() to affect the change.

2.3.3. *Cdt performance.* The token counting application example in the last section was used to compare the performance of various container data type packages. A single program was written with compile time options to switch implementations based on packages. The unordered set container data types to be compared were the Cdt method Dtset, the class Set [15] available in many C++ environments and the STL template hashmap. The ordered set container data types to be compared were the Cdt method Dtoset, the ANSI-C tsearch() function, the C++ class Map [20, 14] and the STL template map. Both map and hashmap [16, 19] were from the distribution ftp://butler.hpl.hp.com/stl, dated 10/31/1995.

File	Size	Tokens	Distinct count	Average token length
ps	1,989K	335,997	11,912	38.00
src	1,169K	149,886	27,964	16.40
kjv	4,441K	822,587	33,916	8.01
mbox	2,701K	419,197	49,903	9.83
city	1,349K	81,206	69,610	18.17
host	2,722K	449,554	102,566	16.71

TABLE 4. Summary of benchmark input files

The resulting executable programs were run using a variety of input files. Table 4 summarizes the statistics of these files: file size in K-bytes, total number of tokens, number of distinct tokens, and average length of a token. Tokens in most input files appear more or less in random order. However, tokens in *host* and *city* are highly ordered. Thus, this set of input data provides a realistic testbed for checking the performance of the various packages.

198   KIEM-PHONG VO

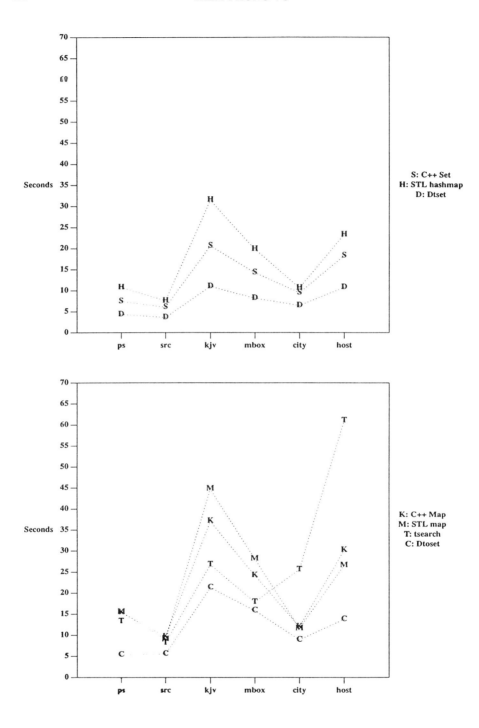

FIGURE 11. Time performance of unordered and ordered set packages

Measurements were done by averaging many runs on a quiescent SPARC-20 running SUN OS5.4. Time values were total CPU and system times. Figure 11 shows the timing results. The Cdt containers were clearly faster than their counterparts. It is interesting to note that Dtoset was faster than both Set and hashmap even though a splay tree operation requires O(logn) amortized time while a hash table operation uses O(1) average time. Since Dtset, another hash table implementation, was clearly faster than Dtoset, this points to problems in either implementation quality or algorithm choice for the non-Cdt packages.

3. Architecture and design principles

Handle	Operations
Resource	Opening/closing handles
States	Resource manipulation
Discipline	**Methods**
Resource definition	Algorithms
Resource acquisition	Management styles
Event handling	Special modes

FIGURE 12. The discipline and method architecture

3.1. The discipline and method library architecture. Sfio, Vmalloc and Cdt show that libraries can be built that are efficient, general and adaptable. A major contributing factor to this success is the *discipline and method library architecture.*[4] Figure 12 shows the four basic interface constructs used in this library architecture:

- *Handle:* A handle holds some collection of resources and states that must be saved across function calls. For example, an Sfio handle is a stream of the type Sfio_t. The stream structure holds buffered data and other state information such as whether the stream type and locking state.
- *Operations:* Applications use the operations provided by a library to create/close handles and to manipulate the resources kept in them. For example, Cdt provides operations such as dtinsert()/dtdelete() to manipulate objects stored in dictionaries.
- *Disciplines:* Each handle is parameterized by a discipline structure that defines resource types, functions to directly manipulate the raw resources, and functions to deal with various applicable events. For example, each Vmalloc region is parameterized by a discipline of type Vmdisc_t which defines a function to obtain raw memory and a function to process exceptional events such as out-of-memory or region initialization.

- *Methods:* When multiple algorithms, heuristics or operational modes can be used to implement the same abstract operations, they should be made available via methods. Then, each handle can select a suitable method and activates the right computational device. For example, each Vmalloc region specifies a particular allocation method, such as general purpose allocation or allocation with debugging.

To summarize, the resources to be manipulated by a library are made abstract by *disciplines* while the operations on them are parameterized by *methods*. In this way, the high level library functions are freed from being bound to low level resource constraints as well as from algorithm selections. Applications can define their own types of resources using disciplines. Once that is done, all algorithms and heuristics implemented in methods are immediately reusable. This should be contrasted, for example, with libraries such as Stdio or Malloc that tightly bind their resources (i.e., data streams and heap memory) with library operations, therefore, making the implemented algorithms unusable for other purposes. Not all libraries require parameterization with methods (e.g., Sfio). However, when applicable, such a parametrization can ease the evolution of a library by allowing addition of new methods or replacement of existing implementation choices without disturbing application code. Further, as the examples in Section 2.3.2 show, methods allow application to select and match proper algorithms and object types with usage contexts so that maximum efficiency can be achieved. For library construction, disciplines and methods form the essence of being general and adaptable.

3.2. Design considerations. There are a number of considerations that have proved useful as guidelines in both the design and implementation of our libraries. These are discussed below:

- *Necessity:* To be reusable, a library must first be used. Many of our libraries are built only when we plan new applications. Such libraries encapsulate the most general aspects of the corresponding applications so that these applications themselves can be implemented as drivers of library code. A good example is `sfdclzw`, the discipline to decompress data produced by the Unix *compress* program.
- *Generality:* Libraries should be designed to be widely applicable. This means unifying seemly separate concepts into single interfaces. The discipline and method architecture discussed above is a way to achieve this.
- *Minimality:* Complex interfaces increase confusion and inconsistency. Our libraries provides only meaningful operations that cannot be missed without losses in efficiency and/or convenience. Examples of gratuitous interfaces are the numerous convenience functions in the Stdio library such as `putchar()` and `getchar()`. Our Sfio avoids such functions.

- *Modularity:* Libraries should be insulated from one another so that the implementation and use of one does not affect that of another. For example, Sfio, Vmalloc and Cdt are all stand-alone libraries. However, as shown in the reference for Cdt,[3] disciplines can be used to combine Cdt with the use of library such as Vmalloc to build dictionaries in shared memory. There are also a number of naming conventions that should be followed to ensure that different libraries work well with each others. These are described in Section 3.3.

- *Evolvability:* A successful library will undergo revisions as its design and implementation are stressed by usage and technological advances. If possible, interface changes should be kept to a minimum to reduce the needs to recode applications. Keeping operations abstract and parameterizing them with methods is a good way to minimize such interface changes. However, there are times when a new interface is required, for example, when Sfio replaces Stdio. Then, compatibility packages should be provided to help existing applications make the transition.

- *Efficiency:* Functions in a library should be based on the best known data structures and algorithms. However, internal efficiency alone is not enough. A library must also allow applications to customize algorithms based on their needs. This means that interface mechanisms should be provided to access certain key resources kept in handles and to alter operational modes. Examples are Sfio functions such as `sfreserve()` and `sfgetr()` for direct buffer accesses and Cdt methods that can be switched dynamically to match computational needs.

- *Robustness:* Widely usable libraries should be tested in a variety of environments to ensure that their implementation does not suffer from artificial resource constraints and they can respond well to unexpected events. For example, all of our libraries are free from constraints such as size of an integers or fixed size arrays. In fact, libraries such as Sfio provide standard functions and types for applications to deal with data portability across platforms. Further, libraries must also provide interfaces for applications to deal with exceptional events that may arise in certain internal computations. For example, if an I/O operation fails due to some equipment failure or resource constraints, applications must be able to deal with that timely. The event handling feature of disciplines is a way to categorize common exceptions so that applications can process them efficiently.

3.3. Naming conventions. An important dimension of library design is to manage the name space so that libraries with similar architectures use similar interface devices. This gives the libraries the same look and feel and helps to ease the learning curve for programmers. The conventions followed by our libraries are discussed below. To be definite, examples shall be taken from the Cdt library.

- A common prefix is used for all library objects. For example, Cdt uses the dt prefix so its function to open a dictionary is named dtopen().
- A public non-function object always starts with a capital letter. For example, the method for set manipulation is Dtset.
- A type object always ends with the suffix _t. For example, the type for a dictionary is Dt_t.
- A value-only object defined by a macro should be in capital letters. For example, the event announcing the change of a method is DT_METHOD.
- A function call should have the affected handle as its first argument. For example, the call to insert an object into a dictionary is dtinsert(dt,obj).
- Discipline and method type names should be modeled after Dtdisc_t and Dtmethod_t.
- Functions to open and close handles and to change disciplines and methods should be modeled after dtopen(), dtclose(), dtdisc() and dtmethod().
- A discipline function should have the respective discipline as its last argument. For example, a Cdt comparator call is (*comparf)(dt,key1,key2,disc).
- Corresponding to the above calls, standard events should be defined similar to DT_OPEN, DT_CLOSE, DT_DISC and DT_METHOD.
- Each event announcement should also give the object causing the event. For example, the call announcing an attempt to change the discipline disc to newdisc is (*eventf)(dt,DT_DISC,newdisc,disc).

In addition to the conventions to reduce name interference across library interfaces, C libraries also need to protect each others from global but private symbols. Our libraries typically gather such private global variables into a single structure so that only a single identifier will be taken from the name space. For example, private global data in Sfio are gathered in a structure _Sfextern. The leading underscore in _Sfextern further emphasizes that it is a private symbol.

3.4. Concrete code image and single source code. A successful software package is often used in diverse applications on a variety of platforms. Many packages [21, 16] use code generation techniques to statically bind resource types to algorithms so that the generated code can operate close to machine performance. The cost in doing this is more complexity in application source code if a separate high level language is used and less flexibility in the ability to dynamically deal with algorithms. Instead, our discipline and method architecture approach emphasizes using dynamic structures to define resource types and specify algorithms. In this way, library code images are concrete and dynamically linkable.

At the implementation level, a successful software library is often ported to a variety of platforms. A major problem is that certain key functionality may not always be available or, if available, not in the same forms. For example, memory mapping is a nice modern facility for efficient I/O. However, not all platforms support memory mapping. In fact, certain platforms only half-heartedly support it and their implementations may be less efficient than the traditional reading and writing system calls. This is a major reason that drives many software packages to use complicated software configuration mechanisms to deal with dependencies on environment-specific features. We use a *feature-based configuration* method to ensure that our code base can remain single source code yet is portable to all relevant environments. This is aided by the use of a small language *iffe* [22] to describe features required by a library and to check if a particular platform supports them. Iffe scripts are automatically run as a part of building a library to generate necessary header files with indicators for available features. The library code uses such indicators to direct different implementations as necessary.

As an example, the below *iffe* specification probes to see if a platform provides either of the poll() or select() system calls:

```
lib poll, select
```

From such a specification, the *iffe* interpreter may produce a header file with data like the below to indicate that select() is available while poll() is not:

```
#define _lib_select  1
```

Then, the library source code can be written to use one or both system calls as below:

```
#if _lib_poll
...
#endif
#if _lib_select
...
#endif
```

Feature-based software configuration is nice in that it identifies exactly what is needed by a library without expressed knowledge of particular platforms. In fact, this can be used to compare alternative features on a given platform. Libraries such as Sfio use *iffe* to run performance tests and identify relevant features to optimally configure library code.

4. Conclusion

Building reusable software libraries is a long-standing tradition in the software engineering community. [21, 23, 24] This paper described Concrete Software Libraries. These highly reusable and portable libraries not only provide unique and useful functionality, they do that with algorithms and data structures that are scalable to extreme application demands. In addition, they emphasize on providing single source code implementations portable to a wide variety of platforms. Three example libraries, Sfio, Vmalloc, and Cdt were described. In our own work, these libraries have been used to build applications that deal with several gigabytes of data and million of objects a day. Performance studies were discussed showing that Sfio, Vmalloc and Cdt were at least as efficient or better than their competitors. Further, they do that with minimal, robust, yet general interfaces. This achievement is due to a discipline and method library architecture and a collection of design and implementation conventions that help to maximize generality and interoperability. A measure of success for these libraries is the fact that they are used in many large applications built by programmers around the world.

Acknowledgement. Many ideas discussed in this paper arose from conversation and collaboration with various colleagues, in particular, Glenn Fowler, David Korn, and Stephen North.

Code availability. The source code for Sfio, Cdt, Vmalloc and other libraries can be obtained at the below website:

    http://www.research.att.com/sw/tools

References

[1] Glenn S. Fowler, David G. Korn, and Kiem-Phong Vo. Sfio: A Buffered I/O Library. *Software—Practice and Experience*, Submitted for publication, 1998.

[2] Kiem-Phong Vo. Vmalloc: A General and Efficient Memory Allocator. *Software—Practice and Experience*, 26:1–18, 1996.

[3] Kiem-Phong Vo. Cdt: A Container Data Type Library. *Software—Practice and Experience*, 27:1177–1197, 1997.

[4] Kiem-Phong Vo. An Architecture for Reusable Libraries. In *Proceedings of the 5th International Conference on Software Reuse*. IEEE, 1998.

[5] Glenn S. Fowler, David G. Korn, and Kiem-Phong Vo. Principles for Writing Reusable Library. In *Proceedings of the ACM SIGSOFT Symposium on Software Reusability*, pages 150–160. ACM Press, 1995.

[6] ANSI. *American National Standard for Information Systems - Programming Language - C*. American National Standards Institute, 1990.

[7] David G. Korn and Kiem-Phong Vo. SFIO: Safe/Fast String/File IO. In *Proceedings of the Summer '91 Usenix Conference*, pages 235–256. USENIX, 1991.

[8] Jeffrey Richter. *Advanced WINDOWS*. Microsoft Press, 1995.

[9] C.J. Stephenson. Fast fits: new methods for dynamic storage allocation. In *Proceedings of the Ninth ACM Symposium on Operating System Principles*, pages 30–32. Bretton Woods, NH, 1983.

[10] D. Lea. libg++, the GNU C++ library. In *Proceedings of the USENIX C++ Conference*, 1988.

[11] H. Boehm and M. Weiser. Garbage collection in an uncooperative environment. *Software—Practice and Experience*, pages 807–820, 1988.

[12] D. Sleator and R.E. Tarjan. Self-Adjusting Binary Search Trees. *Journal of the ACM*, 32:652–686, 1985.

[13] David G. Korn and Kiem-Phong Vo. In Search of a Better Malloc. In *Proceedings of the Summer '85 Usenix Conference*, pages 489–506. USENIX, 1985.

[14] Andrew R. Koenig. Associative Arrays in C++. In *Proceedings of 1988 Summer USENIX Conference*, pages 173–186, 1988.

[15] Unix System Laboratories. *USL C++ Standard Components Programmer's Reference*. AT&T and Unix System Laboratories, Inc., 1990.

[16] David R. Musser and Atul Saini. *STL Tutorial and Reference Guide*. Addison-Wesley, 1995.

[17] D.R. Musser and A.A. Stepanov. Algorithm-oriented generic libraries. *Software—Practice and Experience*, 24(7):623–642, 1994.

[18] Donald E. Knuth. *The Art of Computer Programming, Volume 3: Sorting and Searching*. Addison-Wesley, 1973.

[19] Robert Sedgewick. *Algorithms, 2nd Edition*. Addison-Wesley, 1988.

[20] G.M. Adelson-Velskii and E.M. Landis. An Algorithm for the Organization of Information. *Soviet Mathematik Doklady*, 3:1259–1263, 1962.

[21] D. Batory, V. Singhal, M. Sirkin, and J. Thomas. Scalable Software Libraries. In *Proceedings of the Symposium on the Foundations of Software Engineering*, 1992.

[22] Glenn S. Fowler, David G. Korn, J.J. Snyder, and Kiem-Phong Vo. Feature-Based Portability. In *Proceedings of the Usenix VHLL Conference*, pages 197–207. USENIX, 1994.

[23] T.J. Biggerstaff. The Library Scaling Problem and the Limits of Concrete Component Reuse. In *Proceedings of the 3rd International Conference on Software Reuse*, 1994.

[24] P.C. Clements, D.L. Parnas, and D. Weiss. Negotiated Interfaces for Software Reuse. *IEEE Trans. on Soft. Eng.*, 18(7):646–653, 1992.

AT&T LABORATORIES, 180 PARK AVENUE, FLORHAM PARK, NJ 07932, U.S.A.
E-mail address: kpv@research.att.com

DIMACS Series in Discrete Mathematics
and Theoretical Computer Science
Volume **50**, 1999

S(b)-Tree Library: an Efficient Way of Indexing Data

Konstantin V. Shvachko

ABSTRACT. We present a library for maintaining external dynamic dictionaries
with variable length keys. A new type of balanced trees, called $S(b)$-trees, is
introduced which contrary to the well-known B-trees provide optimal packing
of keys of variable length, while the data access time remains logarithmic,
the same as for B-trees. $S(b)$-trees are implemented as a stand-alone library.
The library functionality includes means for creating, storing, and performing
basic operations on $S(b)$-trees. The library documentation, source code, and
executables are available at `http://namesys.botik.ru/~shv/stree`

Introduction

Indexing is a common mechanism used in data base retrieval. Any data base
programming system provides means for the creation and maintenance of indexes.
Most index implementations are based on balanced trees. The indexing problem
is usually referred to as the problem of *"maintaining dynamic dictionaries"*. A
dynamic dictionary is a data structure for storing dictionary elements, called keys,
together with algorithms for accessing, inserting, and deleting a key.

A number of variants of balanced trees are known. AVL-trees [**AVL62**], 2-3-
trees ([**LD91**], [**Y78**]), red-black-trees [**W93**] and some other variants of balanced
trees with low branching are mostly appropriate for representation of *internal dictionaries*.

From the practical viewpoint, B-trees ([**B72**], [**BM72**], [**K73**], [**TF82**], [**W86**]),
together with all their modifications, like B^+-trees [**W93**], B^*-trees [**M77**], (a,b)-
trees [**LD91**], are considered to be the most common data structure for representing indexes in *external memory*. Unfortunately, B-trees have a serious disadvantage
([**K73**], [**PS92**]) in the case of variable length keys, since in this case they can lead
to an exhaustive waste of memory.

We introduce a new type of balanced trees, called $S(b)$-trees, which contrary
to B-trees provide optimal packing of keys of variable length, while the data access
time remains logarithmic, the same as for B-trees.

1991 *Mathematics Subject Classification*. Primary 68P05, 68P20, 68P15, 68P10, 68Q25,
68N05.

This project was supported in part by the Russian Foundation of Basic Research (grant
#96-01-01005). The project was started within the DIMACS Challenge'96.

$S(b)$-trees are implemented as a stand-alone library. The library functionality includes means for creating, storing, and performing the basic operations for $S(b)$-trees. The implementation is written in C++ and a pure C function interface is provided. The $S(b)$-tree library is designed to be platform independent, and it runs on both UNIX and MS Windows platforms.

1. Common properties of balanced trees

Balanced trees are intended to provide an efficient solution to the problem of maintaining dynamic dictionaries.

A balanced tree stores keys chosen from a finite set of keys K. K is linearly ordered, and the keys are placed in the tree according to the ordering. Each node of the tree contains a number of keys. Leaves don't contain anything else, while internal nodes contain references to other nodes. The number of references in an internal node equals the number of keys in the node plus 1. An internal node S composed of m keys is denoted by

$$S = (S_0, k_0, ..., S_i, k_i, S_{i+1}, ..., k_{m-1}, S_m)$$

Trees satisfying the properties below are called *structured trees*.

- All paths in the tree from the root to a leaf have equal lengths.
- For any node S of the tree, the keys it stores are located in S according to K's linear ordering, that is k_i is less than k_j for $i < j$.
- For any key k_i of an internal tree node $S = (S_0, k_0, ..., S_i, k_i, S_{i+1}, ..., k_{m-1}, S_m)$ all keys located in the sub-tree accessible via the reference S_i to the left of k_i are less than k_i, and all keys located in the sub-tree accessible via the reference S_{i+1}, which is to the right of k_i, are greater than k_i.

The last property is the base of an efficient tree search. The search algorithm is common to all structured trees. Starting from the root, one should look for the given key k in the current node, and either find the key in the node (the searching is finished) or find a pair of adjacent keys k_{i-1}, and k_i such that $k_{i-1} \leq k \leq k_i$. In the latter case searching is continued in the sub-tree referenced by S_i.

2. B-Trees

DEFINITION 2.1. *A B-tree of order q is a structured tree whose nodes except the root contain at least q, and at most $2q$ keys. The root must contain at least 1 key, and at most $2q$.*

The restrictions associated with the tree order q provide high branching of the B-tree internal nodes, and guarantee a fast tree searching.

It is well known that searching, insertion, and deletion in a B-tree can be performed in time proportional to the height of the tree, which is at most $\log_2 n$, where n is the number of nodes in the tree.

A *utilization* $\delta(T)$ of a B-tree T with n nodes is defined to be the ratio of the total number $|T|$ of keys in the tree to the maximal possible number $2qn$ of keys in an n-node B-tree. Since the minimal number of keys is qn, and the maximal is $2qn$, the utilization is bounded by

$$\frac{1}{2} \leq \delta(T) = \frac{|T|}{2qn} \leq 1$$

The lower bound appears to be acceptable, but in practice things look different. Indeed, in an implementation of B-trees it is natural to use fixed size blocks to store the tree nodes – one block per node. According to the definition of B-trees each block must fit $2q$ keys. Suppose that the key set K consists of keys of different lengths, and that their maximal length is l. Then the block size should be at least $2ql$. This means that if a node stores keys of size $l/10$, then it is only $1/10$ full even if the number of keys is maximal. Therefore, actually we cannot guarantee any lower bound greater than 0 for the utilization of the tree when key lengths are taken into account.

3. Definition of $S(b)$-trees

To improve tree utilization a *weight function* μ for the key set K have to be defined. From the practical viewpoint the weight is the number of bytes the key is stored in. The weight is different to the length since sometimes additionally to the key itself it is necessary to store some extra information, like the key length or the ending zero byte. For any key k, $\mu(k)$ denotes the *weight* of k. $\mu_{\max}(K)$ denotes the maximum of key weights in K. For any node S, $\mu(S)$ denotes the node weight, that is the total weight of keys contained in S, while $M(S)$ denotes the total weight of keys contained in the tree rooted at S.

A $S(b)$-tree (read as *sweep b tree*) is characterized by the following three parameters:

1. b – the *locality parameter*,
2. q – the tree *order*, and
3. p – the tree *rank*.

The tree order q specifies the minimal number of keys in a non-root node of the tree. The root must contain at least one key. The tree rank p specifies the maximal total weight of keys in a node. We say that a tree node S is *well-formed* if the number of keys $|S| \geq q$ ($|S| \geq 1$ for the root), and if the total weight of the node $\mu(S) \leq p$.

To provide tight packing of keys an $S(b)$-tree must satisfy an *incompressibility* property.

Let us consider $m + 1$ adjacent nodes at the same level of the tree and m *delimiting keys* such that for any pair of adjacent nodes their references to the nodes in the nodes common parent are separated. Such a collection of nodes, and their delimiting keys is called a *sweep*, and m is defined to be the length of the sweep. A sweep of length m is called *compressible* if one can construct a sweep of smaller length containing all the keys of the initial sweep, and composed of at most m well-formed nodes, and at most $m - 1$ delimiting keys. Otherwise, the sweep is called *incompressible*. Therefore, a *tree is m-incompressible* if any sweep of length m in it is incompressible.

DEFINITION 3.1. *An $S(b)$-tree of order q, and rank p is a structured tree that is incompressible with respect to the locality parameter b, such that*

1. $b \geq 0$
2. $q \geq b$ & $q \geq 1$
3. $p \geq 2q\,\mu_{\max}(K)$

The restrictions above are essential for the correctness proofs.

In [**PS92**], [**S95Obn**], [**S95CMA**] we proved that $S(b)$-trees generalize B-trees, meaning that if the weight function μ equals to 1 on K then each B-tree of order q is a $S(0)$-tree of order q, and rank $2q$.

4. Utilization of $S(b)$-trees

A *utilization* $\Delta(T)$ of an n-node $S(b)$-tree of rank p is the ratio of the total weight of the tree $M(T)$, that is the sum of weights of keys stored in the tree, divided by the maximum possible weight of an n-node $S(b)$-tree of rank p:

$$\Delta(T) = \frac{M(T)}{np}$$

We sketch a proof below of the following key result

THEOREM 4.1. *For any linearly ordered weighted finite key set K, and for any $\varepsilon > 0$ the three parameters $b > 0$, $q \geq b$ and $p \geq 2q\mu_{\max}(K)$ can be chosen in such a way that for any $S(b)$-tree T of order q and rank p, composed of keys from K and containing at least $(b+1)^2$ vertices,*

$$\Delta(T) > 1 - \varepsilon$$

where ε is inversely proportional to b.

Under the assumptions of the theorem, for any n–node $S(b)$-tree we prove that

$$\Delta(T) > \frac{b}{b+1}\frac{q}{q+1} - \frac{b+1}{n}.$$

The proof of the lower bound is based on the following technical lemma.

LEMMA 4.1. *Consider an n–node structured tree T of order q such that $n > q+1$. Then the number of leaves, x, of the tree is bounded by*

$$x > \frac{q(n-1)+1}{q+1}$$

Lemma 4.1 can be proven using induction on the tree height.

To prove the theorem let us consider an n-node $S(b)$-tree T and a sweep σ of T, composed of all leaves of the tree together with their delimiting keys. Note that this sweep includes all keys contained in the tree. Let us partition σ into the maximal number s of disjoint sub-sweeps $\sigma_i = S_0^i, k_1^i, S_1^i, \ldots, k_b^i, S_b^i$ $(i = 1, \ldots, s)$ of length b.

By Definition 3.1 all sweeps of length b in T are incompressible, which implies that $\mu(\sigma_i) > bp$. Therefore, since the chosen sweeps are disjoint we have

$$M(T) = \mu(\sigma) \geq \sum_{i=1}^{s} \mu(\sigma_i) > bps.$$

Let x be the number of leaves of T. Then the number of chosen sub-sweeps is

$$s = \left\lfloor \frac{x}{b+1} \right\rfloor$$

Applying Lemma 4.1 we obtain

$$s > \frac{x}{b+1} - 1 > \frac{1}{b+1}\frac{q(n-1)+1}{q+1} - 1 > n\frac{1}{b+1}\frac{q}{q+1} - \frac{1}{b+1}\frac{q-1}{q+1} - 1$$

Using the two bounds above we get

$$\Delta(T) = \frac{M(T)}{np} > b\frac{s}{n} > \frac{b}{b+1}\frac{q}{q+1} - \frac{1}{n}\left(\frac{b}{b+1}\frac{q-1}{q+1} + b\right)$$

which implies the desired bound.

This lower bound means that $S(b)$-trees provide almost optimal packing of any given finite set of keys. Special cases of this bound are proven in [**PS92**] and [**S95CMA**].

5. The sketch of the algorithms

The algorithm to *search* in an $S(b)$-tree is the same as for all other balanced trees, (see Section 2).

The algorithms for insertion and deletion in an $S(b)$-tree are more complicated than the corresponding ones for B-trees. The difficulty is that in the $S(b)$-tree case, balancing of a modified node S involves additional b neighboring nodes to the left and to the right. For B-trees, insertions and deletions involve at most one neighbor.

It can be proven that an insertion, and a deletion of a key in an n-node $S(b)$-tree can be performed in at most $C \log(n)$ time where C is a constant that is independent of n and proportional to the locality parameter b.

5.1. Insertion. Using the search algorithm an *insertion* first verifies whether the given key k is contained in the given tree T. If k is in T then the insertion is finished. Otherwise the search procedure returns a leaf S, and a key k_i in it, before which the key k must be inserted. k is inserted in its corresponding place in S, and the insertion starts balancing the tree. Balancing is performed by a special procedure, which is a common part of both the insertion and the deletion algorithms. Balancing starting from the enlarged leaf S balances all the nodes that lay on the path from the root of the tree to S.

5.2. Deletion. Similarly to insertion, a *deletion* begins with a search. If the given key k is not found in the given $S(b)$-tree T, then the deletion is finished. Otherwise the search returns a node S, and a key k_i in it, that must be deleted. Note that S can be either an internal node or a leaf. The case when S is not a leaf can be easily reduced to the case of deletion from a leaf. If S is internal, we consider the sub-tree accessible via the reference S_{i+1}, which is to the right of k_i in S. Take the left most leaf S' in the sub-tree, and the smallest key k' in S', and replace k_i by k' in S. Thus the problem is reduced to the deletion of k' from the leaf node S'. Now let S be a leaf, and let k_i be the key that must be deleted from S. The deletion algorithm removes k_i from S and starts balancing the tree with the same balancing algorithm that is used in insertions.

5.3. Balancing. The Balance() procedure is the main common part of the insertion and the deletion algorithms. Balancing starts at the leaf node S given as an input parameter to the procedure. After working on the level of the current node S the procedure takes for balancing the direct parent of S. The process proceeds further up to the tree root. The balanced tree is the result of the procedure Balance().

For any current node S the procedure decides to balance S if one of the following three conditions holds,

1. One of the $b+1$ sweeps of length b, containing S, is not incompressible.

2. $|S| < q$.
3. $\mu(S) > p$.

In the first case the procedure Balance_B() is used for balancing S. If Condition 2 holds for S then Balance_C() is called. And in the case of Condition 3, S is balanced by Balance_W(). When none of the conditions holds, Balance() skips the level.

Each of the three procedures restores the structure of the $S(b)$-tree disturbed locally for one, two or three vertices of the current tree level. While correcting the structure of the tree on the current level, the algorithm changes also the ancestors of S. This can break in turn the balance conditions for the lower level nodes. Such breakdowns are also local, since not more than three lower level nodes can be changed: the direct parent of S, and two its neighbors to the left and to the right of S. Coming to the next level of the tree Balance() merges the modified nodes of the level, and balances them in its entirety.

The computation stops at the tree root. Thus the algorithm examines all the nodes that lay on the path from the root to the modified leaf, and balances them if necessary. Only these nodes and their neighbors (b to the left and b to the right) in the tree can be transformed by the algorithm.

After balancing the structure of the $S(b)$-tree is restored, all tree nodes are well-formed, and all sweeps of the tree are incompressible.

6. The specifications

Let T be a structured tree and let S be its node. Below we present a list of variables and instrumental procedures used to describe the algorithms.

6.1. Denotations and definitions. With respect to the current node S a sweep

$$\langle L_{b-1}, l_{b-1}, \ldots, L_0, l_0, S, r_0, R_0, \ldots, r_{b-1}, R_{b-1} \rangle$$

composed of b neighbors and their delimiting keys to the left and to the right of S is called the *vicinity* of S, where

- S is the current node;
- L_i denotes the i-th left neighbor of S;
- R_i denotes the i-th right neighbor of S;
- l_i is the delimiting key of nodes L_i and L_{i-1};
- r_i is the delimiting key of nodes R_i and R_{i-1}.
- F denotes the direct parent of S.
- FL_i (FR_i) denotes the parent of node L_i (R_i).
- Fl_i (Fr_i) denotes the node containing the delimiting key l_i (r_i).
- $L_i, l_i, R_i, r_i, FL_i, FR_i, Fl_i, Fr_i$ are the local variables of the procedures.
- $WW(S)$ means that $\mu(S) \leq p$.
- $WC(S)$ means that $|S| \geq q$.
- $WF(S)$ means that $WW(S)$ and $WC(S)$ hold.
- $IC(\sigma)$ means that the sweep σ is incompressible
- $WB(\sigma)$ means that $IC(\sigma)$ holds for any sweep σ of length b containing S

- $Sweep_b^m(S)$ denotes the m-th $(m = 0, \dots, b)$ sweep of length b containing node S, namely

$$
\begin{aligned}
Sweep_b^0(S) &= \langle L_{b-1}, l_{b-1}, \dots, L_0, l_0, S \rangle \\
Sweep_b^b(S) &= \langle S, r_0, R_0, \dots, r_{b-1}, R_{b-1} \rangle \\
Sweep_b^m(S) &= \langle L_{b-m-1}, l_{b-m-1}, \dots, L_0, l_0, S, r_0, R_0, \dots, r_{m-1}, R_{m-1} \rangle
\end{aligned}
$$

- $WB(S)$ means that for any sweep, $Sweep_b^m(S)\,(m = 0, \dots, b)\,WB(Sweep_b^m(S))$ hold.

6.2. Instrumental procedures. The following procedures and functions are used for describing the algorithms.

- $Search(T, k)$, given an $S(b)$-tree T and key k, verifies whether the key is contained in the tree. The result of the function is $(IsFound, S, k_i)$. $IsFound$ is a Boolean variable that specifies whether k is found in the tree or not. S is the node that was visited last while searching in the tree, and k_i is the minimal key of S that is greater than or equal to the given key k.
- $MakeVicinity(S)$ initializes the local variables $L_i, l_i, R_i, r_i, FL_i, FR_i, Fl_i, Fr_i$ according to S.
- $Replace(S, P, Q)$, substitutes that portion of the node S that coincides with P, with Q.
- The functions $LeftOf(S, k)$ and $RightOf(S, k)$ specify two portions of S, that are to the left and to the right of key k in S, respectively.
- For each $m = 0, \dots, b$ we define a function $Compress^m$ that is applied to compressible sweeps of length b. The result is an incompressible sweep of length $b - 1$, composed of b well-formed nodes. Namely, if

$$
Compress^m(A_0, a_1, \dots, a_b, A_b) = \langle C_0, c_1, \dots, c_{b-1}, C_{b-1} \rangle
$$

  then the resulting sweep is obtained by distributing the contents of node A_m between the other nodes of the initial sweep in such a way that $a_i < c_i$ for $i = 1, \dots, m$, and $c_i < a_{i+1}$ for $i = m, \dots, b - 1$.
- $ComputeSets(S)$ calculates seven subsets of the set of keys of node S: $MLeft$, $MRight$, $MDelimL$, $MDelimR$, $MDelim$, $\overline{MLeft}$, and $\overline{MRight}$, defined below.
- $ChooseAny(E)$, $ChooseMax(E)$, and $ChooseMin(E)$ for a given key set E return, respectively, an arbitrary, the maximum and the minimum keys of the set.
- $SearchMinimal(S)$ looks for the minimal key in the sub-tree rooted at S, returns (S', k') where k' is the minimal key, and S' is the leaf that contains k'.
- $GlueParents(F)$ takes the parent F of the current node S, the left, and the right neighbors of F, glues them into one node, and returns the result. A more sophisticated variant of this function is to check before gluing whether the neighbors have been modified, and glue to F only the modified ones.
- $CreateNewRoot(S)$ creates the new root of the tree, containing the only reference to the given node S.
- $ReleaseRoot()$. If the root of the tree contains only one reference to a node S, and no keys, then this root is removed, and S is assigned to be the new root of the tree.

- $[P, d, Q]$ creates a new node composed of all keys, and node references (in case of internal nodes), of the given nodes P and Q, with the key d between them. E.g., $[\langle L_0, l_1, L_1\rangle, d, \langle R_0, r_1, R_1\rangle] = \langle L_0, l_1, L_1, k, R_0, r_1, R_1\rangle$

If $k(S)$ denotes the set of all keys of node S, then the definitions of the subsets are as follows.

$$
\begin{aligned}
MLeft &= \{d \in k(S) \mid IC(L_{b-1}, l_{b-1}, \ldots, L_0, l_0, LeftOf(S, d))\} \\
MRight &= \{d \in k(S) \mid IC(RightOf(S, d), r_0, R_0, \ldots, r_{b-1}, R_{b-1})\} \\
MDelimL &= \{d \in k(S) \mid WF(RightOf(S, d))\} \\
MDelimR &= \{d \in k(S) \mid WF(LeftOf(S, d))\} \\
MDelim &= MDelimL \cap MDelimR \\
\overline{MLeft} &= k(S) \setminus MLeft \\
\overline{MRight} &= k(S) \setminus MRight
\end{aligned}
$$

The sets, and the *ComputeSets*() procedure are used in Balance_W() to control the process of computation.

6.3. The balance stages. As we mentioned above, to balance the tree on each tree level one of the three balance procedures is called. In each case balancing is performed by redistributing the keys within the b-vicinity of the current node S in such a way that both $WF(Q)$ and $WB(Q)$ hold for any node Q of the vicinity after the balancing is finished. The explicit algorithms are outlined in the Appendix.

Procedure BALANCE_W is used to balance the input tree when the current weight of vertex S is larger than p ($\neg WW(S)$), while the other two properties $WC(S)$ and $WB(S)$ are satisfied for S. The procedure consists of five stages. The decision on whether a stage should be performed or not is based upon the interrelation of the seven subsets of $k(S)$.

Informally,

- a key d from $k(S)$ belongs to *MLeft* (*MRight*) iff the current vertex S can be split into two parts in such a way that the fragment of the b-vicinity of S which is to the left (right) of d is incompressible;
- a key d from $k(S)$ belongs to *MDelimL* (*MDelimR*) iff the current vertex S can be split into two parts in such a way that the weight of the part of S that is to the right (left) of d is not greater than p, while the number of keys of the part is at least q.

The Compression stage is performed when the intersection $\overline{MLeft} \cap \overline{MRight}$ is not empty, meaning that all keys of S can be distributed between the other nodes of the vicinity and S can be eliminated.

The Move left stage is performed if $\overline{MLeft}$ is not empty, which means that a number of keys of S can be moved to the left part of the vicinity.

The Move right stage is analogous.

If after shifting as many keys as possible from S to the left and right parts of the vicinity, the weight of the keys remaining in S is still greater than p, then the Split stage is performed in order to partition S into two parts at least one of which (actually the left one) is well-formed. If the second node is also well-formed, then the balancing is finished.

Otherwise, the recursion stage is invoked. It balances the remaining not well-formed node by recursive calls of BALANCE_W. It can be shown that the depth of this recursion is constant.

The procedure BALANCE_B is used to balance the input tree when one of the sweeps $Sweep_b^m(S)$ $m = 0, \ldots, b)$ is compressible ($\neg WB(S)$). Balancing in this case is performed by compressing one or two sweeps of the b-vicinity of S. It can be shown that elimination of more than two vertices of the vicinity is impossible. Starting from the left most sweep $Sweep_b^0(S)$ BALANCE_B scans the vicinity and compresses each compressible sweep found. After the second compression it terminates.

Procedure BALANCE_C is called when the number of keys in the current node S is less than q ($\neg WC(S)$). In this case we join S with one of it's neighbors and call BALANCE_W for the resulting node if required.

7. $S(b)$-tree library interface

The $S(b)$-tree library implements the algorithms described above, and provides means for creating, storing, and performing basic operations for $S(b)$-trees. The implementation is written in C++ and a pure C function interface is provided. The $S(b)$-tree library is designed to be platform-independent, and it runs both on UNIX and Windows NT/95 platforms.

7.1. Keys. The *key* type is S_KEY. To create a key an S_CreateKey(Key, String, Length) function is used. Given an array of bytes and its length, it sets the given key to the specified string value.

Note that one can store not only string-valued keys, but keys of an arbitrary structure by providing a conversion of a user defined key to a byte array. Namely, a number can be easily converted to a string using, say, the C run-time library routines itoa(), or ltoa().

A key *weight* equals the length of the byte array plus a constant given by S_EMPTY_KEY_WEIGHT.

7.2. Maintaining the trees. Several functions provide means for creating new $S(b)$-trees, opening existing ones, saving, and closing the tree modification sessions.

In order to create an $S(b)$-tree it is necessary to specify the following parameters.

- FileName is the name of the file where the tree will be stored.
- b is the locality parameter. The better packing you need the greater b should be chosen.
- q is the order of the $S(b)$-tree. It specifies the minimal number of keys in a node, and is intended to provide high branching of the tree internal nodes.
- p is the tree rank. It is the size of the block for storing the tree nodes, meaning that a node weight cannot exceed p.
- MuMax characterizes the key set K, in the way that all keys in K have length not greater than MuMax.

A node weight is the sum of the weights of keys contained in the node plus a constant S_EMPTY_NODE_WEIGHT. In the implementation we need to store some header for each tree node that contains the number of keys in the node, which is required for correct reading of tree nodes. Thus, actually the tree rank according to our definitions in Section 4, is $p - $ S_EMPTY_NODE_WEIGHT.

The restrictions for the parameters that provide correctness of the algorithms are as follows:

1. $b \geq 1$
2. $q \geq b$ & $q \geq 2$
3. $p \geq 2q$ MuMax + S_EMPTY_NODE_WEIGHT

S_CreateTree(ResTreeHandle, FileName, b, q, p, MuMax)
creates an empty tree with the specified parameters, and returns the new tree handle ResTreeHandle, which provides access to the tree from other functions.

Another way to get access to an $S(b)$-tree is to load it, which is performed by S_LoadTree(ResTreeHandle, FileName).

S_CloseTree(TreeHandle) closes the tree, saves it on disk in the file it was created or loaded from, and releases the tree handle.

S_SaveTree(TreeHandle) saves the specified tree on disk in the file it was created or loaded from, but does not close the tree leaving it accessible via the tree handle.

7.3. The basic tree operations. The main algorithms of search, insertion, and deletion are implemented with the following functions
S_Search(TreeHandle, Key, IsFound)
S_Insert(TreeHandle, Key)
S_Delete(TreeHandle, Key).

7.4. Additional operations. Since the set of keys stored in an $S(b)$-tree is linearly ordered, it is natural to provide access to the first, the last key in the dictionary, the next, and the previous keys with respect to the given one. This is achieved with the functions
S_First(TreeHandle, FirstKey, IsFound)
S_Last(TreeHandle, LastKey, IsFound)
S_Next(TreeHandle, InpKey, NextKey, IsFound)
S_Prev(TreeHandle, InpKey, PrevKey, IsFound).
Note that the given key InpKey should not necessarily be contained in the tree. To find the next (the previous) means to find the minimal (maximal) key in the tree that is greater (less) than the input key.

7.5. $S(b)$-tree properties. The rest of the functions return the specified tree parameters: b, q, p, and MuMax, and the tree intrinsic properties:

- Number of nodes in the tree
- Number of keys in the tree
- The tree total weight
- The tree height
- The tree utilization

The prototypes of the functions are:
int S_TreeLocality(TreeHandle)
int S_TreeOrder(TreeHandle)
int S_TreeRank(TreeHandle)
int S_TreeMaxKeyWeight(TreeHandle)
long S_TreeNrNodes(TreeHandle)
long S_TreeNrKeys(TreeHandle)
long S_TreeWeight(TreeHandle)
int S_TreeHeight(TreeHandle)
double S_TreeUtilization(TreeHandle)

8. Conclusions

Balanced trees are the standard data structures for indexing information. This paper introduces a new type of balanced trees, called $S(b)$-trees. $S(b)$-trees generalize B-trees for the case of variable length keys. In Theorem 4.1 we present a lower bound of utilization of an $S(b)$-tree, which shows that $S(b)$-trees provide almost optimal packing of any given finite set of variable length keys. We described logarithmic running time algorithms for the $S(b)$-tree based dictionary search and update operations. $S(b)$-trees are implemented as a stand-alone library. The library functionality includes means for creating, storing, and performing an extended set of operations for $S(b)$-trees.

Besides the library, $S(b)$-trees were implemented in two different software systems with the common feature that both of them were designed to store completely unstructured data collections.

The high-level universal programming language *Starset* [**G94**] was developed to generalize the traditional relational database approach. Starset is based on the set data model, which eliminates restrictions of the relational approach, such as multi-valued attributes, varying arity, null values, etc. The Starset programming language uses $S(1)$-trees as described in [**PS92**] for representing its set data aggregates.

The high-performance Tree File System (TreeFS) was designed to break down a tradeoff common for block oriented file systems where desire to increase the system block size in order to accelerate disk access contradicts the necessity of keeping the size small enough to avoid waste of disk space. A substantially modified variant of $S(1)$-trees is implemented in the *Tree File System*. This is probably the first attempt to represent a whole file system by a balanced tree in a Unix–like operating system. We have conducted experiments that show that most file operations are faster in TreeFS, especially for small files, and that the disk space utilization is higher compared to traditional file systems. TreeFS is implemented as a virtual FS under the Linux operating system.

9. Acknowledgments

I am thankful to Elena A. Zinovieva for helping with the implementation.

References

[AVL62] G.M. Adel'son-Vel'skii, E.M. Landis, *An Algorithm for the Organization of Information*, Soviet Math. Doklady vol. 3, 1972, pp. 1259–1262.

[B72] R. Bayer, *Symmetric binary B-tree: Data Structure and Maintenance Algorithms*, Acta Inf., vol. 1, 4, 1972, pp. 290–306.

[BM72] R. Bayer, E. McCreight, *Organization and Maintenance of Large Ordered Indexes*, Acta Inf., vol. 1, 3, 1972, pp. 173–189.

[G94] M.M. Gilula, *The Set Model for Database and Information Systems*, Addison-Wesley (In Association with ACM Press): Wokingham, 1994.

[K73] D.E. Knuth, *The Art of Computer Programming*, vol. 3 (Sorting and Searching), Addison–Wesley, Reading, MA, 1973.

[LD91] H.R. Lewis, L. Denenberg, *Data Structures and Their Algorithms*, HarperCollins, NY, 1991.

[M77] E.M. McCreight, *Pagination of B^*-Trees with Variable-Length Records*, Commun. ACM, vol. 20, 9, 1977, pp. 670–674.

[PS92] A.P. Pinchuk, K.V. Shvachko, *Maintaining Dictionaries: Space-Saving Modifications of B-Trees*, Lecture Notes in Computer Science, vol. 646, 1992, pp. 421–435.

[S93] K.V. Shvachko, *Space-Saving Modifications of B-Trees*, In Proceedings of Symposium on Computer Systems and Applied Mathematics, St.Petersburg, 1993, p. 214.

[S95Obn] K.V.Shvachko, *Optimal Representation of Dynamic Dictionaries by Balanced Trees*, In Proceedings of XI International Conference on Logic, Methodology, and Philosophy of Science, Obninsk, Russia, vol. 2, 1995, pp. 181-186 (in Russian).

[S95CMA] K.V.Shvachko. *Space Saving Generalization of B-Trees with 2/3 Utilization*, Computers and Mathematics with Applications, vol. 30, No 7, 1995, pp. 47-66.

[S95NN] K.V.Shvachko, *A Hierarchy of Weight Balanced Trees*, In Proceedings of II International Conference on Mathematical Algorithms, N.Novgorod, Russia, 1995.

[TF82] T.J. Teorey, D.P. Fry, *Design of Database Structures*, vol. 2, Prentice-Hall, Englewood Cliffs, NJ, 1982.

[W86] N. Wirth, *Algorithms and Data Structures*, Prentice-Hall, Englewood Cliffs, NJ, 1986.

[W93] D. Wood, *Data Structures, Algorithms, and Performance*, Addison-Wesley Publishing Company, 1993.

[Y78] A.C.-C. Yao, *On Random 2-3 Trees*, Acta Inf., vol. 9, 1978, pp. 159–170.

Appendix A. The main procedures

Procedure Insert(T, k)
$(IsFound, S, k_i) = Search(T, k)$;
if $IsFound = True$ **then return fi**;
/* k haven't been found means that S is a leaf */
$Replace(S, \langle k_i \rangle, \langle k, k_i \rangle)$ /* insert k before k_i */
Balance(S); /* balance T starting from leaf S */
EndProcedure

Procedure Delete(T, k)
$(IsFound, S, k_i) = Search(T, k)$;
if $IsFound = False$ **then return fi**;
if S is not a leaf **then**
    /* S is internal, and contains reference */
    /* S_{i+1}, which is to the right of k_i in S */
    $(S', k') = SearchMinimal(S_{i+1})$;
    $Replace(S, \langle k_i \rangle, \langle k' \rangle)$;
    $S := S'$;
    $k := k'$;
fi
/* S is a leaf */
$Replace(S, \langle k \rangle, \langle \rangle)$ /* delete k from leaf S */
Balance(S); /* balance T starting from leaf S */
EndProcedure

Procedure Balance(S)
while S is not the root **do**
    if $\neg WB(S)$ **then** $F :=$ Balance_B(S);
    else if $\neg WC(S)$ **then** $F :=$ Balance_C(S);
    else if $\neg WW(S)$ **then** $F :=$ Balance_W(S);
    else $S := F$; **continue**;
    fi fi fi
    $S := GlueParents(F)$;
od
/* S is the root now */
if $|S| = 0$ **then**
    $ReleaseRoot()$;
fi
if $\neg WW(S)$ **then**
    $CreateNewRoot(S)$;
    Balance_W(S);
fi
EndProcedure

Procedure Balance_W(S)
$MakeVicinity(S)$;
$ComputeSets(S)$;

/* Compression: */
/* Distribute all keys of S between */
/* the other nodes of the vicinity */
if $\overline{MLeft} \cap \overline{MRight} \neq \emptyset$ **then**
    $d := ChooseAny(\overline{MLeft} \cap \overline{MRight})$;
    $\langle X_{b-1}, x_{b-1}, \dots, x_1, X_0 \rangle := Compress^b(L_{b-1}, l_{b-1}, \dots, L_0, l_0, LeftOf(S, d))$;
    $\langle Y_0, y_1, \dots, y_{b-1}, Y_{b-1} \rangle := Compress^0(RightOf(S, d), r_0, R_0, \dots, r_{b-1}, R_{b-1})$;
    for $i := 1$ **to** $b - 1$ **do**
        $Replace(FL_i, \langle L_i \rangle, \langle X_i \rangle)$;
        $Replace(Fl_i, \langle l_i \rangle, \langle x_i \rangle)$;
        $Replace(FR_i, \langle R_i \rangle, \langle Y_i \rangle)$;
        $Replace(Fr_i, \langle r_i \rangle, \langle y_i \rangle)$;
    od
    $Replace(FR_0, \langle R_0 \rangle, \langle Y_0 \rangle)$;
    $Replace(Fr_0, \langle r_0 \rangle, \langle d \rangle)$;
    if $FL_0 = F$ **then**
        $Replace(F, \langle L_0, l_0, S \rangle, \langle X_0 \rangle)$;
    else
        $Replace(F, \langle S \rangle, \langle X_0 \rangle)$;
        $Replace(Fl_0, \langle l_0 \rangle, \langle x_1 \rangle)$;
        $Replace(FL_0, \langle X_1, x_1, L_0 \rangle, \langle X_1 \rangle)$;
    fi
    return F;
fi

/* Move left: */
/* Move to the left part of the vicinity */
/* as much keys from S as possible */
if $\overline{MLeft} \neq \emptyset$ **then**
    if $MDelimL \cap \overline{MLeft} \neq \emptyset$ **then** $d := ChooseAny(MDelimL \cap \overline{MLeft})$;
    else $\quad\quad\quad\quad\quad\quad\quad\quad\quad\quad d := ChooseMax(\overline{MLeft})$; **fi**
    $\langle X_{b-1}, x_{b-1}, \dots, x_1, X_0 \rangle := Compress^b(L_{b-1}, l_{b-1}, \dots, L_0, l_0, LeftOf(S, d))$;
    $Y := RightOf(S, d)$;
    for $i := 1$ **to** $b - 1$ **do**
        $Replace(FL_i, \langle L_i \rangle, \langle X_i \rangle)$;
        $Replace(Fl_i, \langle l_i \rangle, \langle x_i \rangle)$;
    od
    $Replace(FL_0, \langle L_0 \rangle, \langle X_0 \rangle)$;
    $Replace(Fl_0, \langle l_0 \rangle, \langle d \rangle)$;
    $Replace(F, \langle S \rangle, \langle Y \rangle)$;
    if $WW(Y)$ **then return** F; **fi**
    $S := Y$;
    $MakeVicinity(S)$;
    $ComputeSets(S)$;
fi

/* Move right: */
/* Move to the right part of the vicinity */
/* as much keys from S as possible */
if $\overline{MRight} \neq \emptyset$ **then**
    if $MDelimR \cap \overline{MRight} \neq \emptyset$ **then** $d := ChooseAny(MDelimR \cap \overline{MRight})$;
    else $\quad\quad\quad\quad\quad\quad\quad\quad d := ChooseMin(\overline{MRight})$; **fi**
    $X := LeftOf(S, d)$;
    $\langle Y_0, y_1, \dots, y_{b-1}, Y_{b-1} \rangle := Compress^0(RightOf(S, d), r_0, R_0, \dots, r_{b-1}, R_{b-1})$;

```
    for i := 1 to b - 1 do
        Replace(FR_i, ⟨R_i⟩, ⟨Y_i⟩);
        Replace(Fr_i, ⟨r_i⟩, ⟨y_i⟩);
    od
    Replace(FR_0, ⟨R_0⟩, ⟨Y_0⟩);
    Replace(Fr_0, ⟨r_0⟩, ⟨d⟩);
    Replace(F, ⟨S⟩, ⟨X⟩);
    if WW(X) then return F; fi
    S := X;
    MakeVicinity(S);
    ComputeSets(S);
fi


/* Split: */
/* Even after moving out of S all keys that fit into the */
/* other nodes of the vicinity S is still too large, and */
/* need to be split into two nodes */
if MDelim ≠ ∅ then d := ChooseAny(MDelim);
else                     d := ChooseMax(MDelimR); fi
X := LeftOf(S, d);
Y := RightOf(S, d);
Replace(F, ⟨S⟩, ⟨X, d, Y⟩);
if WW(Y) then return F; fi


/* Recursion: */
/* Even after splitting off a node X the remaining */
/* part Y is still too large, */
/* Balance_W() will be applied to Y again */
F := Balance_W(Y);
return F;
EndProcedure


Procedure Balance_B(S)
MakeVicinity(S);


/* First Compression: */
/* Check which of the first b sweeps of the vicinity is */
/* compressible, and compress the first one found */
for m := 0 to b - 1 do
    if ¬IC(Sweep_b^m(S)) then break; fi
od


if m < b then /* a compressible sweep is found, compress it */
    ⟨X_{b-m-1}, x_{b-m-1}, ..., x_1, X_0, y_0, Y_0, y_1, ..., y_{m-1}, Y_{m-1}⟩ :=
        Compress^{b-m}(Sweep_b^m(S));
    for i := 1 to b - m - 1 do
        Replace(FL_i, ⟨L_i⟩, ⟨X_i⟩);
        Replace(Fl_i, ⟨l_i⟩, ⟨x_i⟩);
    od
    for i := 0 to m - 1 do
        Replace(FR_i, ⟨R_i⟩, ⟨Y_i⟩);
        Replace(Fr_i, ⟨r_i⟩, ⟨y_i⟩);
    od
    if FL_0 = F then
        Replace(F, ⟨L_0, l_0, S⟩, ⟨X_0⟩);
    else
        Replace(F, ⟨S⟩, ⟨X_0⟩);
        Replace(Fl_0, ⟨l_0⟩, ⟨x_1⟩);
```

       $Replace(FL_0, \langle X_1, x_1, L_0 \rangle, \langle X_1 \rangle);$
   fi
   $m := m + 1;$
   $S := X_0;$
   $MakeVicinity(S);$
fi

/* Second Compression: */
/* Among the remaining sweeps of the vicinity check which */
/* one is compressible, and compress the first one found */
for $m := m$ **to** b **do**
   if $\neg IC(Sweep_b^m(S))$ **then break; fi**
od

if $m > b$ **then return** $F;$ **fi**
/* A compressible sweep is found, compress it */
$\langle X_{b-m-1}, x_{b-m-1}, \ldots, x_1, X_0, y_0, Y_0, y_1, \ldots, y_{m-1}, Y_{m-1} \rangle := Compress^{b-m}(Sweep_b^m(S));$
for $i := 0$ **to** $b - m - 1$ **do**
   $Replace(FL_i, \langle L_i \rangle, \langle X_i \rangle);$
   $Replace(Fl_i, \langle l_i \rangle, \langle x_i \rangle);$
od
for $i := 1$ **to** $m - 1$ **do**
   $Replace(FR_i, \langle R_i \rangle, \langle Y_i \rangle);$
   $Replace(Fr_i, \langle r_i \rangle, \langle y_i \rangle);$
od
if $FR_0 = F$ **then**
   $Replace(F, \langle S, r_0, R_0 \rangle, \langle Y_0 \rangle);$
else
   $Replace(F, \langle S \rangle, \langle Y_0 \rangle);$
   $Replace(Fr_0, \langle r_0 \rangle, \langle y_1 \rangle);$
   $Replace(FR_0, \langle R_0, y_1, Y_1 \rangle, \langle Y_1 \rangle);$
fi
/* It is proven that not more than 2 nodes of the */
/* vicinity can be shrunken */
return $F;$
EndProcedure

Procedure Balance_C(S)
$MakeVicinity(S);$

/* If S has too few keys we glue it with one of */
/* the nearest neighbors, and apply Balance_W */
/* to the result if necessary */
if $|F| > 0$ **then** /* at least one key is in F */
  if $FL_0 = F$ **then**
    $X := [L_0, l_0, S];$
    $Replace(F, \langle L_0, l_0, S \rangle, \langle X \rangle);$
  else /* $FR_0 = F$, since F contains at least one key */
    $X := [S, r_0, R_0];$
    $Replace(F, \langle S, r_0, R_0 \rangle, \langle X \rangle);$
  fi
else /* there is no keys in F only the reference to S */
  if $|FL_0| > 1$ **then**
    $X := [L_0, l_0, S];$
    $Replace(F, \langle S \rangle, \langle X \rangle);$
    $Replace(Fl_0, \langle l_0 \rangle, \langle l_1 \rangle);$
    $Replace(FL_0, \langle L_1, l_1, L_0 \rangle, \langle L_1 \rangle);$
  else /* $|FR_0| > 1$ */

$$X := [L_0, l_0, S];$$
$$Replace(F, \langle S \rangle, \langle X \rangle);$$
$$Replace(Fr_0, \langle r_0 \rangle, \langle r_1 \rangle);$$
$$Replace(FR_0, \langle R_0, r_1, R_1 \rangle, \langle R_1 \rangle);$$
  fi
fi

if $\neg WW(X)$ **then**
  $F := \text{Balance\_W}(Y);$
fi
return F;
EndProcedure

RESEARCH CENTER FOR INFORMATION SYSTEMS, PROGRAM SYSTEMS INSTITUTE, PERESLAVL–ZALESSKY, 152140 RUSSIA

Current address: Lesnoj 3–21, Pereslavl–Zalessky, 152140 Russia

E-mail address: `shv@namesys.botik.ru`

DIMACS Series in Discrete Mathematics
and Theoretical Computer Science
Volume **50**, 1999

ASP: Adaptive Online Parallel Disk Scheduling

Mahesh Kallahalla and Peter J. Varman

ABSTRACT. In this work we address the problems of prefetching and I/O scheduling for read-once reference strings in a parallel I/O system. We use the standard parallel disk model with D disks a shared I/O buffer of size M. We design an on-line algorithm ASP (Adaptive Segmented Prefetching) with ML-block lookahead, $L \geq 1$, and compare its performance to the best on-line algorithm with the same lookahead. We show that for any reference string the number of I/Os done by ASP is with a factor $\Theta(C)$, $C = \min\{\sqrt{L}, D^{1/3}\}$, of the number of I/Os done by the optimal algorithm with the same amount of lookahead.

1. Introduction

Continuing advances in processor architecture and technology have resulted in the I/O subsystem becoming the bottleneck in many applications. The problem is exacerbated by the advent of multiprocessing systems that can harness the power of hundreds of processors in speeding up computation. Improvements in I/O technology are unlikely to keep pace with processor-memory speeds, causing many applications to choke on I/O. The increasing availability of cost-effective multiple-disk storage systems [**CLG**$^+$**94**] provides an opportunity to improve the I/O performance through the use of parallelism. However it remains a challenging problem to use the increased disk bandwidth effectively and reduce the I/O latency of an application.

The parallel I/O system is modeled using the intuitive parallel disk model introduced by Vitter and Shriver [**VS94**]: the I/O system consists of D independently-accessible disks and an associated I/O buffer with a capacity of M blocks, shared by all the disks. The data for the computation is stored on the disks in blocks; a block is the unit of access from a disk. In each I/O up to D blocks, at most one from each disk, can be read into the buffer. From the viewpoint of the I/O, the computation is characterized by a *reference string* consisting of the ordered sequence of blocks that the computation accesses. A block should be present in the I/O buffer before it can be accessed by the computation. Serving the reference string requires performing I/O operations to provide the computation with blocks in the order specified by the reference string. The measure of performance of the system is the number of I/Os required to service a given reference string. In this

1991 *Mathematics Subject Classification.* Primary 68Q22; Secondary 68M20.

Supported in part by the National Science Foundation under grant CCR-9704562 and a grant from the Schlumberger Foundation.

work we consider *read-once* reference strings in which each block is read exactly once. Such reference strings arise naturally in database operations such as external merging [**BGV96, PSV94**] (including carrying out several of these concurrently), joins and real-time retrieval and playback of multiple streams of multimedia data.

I/O parallelism is obtained by *prefetching* blocks from the idle disks in parallel with the block currently requested by the computation. These prefetched blocks are buffered until required. In order to prefetch accurately, the I/O scheduling algorithm needs to have some knowledge regarding future accesses. [**BKVV97**] introduced the notion of M-*block lookahead* to model this information. An algorithm having this form of lookahead knows the sequence of next M blocks beyond the currently referenced block. [**BKVV97**] showed that any algorithm having M-block lookahead can require $\Omega(\sqrt{D})$ times as many I/Os as the optimal *off-line* algorithm. They also presented a simple algorithm, NOM, which achieves this bound.

We are interested in designing on-line algorithms which can use ML-block lookahead, where $L \geq 1$. One straightforward approach to use ML-block lookahead is to be greedy: on every I/O fetch, from each disk, the next block in the lookahead not present in the buffer. However, such an aggressive policy can require $\Omega(D)$ times more I/Os than the optimal on-line algorithm (when $L \geq 2$). On the other hand NOM, which is greedy only within M requests, can be shown to require $\Theta(C)$, $C = min\{L, D^{1/2}\}$, times as many I/Os as the optimal on-line algorithm using ML-block lookahead. The fact that NOM uses only the next M requests is intrinsic to the algorithm and there does not seem to be any way to generalize it to make use of the additional lookahead.

The main result of this paper is a new prefetching and scheduling algorithm called ASP (Adaptive Segmented Prefetching), with improved on-line performance. ASP uses ML-block lookahead to schedule I/Os. The number of I/Os performed by ASP is within a factor $\Theta(C)$, $C = min\{\sqrt{L}, D^{1/3}\}$, of the optimal on-line algorithm with ML-block lookahead. The only other result we are aware of for scheduling read-once reference strings in the parallel disk model is the algorithm RBP whose competitive ratio was shown to be $\Theta(C)$, $C = max\{\sqrt{D/L}, D^{1/3}\}$ [**KV98**]. The relation between RBP and ASP is further discussed in Section 4.2.

Classical buffer management which primarily deals with optimizing buffer evictions has been studied extensively in sequential I/O models [**Bel66, ST85, BGV95**]. In the parallel disk model of this paper, a randomized caching and scheduling algorithm using M-block lookahead was presented in [**Var98**]. Using a distributed buffer configuration, in which each disk has its own private buffer, [**VV96**] presented an optimal off-line I/O scheduling algorithm. An interesting alternative measure of performance is the elapsed or stall time that includes the time required to consume a block as an explicit parameter. Off-line approximation algorithms for a single-disk and multiple-disk systems in this model were addressed in [**CFKL95**] and [**KK96**] respectively. Recently a polynomial time optimal algorithm for the single disk case was presented in [**AGL98**].

The rest of this paper is organized as follows. In Section 2 we introduce some notation and definitions. In Section 3 we present the algorithm ASP. The analysis is done in two parts: in Section 4.1 we bound its performance for lookahead $L \leq D^{2/3}$, while the case of $L > D^{2/3}$ is analyzed in Section 4.2.

2. Definitions

The I/O system is modeled by the Parallel Disk Model [**VS94**] with D disks and a buffer of capacity M blocks, $M \geq 2D$. The sequence of accesses to the I/O system is modeled by a reference string which is the ordered sequence of blocks accessed by the computation. The reference string is partitioned into *phases*, where each phase corresponds to a buffer-load of I/O requests. We quantify the performance of an on-line algorithm by comparing it with the optimal on-line algorithm that has the same amount of lookahead. This motivates the definition of *on-line ratio*, which is a measure of how effectively a given algorithm uses the lookahead available to it.

DEFINITION 2.1. Let the reference string be $\Sigma = \langle r_0, r_1, \cdots, r_{N-1} \rangle$.

- The ith phase, $i \geq 0$, $phase(i)$, is the substring $\langle r_{iM}, \cdots, r_{(i+1)M-1} \rangle$.
- If the last block referenced is r_i, then an on-line scheduling algorithm has ML-*block lookahead* if it knows the substring $\langle r_{i+1}, \ldots, r_{i+ML} \rangle$.
- Let $\mathcal{C}_L$ be the set of all algorithms with ML-block lookahead. For any algorithm $A \in \mathcal{C}_L$, let $T_A(\Sigma)$ denote the number of I/Os needed by A to service Σ. The *optimal on-line algorithm* with ML-block lookahead is the algorithm $OPT \in \mathcal{C}_L$, such that $T_{\text{OPT}}(\Sigma) \leq T_A(\Sigma)$, for every Σ.
- An on-line scheduling algorithm $\mathcal{A} \in \mathcal{C}_L$ has an *on-line ratio* of C_A if there is a constant b such that for any reference string Σ, $T_A(\Sigma) \leq C_A T_{\text{OPT}}(\Sigma) + b$.

The performance of on-line algorithms has traditionally been studied through competitive analysis [**ST85**]. The competitive ratio, the usual measure of performance in this framework, is the worst case ratio of the number of I/Os needed by the on-line algorithm to that required by the optimal off-line algorithm to service a reference string. The competitive ratio of an algorithm with a certain amount of lookahead is influenced by two factors: unknown information about the reference string beyond the lookahead, and the inability of the algorithm to effectively exploit information available in the lookahead. The competitive ratio does not differentiate between these two factors and hence cannot distinguish algorithms for which the contribution of the former factor dominates. A complementary measure of performance, the comparative ratio, was introduced in [**KP94**]. This measure tries to quantify the performance loss of an on-line algorithms due to the unknown portion of the reference string beyond the lookahead. In contrast the on-line ratio measures how well the on-line algorithm exploits the information available in the lookahead.

3. Adaptive Segmented Prefetching

In this section we present the algorithm ASP, that constructs a schedule for a read-once reference string. ASP partitions the reference string into *segments*: a segment is a sequence of contiguous phases. The I/O schedule for each segment is generated by an algorithm THIN. The schedule for the overall reference string is obtained by concatenating the individual schedules. ASP uses a dynamic programming method to adaptively partition the reference string into segments. This procedure is presented later. We first present the algorithm THIN that generates the schedule for a given segment.

THIN colors each block of a segment either red or black. The buffer is also partitioned into a red buffer and a black buffer, each of size $M/2$. When a requested block is not present in the buffer, a batched I/O is initiated. If the block is red then the next $M/2$ red blocks are fetched into the red buffer. Similar action is taken when the block is black. The coloring of the blocks by THIN is based on the following definitions.

DEFINITION 3.1. Within a phase, a block on some disk has a *depth* k if there are $k-1$ blocks from that disk referenced before it in that phase. The set of blocks in a phase with the same depth is called a *stripe*. The *width* of a stripe is the number of blocks in that stripe. The maximum depth of any block in a phase is called the *max-depth* of that phase.

Figure 1 illustrates the definitions above. Note that there can be at most D blocks in a stripe. The max-depth of $phase(i)$ is the minimum number of I/Os required for $phase(i)$ if no blocks of that phase have been fetched prior to the start of $phase(i)$. Each stripe present in the buffer at the start of a phase guarantees that the phase can be serviced in one I/O less than its max-depth. The width of a stripe indicates how much buffer space needs to be allocated to reduce the number of I/Os by one: the narrower a stripe is the lesser space it needs for the same benefit. We use $\sum_{\eta}$ to denote the sum over all k, such that $phase(k)$ belongs to a segment η.

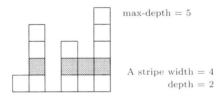

FIGURE 1. Illustration of a stripe

The blocks of a segment are classified as red or black depending on the widths of the stripes in the segment. Red blocks belong to stripes that are no wider that the stripe of any black block. Thus the red blocks of a phase span a small number of disks while the black blocks span a larger number. The details of algorithm THIN are presented in Figure 2. THIN is a building block which is used by ASP to generate the overall I/O schedule.

DEFINITION 3.2. If h_k is the maximum number of red blocks from a single disk in $phase(k)$ and R the maximum number of red blocks from a single disk in segment η, then the *benefit* of THIN in that segment is defined as $B_{\text{THIN}}(\eta) = \sum_{\eta} h_k - R$.

If the red blocks of η were fetched on a phase-by-phase basis, the number of I/Os required would be at least $\sum_{\eta} h_k$. The number of I/Os done by THIN to fetch these blocks in a batched manner is proportional to R.

LEMMA 3.3. *The number of I/Os done by THIN in a segment η is at most* $T_{\text{THIN}}(\eta) \leq 3(\sum_{\eta} d_k - B_{\text{THIN}}(\eta))$, *where d_k is max-depth of $phase(k)$.*

PROOF. Since a stripe has at most D blocks, the total number of red blocks in a segment is at most $M + D - 1$. In each batched-I/O THIN fetches $M/2$ red blocks; hence at most 3 batched-I/Os are required to fetch all the red blocks of a segment. Since the maximum number of red blocks from any single disk in the segment is R, in each batched-I/O operation for red blocks THIN performs no more than R I/Os.

Algorithm THIN takes as input a segment $\eta = \langle phase(i) \ldots phase(i') \rangle$, and generates an I/O schedule to service this segment.

- Partition the I/O buffer into two parts, *red* and *black*, each of size $M/2$. Each half of the buffer will only be used to hold blocks of that color.
- Order the stripes in a segment in increasing order of their width, breaking ties by giving priority to stripes which occur in earlier phases and, within a phase to a stripe with a larger depth. Choose the minimal number of stripes using the above ordering such that the total number of blocks in these stripes is at least M: color all these blocks red. All other blocks are colored black.
- On a request for block b, one of the following actions is taken:
  - If b is present in either the red or the black buffer, service the request and evict block b from the corresponding buffer.
  - If b is not present in either buffer then
    * Begin a batched-I/O as follows: If b is red (black), fetch the next $M/2$ red (respectively black) blocks beginning with b in order of reference, into the red (respectively black) buffer. These $M/2$ blocks are fetched with maximal parallelism.
    * Service the request and evict block b from the buffer.

FIGURE 2. Algorithm THIN

By a similar argument at most 2 batched-I/Os are required to fetch the black blocks of a phase. Since the maximum number of red blocks from a single disk in $phase(k)$ is h_k, then the maximum number of black blocks from a single disk in $phase(k)$ is $d_k - h_k$. Hence the maximum number of I/Os performed in each of the two batched-I/Os is at most $d_k - h_k$. Therefore, the total number of I/Os done by THIN in a segment is $T_{\text{THIN}}(\eta) \leq 2\sum_\eta (d_k - h_k) + 3R$. The desired inequality now follows from the definition of B_{THIN}. □

Algorithm ASP partitions the reference string into segments and schedules each segment independently using THIN. The total number of I/Os done by ASP is therefore $\sum T_{\text{THIN}}(\eta)$, where the sum is taken over all segments η in Σ. From Lemma 3.3 minimizing $\sum T_{\text{THIN}}(\eta)$ translates to maximizing $\sum B_{\text{THIN}}(\eta)$ as $\sum d_k$ over all phases in the reference string is independent of the partitioning.

DEFINITION 3.4. Algorithm ASP partitions the reference string into disjoint segments $\langle \eta_1, \ldots, \eta_n \rangle$ such that $\sum_{j=1}^n B_{\text{THIN}}(\eta_j)$, is maximized over all possible partitions of the reference string. The I/O schedule is then constructed by concatenating the schedules generated by THIN for each segment η_j.

A dynamic programming approach can be used to find the partitioning. Consider a graph where node (i,j), $0 \leq i \leq j < L$, denotes that $phase(j)$ is the last phase of the ith segment. An edge from node $(i-1,k)$ to node (i,j) indicates that the ith segment is from $phase(k+1)$ to $phase(j)$. The weight of this edge, $w_{i,j,k}$, is the benefit of THIN for this segment. Associated with node (i,j) is $b_{i,j}$ where $b_{i,j} = \max_{i<k<j}\{b_{i-1,k} + w_{i,j,k}\}$. The entries $b_{i,L-1}$, give the maximum benefit possible for a partitioning into $i+1$ segments. Hence the maximum benefit is $\max_{0 \leq i < L}\{b_{i,L-1}\}$. A straightforward implementation of the above dynamic programming algorithm has complexity $O(ML^3 + ML\log(ML))$. This can be improved to $O(ML + DL^2 + L^3)$ by working with a concise representation of each phase consisting of D buckets, one for each possible width. We omit the details of the scheme for brevity.

4. Analysis of algorithm ASP

In this section we present tight bounds on the on-line ratio of ASP. Let OPT denote the optimal on-line algorithm with ML-block lookahead. First, note that any I/O schedule can be transformed into another schedule of the same length, or less, in which a block is never evicted before it has been referenced. Hence we shall implicitly assume this property for all the schedules considered in this paper.

DEFINITION 4.1. An I/O is said to be performed in $phase(i)$ if the next block to be referenced is in $phase(i)$. The $start$ of a phase (respectively segment) refers to the first reference of that phase (segment). An $inter\text{-}segment$ block is one, which is fetched in a segment different from the one in which it is referenced. Similarly an $inter\text{-}phase$ block is one which is fetched in a phase different from the one in which it is referenced.

The bounds of ASP are presented in two ranges: $L \leq D^{2/3}$ and $L > D^{2/3}$. In the lower range the on-line ratio is shown to be $\Theta(\sqrt{L})$ in Theorem 4.9. An on-line ratio of $\Theta(D^{1/3})$ is shown in Theorem 4.11 for $L > D^{2/3}$.

THEOREM 4.2. *The on-line ratio of ASP is* $\Theta(\sqrt{L})$ *when* $L \leq D^{2/3}$, *and* $\Theta(D^{1/3})$ *when* $L > D^{2/3}$.

To simplify the analysis, in the subsequent sections we implicitly assume that the length of the reference string is ML. This is justified because of the following observations. Let the reference string $\Sigma = \langle l_1, l_2, \cdots, l_n \rangle$, where each l_i is a substring of Σ with ML requests. Since OPT has ML-block lookahead, during l_i OPT might prefetch a block from l_{i+1}, but never beyond l_{i+1}. In contrast, consider a schedule generated by independently scheduling each l_i optimally. It is easy to show that such a schedule is within a factor of 2 of OPT. Hence the on-line ratio of an algorithm A for arbitrary reference strings, can be computed to within a factor 2 by computing the on-line ratio of A over reference strings of length ML.

4.1. Analysis of ASP for L ≤ D²/³. In the range $L \leq D^{2/3}$ we proceed as follows. We define a weaker form of ASP, called Fixed Segment Prefetching (FSP). FSP *statically* partitions the reference string into fixed-size segments and, like ASP, services each segment independently using THIN. Since the segments used by FSP is just one of the many partitions considered by ASP, the number of I/Os required by ASP is no more than that required by FSP. Hence we bound the on-line ratio of ASP by bounding the on-line ratio of FSP.

DEFINITION 4.3. Algorithm FSP partitions the reference string Σ into $\sqrt{L}$ segments of equal length, $\Sigma = \langle \eta_0, \ldots, \eta_{\sqrt{L}-1} \rangle$. Each η_j thus consists of $\sqrt{L}$ consecutive phases. The I/O schedule for Σ is constructed by concatenating the schedules generated by THIN for each segment η_j.

To bound the on-line ratio of FSP we proceed in two steps. We first define a weaker form of OPT, called OPT*, which performs $O(\sqrt{L})$ times as many I/Os as OPT. We then show that FSP does $O(1)$ times as many I/Os as OPT*. The construction of OPT* is described in Figure 3. OPT* is constructed from OPT by splitting each I/O done by OPT into I/Os performed currently for the present phase, I/Os performed at the start of other phases of the same segment, and I/Os performed at the start of other segments. By assumption, no block fetched by OPT is evicted before it is referenced. Hence we can safely delay an I/O for a block without overflowing the buffer at the new time it is fetched. The total number of segments is $\sqrt{L}$ and each segment has $\sqrt{L}$ phases. Hence, each I/O of OPT is dilated to at most $2\sqrt{L}$ I/Os in OPT*.

LEMMA 4.4. *The number of I/Os done by OPT* to service Σ is within a factor $2\sqrt{L}$ of the number of I/Os done by OPT to service Σ.*

- Partition the reference string into $\sqrt{L}$ segments $\langle \eta_1, \ldots, \eta_{\sqrt{L}} \rangle$, such that each segment has $\sqrt{L}$ contiguous phases. Consider the blocks fetched by OPT in some I/O in $phase(i)$ of segment η_j.
- For blocks belonging to $phase(i)$, an I/O is performed by OPT* at the current time.
- For blocks belonging to $phase(i')$ in η_j, $i \neq i'$, an I/O is performed by OPT* just prior to the start of $phase(i')$.
- For blocks belonging to $\eta_{j'}$, $j \neq j'$, an I/O is performed by OPT* just prior to the start of $\eta_{j'}$.

FIGURE 3. Construction of OPT* from OPT

THEOREM 4.5. *The on-line ratio of FSP is $O(\sqrt{L})$.*

PROOF. The schedule OPT* has the following properties

1. There are no inter-segment blocks.
2. There are at most M inter-phase blocks within any segment, and all of them are fetched at the start of the segment.

By construction, the segments of OPT* match those of FSP. Since FSP uses THIN to schedule each segment the number of I/Os done by FSP and THIN in a segment are equal. In Lemma 4.8 we use property 2 to show that the number of I/Os performed by THIN in any segment is within a constant factor of the number of I/Os performed by OPT* in the same segment. By property 1 the number of I/Os done by OPT* is the sum of the number of I/Os done by it in each segment. Hence the number of I/Os needed by FSP is within $O(1)$ of OPT*. The theorem then follows from Lemma 4.4. □

We next define the notion of *useful blocks*, which is a measure of the value of the blocks that OPT* prefetches at the start of the segment for some phase in that segment. This will be used in comparing the number of I/Os done by OPT* and FSP in a segment.

DEFINITION 4.6. A total of u_i *useful blocks* are said to be prefetched by OPT* for $phase(i)$ in segment η_j, if after the prefetches at the start of η_j, the maximum number of blocks of $phase(i)$ that are yet to be fetched from a single disk is $d_i - u_i$, where d_i is the max-depth of $phase(i)$.

Note that if no useful blocks were fetched by OPT* for $phase(i)$, then OPT* *must* do at least as many I/Os as the max-depth of that phase. Hence the number of useful blocks fetched for a phase is the reduction in the number of I/Os that OPT* needs to do in that phase. In addition if u_i useful blocks are fetched for a phase, the buffer must contain at least as many blocks as the u_i thinnest stripes of that phase. This gives a handle on the buffer space occupied by u_i useful blocks.

CLAIM 4.7. *The maximum number of useful blocks prefetched in a segment equals the maximum number of stripes in that segment such that the total number of blocks in them is at most M.*

LEMMA 4.8. *The number of I/Os done by THIN in segment η_j, $0 \leq j < \sqrt{L}$, is within $O(1)$ of the number of I/Os done by OPT* for η_j.*

PROOF. Let the number of I/Os performed by OPT* at the start of segment η_j, to fetch the useful blocks, be I_{OPT^*}. Let the total number of useful blocks fetched by OPT* for phases in η_j be U_{OPT^*}. Let $H = \sum_{\eta_j} d_i$, be the sum of the max-depths of the phases of η_j. By Definition 4.6, the total number of I/Os done by OPT* to service η_j is

$$(4.1) \qquad T_{\mathrm{OPT}^*} = H - U_{\mathrm{OPT}^*} + I_{\mathrm{OPT}^*}$$

By Claim 4.7 U_{OPT^*} is less than the maximum number of stripes in that segment such that the total number of blocks in them is at most M. From the definition of THIN, red blocks belong to stripes which have the least width and number at least M. Hence U_{OPT^*} is bounded by the total number of stripes containing red blocks.

$$(4.2) \qquad U_{\mathrm{OPT}^*} \leq \sum_{\eta_j} h_i$$

where h_i is the maximum number of red blocks from a single disk in $phase(i)$. By Definition 3.2 if R is the maximum number of red blocks from a single disk in η_j, and the benefit of THIN in the segment is $B_{\mathrm{THIN}}(\eta_j)$,

$$(4.3) \qquad \sum_{\eta_j} h_i = B_{\mathrm{THIN}}(\eta_j) + R$$

Also, in η_j there are at least R (red) blocks required from a single disk. Hence,

$$(4.4) \qquad T_{\mathrm{OPT}^*} \geq R$$

From Lemma 3.3 and Equations 4.1-4.4, the total number of I/Os done by THIN in η_j is at most

$$T_{\mathrm{THIN}}(\eta_j)/3 \ \leq \ T_{\mathrm{OPT}^*} + U_{\mathrm{OPT}^*} - B_{\mathrm{THIN}}(\eta_j) \ \leq \ T_{\mathrm{OPT}^*} + R \ \leq \ 2T_{\mathrm{OPT}^*}$$

$\square$

The on-line ratio of ASP follows from Theorem 4.5 and the fact that the number of I/Os done by ASP is no more than the number of I/Os done by FSP. We can show that the bound is tight by constructing a reference string of length ML such that the bound is achieved. The proof of the lower bound is omitted for brevity.

THEOREM 4.9. *The on-line ratio of ASP with ML-block lookahead is $\Theta(\sqrt{L})$.*

The analysis above may suggest that FSP is comparable in performance to ASP. However with $L > D^{2/3}$, it will be shown in Section 4.2 that the on-line ratio of ASP is $\Theta(D^{1/3})$, while that of FSP can be shown to be $\Omega(\sqrt{L})$, for $D^{2/3} < L \leq D$. Even in the range $L \leq D^{2/3}$ there are reference strings for which ASP performs $\Omega(\sqrt{L})$ times less I/Os than FSP. The details of these constructions are omitted in this abstract.

4.2. Analysis of ASP for L > D$^{2/3}$. In this range of L, we shall bound the on-line ratio of ASP by comparing it to the algorithm RBP presented in [**KV98**]. We summarize the algorithm and its analysis in the current context in Figure 4. The main difference between ASP and RBP is the scheme used by RBP to color blocks. RBP uses a fixed threshold width to decide which blocks to color red, while THIN adapts to the structure of the reference string.

Partition the I/O buffer into two parts, *red buffer* and *black buffer*, each of size $M/2$. Each part will only be used to hold blocks of that color.

The blocks of all stripes in the reference string with a width smaller than $D^{1/3}$ are colored red. All other blocks are colored black. Partition the reference string into segments of maximal length such that the total number of red blocks in each segment is at most M.

On a request for block B, one of the following actions is taken:

- If B is present in either the red or the black buffer, service the request and evict block B from the corresponding buffer.
- If B is not present in either buffer then
    - Begin a batched-I/O operation as follows: (a) If B is red, fetch up to the next $M/2$ red blocks in the segment beginning with B in order of reference, into the red buffer. (b) If B is black, fetch up to the next $M/2$ black blocks in the phase beginning with B in order of reference, into the black buffer. These blocks are fetched with maximal parallelism.
    - Service the request and evict block B from the buffer.

FIGURE 4. Algorithm RBP

THEOREM 4.10 ([**KV98**]). *The ratio of the number of I/Os done by RBP to that done by OPT is* $\Theta(D^{1/3})$, *for* $L \geq D^{2/3}$.

To show that the on-line ratio of ASP is $O(D^{1/3})$ it is sufficient to show that the number of I/Os done by ASP is within a factor of $O(1)$ of the number of I/Os done by RBP. This follows from the way RBP partitions the reference string: At most M blocks are prefetched for phases in a segment, and in any I/O done in a segment no block is prefetched for a different segment. By a proof similar to that used in Lemma 4.8 we can show that in any segment the number of I/Os done by THIN is within $\Theta(1)$ of the number of I/Os done by RBP. As ASP partitions the reference string into segments to minimize the total cost of THIN, the number of I/Os done by ASP is within $\Theta(1)$ of the number of I/Os done by RBP. Like in the previous range of L, the bound above is tight. The lower bound can be shown by constructing a reference string for which the equality holds.

THEOREM 4.11. *The on-line ratio of ASP is* $\Theta(D^{1/3})$.

We can show that there are reference strings for which ASP performs $\Omega(D^{1/3})$ times fewer I/Os than RBP. For the range $L \leq D^{2/3}$, it can be shown that RBP performs significantly more I/Os than ASP. Even if the best threshold width is chosen for a given L, we can show that there are reference strings of length ML, $L < D^{2/3}$, for which the number of I/Os done by RBP is at least $\Omega(C)$ times the number of I/Os done by OPT, where $C = \min\{L, (DL)^{1/5}\}$.

5. Conclusions

In this paper we addressed the problem of on-line scheduling of read-once reference strings in the parallel disk model using bounded ML-block lookahead. We presented a novel algorithm ASP and compared its performance with the best online algorithm with the same amount of lookahead, using the on-line ratio as the performance measure.

ASP partitions the reference string into segments of contiguous phases, and performs all prefetching within a segment only. Two auxiliary algorithms RBP and FSP, which also schedule I/Os on a segment-by-segment basis, were defined for bounding the performance of ASP. RBP defines segments based on a predetermined threshold width, while FSP defines segments of fixed length. In contrast, ASP adaptively chooses its segments to minimize the total cost of servicing them and is thereby able to match the best performance of both FSP and RBP. We showed that ASP has an online ratio of $\Theta(\sqrt{L})$ when $L \leq D^{2/3}$ and $\Theta(D^{1/3})$ when $L > D^{2/3}$.

References

[AGL98]   S. Albers, N. Garg, and S. Leonardi, *Minimizing Stall Time in Single and Parallel Disk Systems*, Proc. of STOC'98, 1998, pp. 454–462.

[Bel66]   L. A. Belady, *A Study of Replacement Algorithms for a Virtual Storage Computer*, IBM Systems Journal **5** (1966), no. 2, 78–101.

[BGV95]   R. D. Barve, E. F. Grove, and J. S. Vitter, *Application-Controlled Paging for a Shared Cache*, Proc. of FOCS'95, 1995, pp. 204–213.

[BGV96]   R. D. Barve, E. F. Grove, and J. S. Vitter, *Simple Randomized Mergesort on Parallel Disks*, Parallel Computing **23** (1996), no. 4, 601–631.

[BKVV97]   R. D. Barve, M. Kallahalla, P. J. Varman, and J. S. Vitter, *Competitive Parallel Disk Prefetching and Buffer Management*, Proc. of IOPADS'97, 1997, pp. 47–56.

[CFKL95]   P. Cao, E. W. Felten, A. R. Karlin, and K. Li, *A Study of Integrated Prefetching and Caching Strategies*, Proceedings of the Joint International Conference on Measurement and Modeling of Computer Systems, ACM, May 1995, pp. 188–197.

[CLG+94]   P. M. Chen, E. K. Lee, G. A. Gibson, R. H. Katz, and D. A. Patterson, *RAID: High Performance Reliable Secondary Storage*, ACM Computing Surveys **26** (1994), no. 2, 145–185.

[KK96]   T. Kimbrel and A. R. Karlin, *Near-Optimal Parallel Prefetching and Caching*, Proc. of FOCS'96, 1996, pp. 540–549.

[KP94]   E. Koustoupias and C. H. Papadimitriou, *Beyond Competitive Analysis*, Proc. of STOC'94, 1994, pp. 394–400.

[KV98]   M. Kallahalla and P. J. Varman, *Red-Black Prefetching: An Approximation Algorithm for Parallel Disk Scheduling*, Proc. of FST&TCS'98, 1998, to appear.

[PSV94]   V. S. Pai, A. A. Schäffer, and P. J. Varman, *Markov Analysis of Multiple-Disk Prefetching Strategies for External Merging*, Theoretical Computer Science **128** (1994), no. 1–2, 211–239.

[ST85]   D. D. Sleator and R. E. Tarjan, *Amortized Efficiency of List Update and Paging Rules*, Communications of the ACM **28** (1985), no. 2, 202–208.

[Var98]   P. J. Varman, *Randomized Parallel Prefetching and Buffer Management*, Parallel and Distributed Computing (1998), 363–372.

[VS94]   J. S. Vitter and E. A. M. Shriver, *Optimal Algorithms for Parallel Memory, I: Two-Level Memories*, Algorithmica **12** (1994), no. 2–3, 110–147.

[VV96]   P. J. Varman and R. M. Verma, *Tight Bounds for Prefetching and Buffer Management Algorithms for Parallel I/O Systems*, Proc. of FST&TCS'96, vol. 16, 1996, pp. 454–462.

DEPT. OF ECE, RICE UNIVERSITY, HOUSTON TEXAS 77251
E-mail address: {kalla,pjv}@rice.edu

DIMACS Series in Discrete Mathematics
and Theoretical Computer Science
Volume **50**, 1999

Efficient Schemes for Distributing Data on Parallel Memory Systems

Sajal K. Das and M. Cristina Pinotti

ABSTRACT. A *perfect* memory system should supply immediately any datum requested by the CPU. However, since such an ideal memory is not practically feasible, this goal is commonly approximated by organizing the memory hierarchically and using a parallel memory system consisting of several modules (e.g. main memory banks and/or I/O disks) at each level of the hierarchy. In this paper, we develop efficient data distribution schemes as a technique for improving the performance of a parallel memory system. The objective is to fetch specific subsets of nodes (called *templates*) of a given data structure called a *host*, in as few memory cycles (ideally one) as possible. If a template to be retrieved is stored in a single memory module, its nodes can only be fetched sequentially, one after the other, thus incurring a high access latency. This is independent of the number of memory modules, and hence the potential bandwidth, available in the parallel memory system. Therefore, our goal is to evenly distribute the nodes of a host data structure among the memory modules in such a way that the number of nodes of a given template assigned to the same module is minimized. Interestingly, to retrieve a template in just one cycle (i.e., without conflicts), the lower bound on the number of memory modules required is often greater than the template size (in terms of number of nodes).

After defining the data access problem and the optimality criteria, we propose efficient schemes for conflict-free access to path templates and subtrees in k-ary trees and binomial trees. The proposed assignments are based on simple node-indexing, modular arithmetic, or graph-theoretic techniques. Our assignments are optimal in terms of the number of memory modules required and also balance the memory load among the modules.

1. Introduction

A *perfect* memory system should supply almost immediately any datum requested by the CPU. However, this concept of ideal memory is not practically feasible because of the contradictory requirements of memory capacity, speed and cost, among others. For example, a faster memory is expensive and hence has a limited capacity, while a cheaper memory is large but slow. In practice, a perfect memory system is approximated by organizing the memory hierarchically, that is placing smaller and faster memories in front of larger, slower, and cheaper memories, and then using parallel memory systems consisting of several *modules* (e.g. main memory banks and/or I/O disks) at each level of the hierarchy.

1991 *Mathematics Subject Classification*. Primary 68Q22, 68P05, 05C05, 68R05, 68M10, 68M99.

This work is supported by Texas Advanced Research Program grant TARP-003594-013.

The performance of a memory can be measured in terms of latency and bandwidth. The *access latency* of a memory request is the elapsed time between the request and the fetching of the data, and can be improved by a hierarchical memory organization. On the other hand, the memory *bandwidth* is the rate at which data are fetched, which can be improved by using a parallel memory system. In fact, when a module is accessed, it remains busy for a period of time, during which the processors may not make any other accesses to that module. By increasing the number of modules and with the help of efficient mechanisms for data distribution, the probability that the processors issue two conflicting requests to the same module can be reduced.

Let us consider parallelizing compilers as an application. The data dependency analysis and the consequent data distribution schemes have been topics of significant research for automatic detection of implicit parallelism in programs written in conventional sequential languages, such as Fortran and C. A derived main memory scheme, called *memory interleaving*, maps consecutive addresses into different memory modules. However, this scheme guarantees conflict-free access only when the strides associated with successive references is relatively prime to the number of modules [L75]. Moreover, most of the proposed work in parallelizing compilers focus only on the loop statements with references to arrays and matrices. For recent results, refer to [PD96, GAG96, PS96, MA97].

Relatively less well understood and investigated is the access to arbitrary data patterns represented by graphs, say, which model many real-life applications. Designing efficient distribution schemes to minimize the number of conflicts for arbitrary sets of data patterns is a very important and challenging problem. This is known as the *data access problem* which is formally defined in the next section.

2. Data Access Problem and Optimality Criteria

Let G be a data structure stored, without replications, in a distributed fashion into M memory modules. The processor(s) read(s) into or write(s) from the memory modules through a shared bus or an interconnection network topology depending on whether it is a shared-memory multiprocessor architecture or a distributed-memory multicomputer system.

A *conflict* occurs when several requests by different processors attempt to access the same memory module at the same time to retrieve different nodes of the data structure.

A *template type* (or briefly, a *template*), T, is defined as a subgraph of the data structure G that is required to be fetched concurrently. The graph G may have several instances (overlapping allowed) of the template type T. We say that a template has k conflicts if at most $k + 1$ nodes of the template belong to the same memory module.

The *data access problem* is to map the nodes of G onto M memory modules such that all instances I of T within G can be accessed with the minimum number of conflicts. This problem can be viewed as a restricted graph *coloring* as follows. The distribution of nodes into memory modules is equivalent to coloring the nodes with the color-set $\{0, 1, 2, \ldots, M - 1\}$ in such a way that all the nodes within a template instance have distinct colors. In fact, this can be formulated as a hypergraph coloring problem [KJD97] and hence NP-hard. Therefore, a first step to study this problem and get a good handle on it is to restrict the data structure graph G to known and more structured topologies which will lead to faster solutions. This motivates our work.

Earlier, Das and Sarkar [**DS94**] formulated the conflict-free data access problem from two different angles: (i) given a processor architecture, the number **M** of memory modules and an access distribution of template instances, find out a mapping or data distribution scheme which minimizes the average number of conflicts; and (ii) develop a mapping svheme under which all the template instances of a given type can be accessed conflict-free and also derive the required number **M** of memory modules. Let us formally define the problem as follows.

Let $U(\mathtt{G},\mathtt{T},\mathtt{M})$ denote a mapping of **G** for accessing a template type **T** employing **M** memory modules, and let $\mathtt{I} \subset \mathtt{G}$ be an instance of **T**. We define the *cost* of the mapping $U(\mathtt{G},\mathtt{T},\mathtt{M})$ as the *maximum number of conflicts* incurred by $U(\mathtt{G},\mathtt{T},\mathtt{M})$ over the set of all possible instances $I \subset \mathtt{G}$.

An **M**-*optimal* mapping $U^*(\mathtt{G},\mathtt{T},\mathtt{M})$ is defined as a mapping with the *minimum cost* over all possible mappings $U(\mathtt{G},\mathtt{T},\mathtt{M})$. Necessarily, for a template of size S, an **M**-*optimal* mapping has a minimum cost of $\lceil S/\mathtt{M} \rceil - 1$. A *conflict-free* mapping $U^0(\mathtt{G},\mathtt{T},\mathtt{M})$ is one that results in zero cost. An *optimal* mapping $U^*(\mathtt{G},\mathtt{T})$ is a conflict-free mapping that requires the *minimum* number of memory modules over all possible conflict-free mappings. Since at least S memory modules are necessary to guarantee a conflict-free mapping for a template of size S, we say that the template size S is the *absolute minimum* number of modules required for its conflict-free access. However, as will be seen later, S modules are often not sufficient even for simple template types. Hence, if there exists a conflict-free mapping $U^0(\mathtt{G},\mathtt{T},\mathtt{S})$ for a template of size S, it holds $U^*(\mathtt{G},\mathtt{T}) = U^0(\mathtt{G},\mathtt{T},\mathtt{S})$.

A data mapping is also desired to satisfy the following properties.

(i) *balanced*: the load (i.e., the number of data items stored) on each memory module is almost the same;

(ii) *direct*: an analytical function determines the memory module to which a given item is assigned,

(iii) *scalable*: the mapping preserves its properties without reassignment of the existing nodes when the template or host size changes.

The definition of the data access problem is not affected by the fact that the modules of the parallel memory system can be either main memory banks or I/O disks. In fact, for our abstraction purpose, these two kinds of elementary memory modules are distinguished only for the grain used in retrieving the data. For instance, the fine grain applies to the main memory banks while the coarse grain to the I/O disks. Thus, we implicitly assume that each single node of the host data structure consists of as many records as can be stored into a disk page when an I/O system is considered, and a single record otherwise.

Note also that we do not take into account the ratio between the main memory capacity and the disk page size since we are not interested in any specific processor configuration. We assume that the CPU is able to work on the entire template data, and this is fair since the CPU under consideration could be a multiprocessor system. Efficient data distribution schemes are very important for both shared-memory and distributed-memory multiprocessors whose local and/or remote memories constitute a natural parallel memory system. The peculiarity of the data access problem in networks of processors like a hypercube is addressed in [**DPS96, DP98**], where the data locality is exploited to reduce the communication overhead.

2.1. Our Contributions. Let us highlight the results presented in this paper which addresses the data access problem for tree-like structures.

We present optimal, direct, balanced and scalable mapping schemes for accessing path templates and t-ary subtrees in k-ary host trees, where $2 \leq t \leq k$. In addition, given a certain number N of memory modules, we discuss an N-optimal mapping for k-ary subtrees. We also investigate the access of subtrees in binomial trees which nicely illustrates the role played by the overlapping of the template instances in determining the minimum number of memory modules for an optimal mapping. For both k-ary and binomial trees, we define the notion of *oriented* subtree templates that can be optimally accessed without conflicts, employing the absolute minimum number of memory modules (the same as the template size).

3. Previous Work

Similar to the arrays and matrices, trees are perhaps the other most frequently used data structures with numerous applications, particularly in non-numeric problems. The most commonly used templates of trees are subtrees of a given height, nodes at any level, paths from the root to the leaves, or any combination of such elementary templates. Previous results on the conflict-free subtree access are summarized in Table 1.

The recursively linear scheme due to Creutzburg et al. [C86] is somewhat elegant but it uses a non-optimal number of memory modules. Wijshoff [Wij89] provided a general framework, although the structural properties of a complete k-ary tree were not exploited to reach a simpler solution. The interleaved storage scheme suggested in [B92] may cause more conflicts than necessary for a given number of memory modules. Das and Sarkar [DS94] proposed a simple solution to the conflict-free subtree access in k-ary trees and the nodes at any level. One algorithm is recursive and the other non-recursive. This indirect approach uses an optimal number of memory modules.

For accessing path templates in trees, we have recently proposed two optimal[1] and load balanced assignments [DSP95, PDS97]. For the sake of completeness, let us review the main ideas behind these path access schemes.

TABLE 1. **Conflict-free subtree access**

# Modules	Templates Accessed	Solution Scheme	Comments/References
$2^{h+1} - 1$	Complete binary subtree of height h	Regular skewing	Difficult to implement [Wij85]
$hk^h + k^{h-1}$	Complete k-ary subtree of height h	Recursively linear	Inefficient use of modules [C86]
k^h	Complete k-ary subtree of height h and	Interleaved storage	As many as h conflicts [B92]
$\frac{k^{h+1}-1}{k-1}$	Complete k-ary subtree of height h	Not Direct	Optimal number of modules [DS94]
$h + 1$	Path of length $h + 1$ in k-ary tree	Balanced and Direct	Optimal [PDS97]

---

[1]As presented in [PDS97], the second algorithm uses one extra memory module, but can be made optimal at the expense of a more cumbersome assignment/retrieval function.

3.1. Path Templates in Trees. Let P_h denote a path template of h nodes in a k-ary tree host, H_w^k, of height w, where $k \geq 2$ and $w \geq h$. The recursive algorithm is balanced but not direct. For simplicity, we outline the case for binary trees (i.e., $k = 2$).

Given a coloring of a complete binary tree H_{h-1}^2, the host tree H_h^2 is recursively colored by decomposing it into three parts – the root of H_h^2, the left subtree LH_{h-1}^2, and the right subtree RH_{h-1}^2. The mapping of LH_{h-1}^2 is the same as that of H_{h-1}^2 using the color-set $\{0, 1, \ldots, h-1\}$, and the root of H_h^2 is assigned a new color, h. Now, to color RH_{h-1}^2, we first count how many times each color has been used in LH_{h-1}^2, and assign the colors $\{0, 1, \ldots, h-1\}$ to the nodes of RH_{h-1}^2 with the help of a reverse mapping such that the least frequently used color in LH_{h-1}^2 is replaced with the most frequently used color, the second least frequently used color with the second most frequently used color, and so on. See Figure 1 for an illustration. By induction, it can be proved that each path can be accessed without conflicts. Also, it is easy to show that the memory load is balanced. Indeed, the ratio between the number of nodes stored, respectively, in the most and the least frequently used memory modules (i.e., colors) is at most 2. This construction can be extended to binary trees of arbitrary height $w \geq h$.

LEMMA 3.1. [**DSP95**] *There exists a conflict-free and optimal mapping*

$$U^*(H_w^2, P_h) = U^0(H_w^2, P_h, h).$$

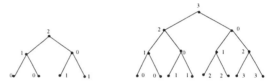

FIGURE 1.   a) Coloring of H_2^2. b) Recursive coloring of H_3^2.

Our non-recursive algorithm for path access is also balanced but direct. It partitions the host tree H_h^k (except the root) into k parts horizontally as well as vertically. This divides the tree into k^2 cells, organized as a $k \times k$ matrix, LS. Then using the colors $\{0, \ldots, h-1\}$, we form k groups G_i, for $0 \leq i \leq k-1$, each of $\frac{h}{k}$ colors. Roughly speaking, every entry $E(r, s)$ of LS, where $1 \leq r, s \leq k$, is assigned one of the groups of colors G_i. Precisely, if the first row of the matrix is assigned the groups of colors in the order $\{G_0, G_1, \ldots, G_{k-1}\}$, every other row is formed by shifting the previous row cyclically one position to the right. Subsequently and in some way recursively, the nodes in each entry $E(r, s)$ of the matrix LS, for $1 \leq r, s \leq k$, are partitioned in a matrix of size $\frac{h}{k} \times \frac{h}{k}$ which is colored as a Latin square using the group of colors previously assigned to $E(r, s)$. Figure 2 sketches the algorithm. It has been proved in [**PDS97**] that this mapping is conflict-free for accessing paths P_h of h nodes. It uses $\mathtt{M} = h + 1$ memory modules for coloring a tree of height h, thus using one extra memory module with respect to optimality. That is,

LEMMA 3.2. [**PDS97**] *The non-recursive algorithm leads to a conflict-free mapping* $U^0(H_w^k, P_h, h+1)$.

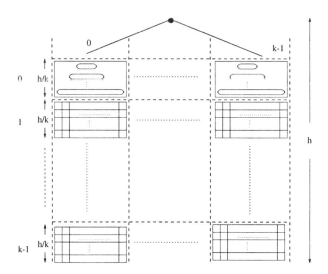

FIGURE 2. Sketch of the non-recursive algorithm for path access in k-ary trees.

This scheme is perfectly load balanced and also direct because the memory module to which a node is assigned can be derived analytically [**DSP95, PDS97**]. With careful attention to preserve the *locality* of data references and minimize the communication overhead, this mapping also led to the first cost-optimal implementation of heaps on hypercube architectures [**DPS96, DP98**].

4. t-ary Subtree Templates in k-ary Trees

In this section, we present an optimal, balanced, direct and scalable mapping to access complete subtrees, T_v^t, of arity t ($2 \leq t \leq k$) and height v ($0 \leq v \leq w$) within complete k-ary trees H_w^k of height $w \geq v$.

THEOREM 4.1. *There exists an optimal mapping*

$$U^*(H_w^k, T_v^t) = U^0(H_w^k, T_v^t, \frac{k^{v+1} - 1}{k - 1}), \text{ for } 2 \leq t \leq k.$$

PROOF. First we prove that a conflict-free access to any occurrence of the template T_v^t in the host H_w^k, where $2 \leq t \leq k$ and $0 \leq v \leq w$, requires $\mathtt{M} \geq \dfrac{k^{v+1} - 1}{k - 1}$ memory modules.

Two nodes of the host can be assigned to the same module if and only if their lowest common ancestor is at a distance $\geq v + 1$ from at least one of them. Therefore, all the nodes of the host at distance $\leq v$ must be assigned to different memory modules, independent of the arity t of the templates considered. Thus, depending only on the height v of the template and the arity k of the host, the minimum number of modules required to achieve a conflict-free mapping is given by $\mathtt{M} \geq \dfrac{k^{v+1} - 1}{k - 1}$. Since $\mathtt{M}$ is independent of the arity of the subtree template, if such a mapping $U^0(H_w^k, T_v^t, \frac{k^{v+1}-1}{k-1})$ exists, it is also optimal for any t, for $2 \leq t \leq k$.

Figure 3 depicts an optimal memory assignment $U^0(H_3^3, T_2^t, 13)$ by the following procedure, where $2 \leq t \leq 3$. In this procedure, $M[(i, j)]$ denotes the memory

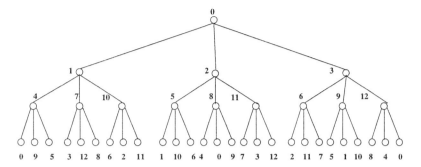

FIGURE 3.   Mapping of H_3^3 for conflict-free access to any subtree T_2^t of height 2, for $t \leq 3$. For each node, the label indicates the memory module to which it is assigned.

module assigned to the jth node at level i of the host, counting the nodes at each level from left to right (from 0) and assuming that the root of the host is at level 0.

Procedure $U^0(H_w^k, T_v^k, \frac{k^{v+1}-1}{k-1})$

- $M[(0,0)] = 0$

- $M[(i,j)] = \left(\sum_{r=0}^{i-1} (j_r + 1) k^{i-1-r} \right) \ \text{mod} \ \frac{k^{v+1} - 1}{k - 1},$

  for all $(i,j) \in H_w^k$ where $j = \sum_{r=0}^{i-1} j_r k^r$ and $i \geq 1$.

We argue that this mapping is conflict-free. Counting from left to right, let us label by s the edge from a node to its sth child, where $0 \leq s \leq k - 1$. Thus, letting $(j_i, j_{i-1}, \ldots, j_0)$ be the k-weighted representation of the index j of the node (i,j) at level i, the digit j_r implies that the edge labeled j_r at level r belongs to the path from the root to the node (i,j) itself. In other words, the path from the root to (i,j) is uniquely identified by the k-weighted representation $(j_i, j_{i-1}, \ldots, j_0)$.

Similarly, the path $A_{(m,n)}(i,j)$ from the host node (i,j) and its ancestor (m,n) is identified by the $i-m$ rightmost digits of the k-weighted representation of j. That is, $A_{(m,n)}(i,j) = j \ \text{mod} \ k^{i-m}$. The path from the root to the ancestor (m,n) is given by the leftmost m digits of the representation of j, that is $\lfloor \frac{j}{k^{i-m}} \rfloor$. Therefore, the set of k-weighted representations of j for all the nodes (i,j) belonging to the instance I of the template T_v^k rooted at node (m,n) yields the following properties: (a) the leftmost m digits of all the k-weighted representations in the set are identical, and (b) the digits $A_{(m,n)}(i,j)$ span all the configurations in the set $\{0, \ldots, (k^{v+1} - 1)/(k-1)\}$.

Now, back to the assignment function used in the procedure $U^0(H_w^k, T_v^t, \frac{k^{v+1}-1}{k-1})$, the node $(i,j) \in$ I is assigned to the memory module

$$M[(i,j)] = \left(\sum_{r=0}^{i-m-1} (j_r + 1)k^{i-1-r} + \sum_{r=i-m}^{i-1} (j_r + 1)k^{i-1-r} \right) \bmod \mathtt{M}$$

$$- \left(\sum_{r=0}^{i-m-1} (J_r + 1)k^{i-1-r} + M[(m,n)] \right) \bmod \mathtt{M}$$

$$= \left(M[(m,n)] + k^m \sum_{r=0}^{i-m-1} (j_r + 1)k^{i-1-r-m} \right) \bmod \mathtt{M}$$

$$= \left(M[(m,n)] + k^m A_{(m,n)}(i,j) \right) \bmod \mathtt{M}.$$

Recall that $A_{(m,n)}(i,j)$ assumes all values in $[0,\mathtt{M})$ in correspondence with all the nodes $(i,j) \in \mathtt{I}$. Observing that k and $\mathtt{M}$ are coprime (i.e., there exists the multiplicative inverse of $k^m \bmod \mathtt{M}$), then for any $0 \le x \le \mathtt{M}$, there exists $(i,j) \in \mathtt{I}$ such that $M[(i,j)] = x$. Therefore, the proposed mapping is conflict-free.

The mapping $U^0(H_w^k, T_v^k, \mathtt{M})$ is also optimal for the family of templates T_t^v, for any $2 \le t \le k$ and $0 \le v \le w$. $\square$

By definition, the mapping presented in Theorem 4.1 is also direct. Observing that the host H_w^k can be partitioned into disjoint instances of the template T_v^t and that exactly one node of each instance is assigned to one memory module, the proposed mapping is balanced as well.

With respect to scalability, this mapping preserves its properties for any host of larger height. If the same mapping is applied to access subtrees of size S of height larger than v, the mapping in Theorem 4.1 will be $\mathtt{M}$-optimal resulting in at most $\lceil \frac{S}{M} \rceil - 1$ conflicts.

Now, fixing $\mathtt{N} = \frac{t^{v+1}-1}{t-1}$ where $t < k$, let us design an $\mathtt{N}$-optimal mapping $U^*(H_w^k, T_v^k, \mathtt{N})$. According to the proof of Theorem 4.1, if the logical addresses of all nodes (i,j) of the instances $\mathtt{I}$ span all possible t-weighted representations corresponding to the integers in $[0, \mathtt{N})$, we can access such instances with the minimum number of conflicts. This can be obtained by forcing every digit of the k-weighted representation of j to belong to the interval $[0, t-1]$ by a "mod t" operation as follows.

Procedure $U^*(H_w^k, T_v^k, \frac{t^{v+1}-1}{t-1})$ **where** $t < k$

- $M[(0,0)] = 0$

- $M[(i,j)] = \sum_{r=0}^{i-1} [(j_r \bmod t) + 1]t^{i-1-r} \bmod \frac{t^{v+1} - 1}{t - 1}$,

  for all $(i,j) \in H_w^k$, with $j = \sum_{r=0}^{i-1} j_r k^r$ and $i \ge 1$.

This mapping is $\mathtt{N}$-optimal since accessing any subtree T_v^k of size S yields $\lceil S/\mathtt{N} \rceil - 1$ conflicts. Recalling that at least $\mathtt{M}$ modules are necessary to achieve an optimal mapping for the family of templates T_v^t, for $2 \le t \le k$, this mapping is also $\mathtt{N}$-optimal for any template belonging to this family.

Employing the absolute minimum number of memory modules (i.e., the template size), the same mapping can optimally access a special family of templates which we call the *oriented templates*. Labeling the children of a host node from 0 to $k - 1$ and assigning the index $i \bmod t$ to the ith child, an *oriented template* O_v^t is a t-ary subtree of height v such that for each node, the indices of its children assume all the values in the closed interval $[0, \dots, t-1]$. Again, the optimality can be proved along the lines of the proof of Theorem 4.1.

5. Subtree Templates in Binomial Trees

In this section, we study mappings of binomial subtrees, which offer a concrete example of the role played by the overlapping of the template instances in determining the minimum number of modules required for an optimal mapping.

A *binomial tree* B_n of order (or, height) n is an ordered tree, rooted at the node r and defined recursively as follows: (i) B_0 has a single node, (ii) B_n consists of two binomial trees, B_{n-1}, linked together such that the root of one is the leftmost child of the root of the other. B_n has n levels with the root at level 0, and $\frac{n!}{i!(n-i)!}$ nodes at level $i \geq 0$. The root r has degree $\delta(r) = n$; its ith (from right to left) child is denoted as r_i which is also the root of a binomial subtree termed $B_i^{r_i}$ of order $\delta(r_i) = i$, where $0 \leq i \leq n - 1$. Here is the first result.

THEOREM 5.1. *Let* $\mathbf{M} = \displaystyle\sum_{i=0}^{t-1} \frac{n!}{i!(n-i)!} + \frac{(n-1)!}{(t-1)!((n-1)-(t-1))!}$. *For fixed* $t \leq n$, *there exists an optimal mapping* $U^*(B_n, B_t) = U^0(B_n, B_t, \mathbf{M})$.

PROOF. First, in order to compute $\mathbf{M}$, the minimum number of memory modules required for conflict-free access to any instance of a binomial subtree $B_t \subset B_n$, for some $t \leq n$, we show the following. Given two nodes x and y in the uppermost t levels of the host B_n rooted at r, we can extract a subtree B_t^r of order t and also rooted at r, that contains both x and y.

There are no restrictions on the choice of nodes x and y in the sense that they can belong to the same path or one of them can be the root r itself. Let r_x and r_y (not necessarily distinct) be the children of r on the paths $x \to r$ and $y \to r$, respectively. Clearly, $\delta(r_x) \geq l(x) - 1$ and $\delta(r_y) \geq l(y) - 1$, where $l(x)$ and $l(y)$ respectively stand for the levels of x and y in B_n. W.l.o.g., suppose $\delta(r_y) \geq \delta(r_x)$. Then, the subtree B_t^r containing nodes x and y (not necessarily distinct) is recursively constructed as follows:

- $t - 1 \geq \delta(r_x) \geq \delta(r_y) \geq 0$ (terminal case):
  B_t^r consists of r and the subtrees rooted at its rightmost t children;
- $\delta(r_y) > t - 1 \geq \delta(r_x) \geq 0$:
  B_t^r consists of r, the subtrees rooted at its rightmost $t - 1$ children, and the recursively built template $B_{t-1}^{r_y}$ containing node y;
- $\delta(r_y) \geq \delta(r_x) > t - 1$:
  B_t^r consists of r, the subtrees rooted at its $t - 2$ rightmost children, and the templates $B_{t-2}^{r_x}$ and $B_{t-1}^{r_y}$ containing nodes x and y respectively.

Therefore, all the $\mathbf{M} = \displaystyle\sum_{i=0}^{t-1} \frac{n!}{i!(n-i)!} + \frac{(n-1)!}{(t-1)!((n-1)-(t-1))!}$ nodes of the uppermost t levels of the host tree B_n must be assigned to distinct memory modules.

Procedure $U^0(B_n, B_t, \mathtt{M})$.

- **Step 1**
  Assign a new memory module to each node of the uppermost t levels of of the host binomial tree, B_n^{r}.
- **Step 2**
  Consider the nodes at level $t-1$ of the tree $B_{n-1}^{r_{n-1}}$. Assign a new module to each of these nodes.
- **Step 3**
  For $0 \le i \le n-2$, **do**
  Assign the nodes at level $t-1$ of the tree $B_i^{r_i}$ to the same memory modules to which are assigned the leftmost $\frac{i!}{(t-1)!(i-(t-1))!}$ nodes in Step 2.
- **Step 4**
  For each level l, $2 \le l \le n-t+1$, **do**
  For each node (l, i), $0 \le i \le \frac{n!}{l!(n-l)!} - 1$, **do**
  Assign the nodes at level t of the binomial tree rooted at (l, i), to the same set of modules as are assigned to the uppermost t levels of the tree rooted at $(l-2, p(p(i)))$ but not to the uppermost $t-1$ levels of the tree rooted at the leftmost child $(l-1, p(i))$ of node $(l-2, p(p(i)))$. Refer to Figure 4 for clarification.

Next, we devise the mapping procedure $U^0(B_n, B_t, \mathtt{M})$ that uses $\mathtt{M}$ memory modules for conflict-free access to the subtree $B_t \subset B_n$, where $t \le n$. From now on, let $(l-1, p(i))$ denote the parent of the host node (l, i) in the host.

For a given node (l, i), Figure 4 depicts the subtrees mentioned in Step 4 of the above procedure. All the $\mathtt{M} = \sum_{i=0}^{t-1} \frac{n!}{i!(n-i)!} + \frac{(n-1)!}{(t-1)!((n-1)-(t-1))!}$ memory modules are assigned the first time in Steps 1-2 and then reused in Steps 3-4.

Finally, since the lowest common ancestor of two nodes x and y assigned to the same memory module by the procedure $U^0(B_n, B_t, \mathtt{M})$ is at a distance $\ge t+1$ from at least one of x and y, the proposed mapping is conflict-free. Therefore, recalling that $\mathtt{M}$ is the minimum number of memory modules required to achieve a conflict-free mapping, we have proved that $U^*(B_n, B_t) = U^0(B_n, B_t, \mathtt{M})$. $\qquad\square$

Although this assignment is optimal, it is not direct and hence it is not easy to compute the load on each module. Moreover, the procedure requires a very large number of memory modules, which is independent of the template size. Therefore, we characterize further the templates of binomial trees with the notion of what we call *orientation*.

5.1. Oriented Templates. Let us define an *oriented tuple* $\Delta = (i_{t-1}, \ldots, i_0)$ from any t integers in the range $[0, n-1]$ such that

(1) $i_j > i_{j-1}$, for $1 \le j \le t-1$, and
(2) $\forall z \in [0, \ldots, t-1]$, $\exists ! \, i_j : (i_j \mod t) = z$, where $0 \le j \le t-1$.

Let $\Delta_j = (i_{j-1}, \ldots, i_0)$ be the rightmost j indices in Δ. Then, $\Delta_t = \Delta$. In other words, an oriented tuple Δ consists of simply t integers in the decreasing order which cover the interval $[0, t-1]$ when mapped by a suitable function.

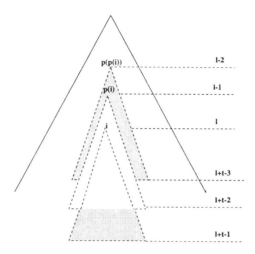

FIGURE 4.   Illustrating Step 4 of the Procedure $U^0(B_n, B_t, \mathbf{M})$.

An *oriented template*, O_t^Δ, of order t and guided by the oriented tuple Δ is an instance of the binomial subtree $B_n \subset B_n$, consisting of only the children of the host node enumerated in Δ.

Let ξ_{i_j} be the i_jth (from left to right) child of a node ξ. Then the instance of the oriented template $O_t^{\xi,\Delta} \subset B_n$, rooted at ξ and based on the oriented tuple Δ, is constructed as follows:

1. The t children of the root ξ in $O_t^{\xi,\Delta}$ are $\{\xi_{i_j} | i_j \in \Delta\}$. Thus, $\delta(\xi) \geq i_{t-1}$.
2. Recursively, $\{\eta_{i_s} | i_s \in \Delta_j\}$ are the j children of a node η and belonging to $O_t^{\xi,\Delta}$ such that $\delta(\eta) = j$.

In Figure 5, the oriented template $O_3^{r,\Delta}$ is drawn in bold lines for $\Delta = (3, 2, 1)$.

THEOREM 5.2.   *There exists an optimal mapping* $U^*(B_n, O_t^\Delta) = U^0(B_n, O_t^\Delta, 2^t)$ *for any oriented t-tuple,* Δ.

PROOF.   With a goal of describing the optimal mapping, let us review a known scheme of assigning *binary addresses* to the nodes of a binomial tree [**JH89**].

Assigning the address $(0, 0, \ldots, 0)$ to the root, let p be such that $c_p = 1$ and $c_m = 0 \; \forall m \in \{p+1, p+2, \ldots, n-1\}$, and let $p = -1$ if $c = 0$. Then, the children of the nodes $i = (0, 0, \ldots, 0, c_i = 1, b_{i-1}, \ldots, b_0)$, where $b_j = \{0, 1\}, 0 \leq j \leq i-1$, are the nodes $(0, 0, \ldots, c_m = 1, 0, \ldots, c_i = 1, b_{i-1}, \ldots b_0)$, for $m \in \{i+1, i+2, \ldots, n-1\}$. In other words, the children of a node whose leading 1 is at position i are the nodes whose binary address is generated by complementing one by one the leading zeros of the binary encoding of the address of their parent (see Fig. 5). The procedure is given below.

Procedure $U^0(B_n, O_t^\Delta, 2^t)$.

- $M[x] = \left(\displaystyle\sum_{i=0}^{n-1} b_i 2^{i \bmod t} \right) \bmod 2^t$, for all $(b_{n-1} \ldots b_0) \in B_n$.

Given Δ and the root of the template ξ, the addresses of the subset of nodes belonging to $O_t^{\xi,\Delta}$ span all the 2^t possible configurations of the dimensions b_i,

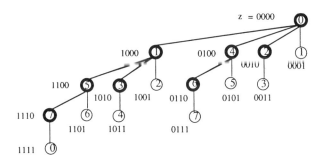

FIGURE 5. For the host binomial tree B_4, the label of each node is shown along with the memory module assigned to it in order to guarantee conflict-free access to the oriented templates of order 3 when the oriented tuple is $\Delta = (3, 2, 1)$.

where $i \in \Delta$. Thus, any oriented template $O_t^{\xi, \Delta}$ can be accessed conflict-free using a number of memory modules equal to the template size, which is optimal.   □

The mapping procedure for the oriented templates is highly scalable with respect to the host size. Indeed, a binomial tree of higher order can be built from a smaller binomial tree without affecting its memory module destination, assuming that the labels of the original binomial tree are augmented with zeros in the leftmost positions.

6. Conclusions

Based on simple indexing and modular arithmetic techniques, we have shown how to map the nodes of k-ary trees into a parallel memory system such that any occurrence of t-ary subtrees, for $t \leq k$, can be retrieved in a single memory access. We have also studied the conflict-free access to binomial subtrees with the help of graph-theoretic or topological properties of binomial trees. The lower and upper bounds on the number of memory modules required for conflict-free, scalable (w.r.t. the host size) and load-balanced assignments are derived.

An open problem for k-ary trees is how to preserve the conflict-free assignment while the host tree grows dynamically from the top or while it changes its arity. This could be useful in applications involving structures based on B-trees where insertions (or deletions) of keys can cause the split (or merger) of a node, which in turn can increase (or decrease) the height of the tree by one.

In this paper, we have discussed the data mapping for conflict-free access to templates of a specific type, e.g. paths or subtrees. For recent work on a unified scheme which allow simultaneous access to multiple templates of different types, refer to [**ADPS98**]. Additionally, in [**DP97**] we proposed a novel approach for conflict-free subcube access in hypercubes, by showing the similarities between this problem and that of generating block linear codes. More precisely, given a linear code $C = [n, k, d]$, we have shown how to partition the nodes of an n-dimensional binary hypercube into 2^{n-k} subsets, each with 2^k nodes, so that any $(d-1)$-dimensional subcube template can be accessed without conflicts.

References

[ADPS98] V. Auletta, S. K. Das, M. C. Pinotti, and V. Scarano, "Toward a Universal Mapping Algorithm for Accessing Trees in Parallel Memory Systems", *Proc. IEEE Int'l Parallel Processing Symp.*, Orlando, Apr 1998, pp. 447-454.

[B92] R. V. Boppana, "On the Effectiveness of Interleaved Memories for Binary Trees", *Proc. of Architectures and Compilation Techniques for Fine and Medium Grain Parallelism*, Elsevier Science Pub. B.V., 1993, pp. 203-214.

[CH92] C. J. Colbourn, and K. Heinrich, "Conflict-Free Access to Parallel Memories", *Journal of Parallel and Distributed Computing*, Vol. 14, 1992, pp. 193-200.

[C86] R. Creutzburg, "Parallel Optimal Subtree Access with Recursively Linear Memory Functions", *Proc. Int'l Workshop on Parallel Processing by Cellular Automata and Arrays*, Berlin, Sept 1986, pp. 203-209.

[DP97] S. K. Das and M. C. Pinotti, "Conflict-Free Access to Templates of Trees and Hypercubes," *Proc. 3rd Annu. Int'l Conf. on Computing and Combinatorics* (COCOON'97), Shanghai, China, pp. 1-10, Aug 1997.

[DP98] S.K. Das and M.C. Pinotti, " $O(\log \log N)$ Time Algorithms for Hamiltonian-Suffix and Min-Max-Pair Heap Operations on the Hypercube", *Journal of Parallel and Distributed Computing*, Vol. 48, No. 2, 1998, pp. 200-211.

[DPS96] S. K. Das, M. C. Pinotti, and F. Sarkar, "Parallel Priority Queues in Distributed Memory Hypercubes", *IEEE Trans. Parallel and Distributed Systems*, Vol. 7, No. 6, June 1996, pp. 555-564.

[DS94] S. K. Das and F. Sarkar, "Conflict-Free Data Access of Arrays and Trees in Parallel Memory Systems", *Proc. 6th IEEE Symp. Parallel and Distributed Processing*, Dallas, Texas, Oct 1994, pp. 377-384.

[DSP95] S. K. Das, F. Sarkar, and M. C. Pinotti, "Conflict-Free Access of Trees in Parallel Memory Systems and Its Generalization with Applications to Distributed Heap Implementation", *Proc. Int'l Conf. on Parallel Processing*, Oconomowoc, Wisconsin, Aug 1995, Vol. III, pp. 164-167.

[GAG96] R. Govindarajan, E.R. Altman, and G.R. Gao, "A Framework for Resource-Constrained Rate-Optimal Software Pipelining", *IEEE Trans. Parallel and Distributed Systems*. Vol. 7, No. 11, Nov 1996, pp. 1133-1149.

[JH89] S. L. Johnsson, and C. Ho, "Optimum Broadcasting and Personalized Communication in Hypercubes", *IEEE Trans. Computers*, Vol. 38, No. 9, 1989, pp. 1249-1268.

[KJD97] D. Kaznachey, A. Jagota and S. K. Das, "Primal-Target Neural Net Heuristics for the Hypergraph k-Coloring Problem," *Proc. Int'l Conf. on Neural Networks*, Houston, Texas, pp. 1251-1255, June 1997.

[L75] D. H. Lawrie, "Access and Alignment of Data in an Array Processor", *IEEE Trans. Computers*, Vol. c-24, No. 12, Dec 1975, pp. 1145-1155.

[KP93] K. Kim, and V. K. Prasanna Kumar, "Latin Squares for Parallel Array Access", *IEEE Trans. Parallel and Distributed Systems*, Vol. 4, No. 4, Apr 1993, pp. 361-370.

[MA97] N. Manjikian and T.S. Abdelrahman, "Fusion of Loops for Parallelism and Locality", *IEEE Trans. Parallel and Distributed Systems*. Vol. 8, No. 2, Feb 1997, pp. 193-209.

[PS96] N.L. Passos and E.H.-M. Sha, "Achieving Full Parallelism Using Multidimensional Retiming", *IEEE Trans. Parallel and Distributed Systems*. Vol. 7, No. 11, Nov 1996, pp. 1150-1163.

[PD96] P. M. Peterson, and D. Padua, "Static and Dynamic Evaluation of Data Dependence Analysis Techniques",*IEEE Trans. Parallel and Distributed Systems*. Vol. 7, No. 11, Nov 1996, pp. 1121-1132.

[PDS97] M. C. Pinotti, S. K. Das, and F. Sarkar, "Conflict-Free Template Access in k-ary and Binomial Trees", *Proc. ACM Int'l Conf. Supercomputing*, Austria, July 1997, pp. 237-244.

[Wij85] H. A. G. Wijshoff, "Storing Trees into Parallel Memories", *Parallel Computing 85*, (Ed. Feilmeier), Elsevier Science Pub., 1986, pp. 253-261.

[Wij89] H. G. Wijshoff, *Data Organization in Parallel Memories*, Kluwer Academic Pub., 1989.

DEPT OF COMPUTER SCIENCE, UNIVERSITY OF NORTH TEXAS, DENTON, TX 76203, USA.
E-mail address: das@cs.unt.edu

I.E.I., NATIONAL COUNCIL OF RESEARCH, VIA S. MARIA, 46, 56126 PISA, ITALY.
E-mail address: pinotti@iei.pi.cnr.it

DIMACS Series in Discrete Mathematics
and Theoretical Computer Science
Volume **50**, 1999

External Memory Techniques for Isosurface Extraction in Scientific Visualization

Yi-Jen Chiang and Cláudio T. Silva

ABSTRACT. Isosurface extraction is one of the most effective and powerful techniques for the investigation of volume datasets in scientific visualization. Previous isosurface techniques are all main-memory algorithms, often not applicable to large scientific visualization applications. In this paper we survey our recent work that gives the first external memory techniques for isosurface extraction. The first technique, *I/O-filter*, uses the existing I/O-optimal interval tree as the indexing data structure (where the corner structure is not implemented), together with the isosurface engine of Vtk (one of the currently best visualization packages). The second technique improves the first version of *I/O-filter* by replacing the I/O interval tree with the metablock tree (whose corner structure is not implemented). The third method further improves the first two, by using a *two-level indexing* scheme, together with a new *meta-cell* technique and a new I/O-optimal indexing data structure (the *binary-blocked I/O interval tree*) that is simpler and more space-efficient in practice (whose corner structure is not implemented). The experiments show that the first two methods perform isosurface queries faster than Vtk by a factor of two orders of magnitude for datasets larger than main memory. The third method further reduces the disk space requirement from 7.2–7.7 times the original dataset size to 1.1–1.5 times, at the cost of slightly increasing the query time; this method also exhibits a smooth trade-off between disk space and query time.

1. Introduction

The field of computer graphics can be roughly classified into two subfields: *surface graphics*, in which objects are defined by surfaces, and *volume graphics* [**17, 18**], in which objects are given by datasets consisting of 3D sample points over their volume. In volume graphics, objects are usually modeled as *fuzzy* entities. This representation leads to greater freedom, and also makes it possible to visualize the *interior* of an object. Notice that this is almost impossible for traditional surface-graphics objects. The ability to visualize the interior of an object is particularly

1991 *Mathematics Subject Classification.* 65Y25, 68U05, 68P05, 68Q25.

Key words and phrases. Computer Graphics, Scientific Visualization, External Memory, Design and Analysis of Algorithms and Data Structures, Experimentation.

The first author was supported in part by NSF Grant DMS-9312098 and by Sandia National Labs.

The second author was partially supported by Sandia National Labs and the Dept. of Energy Mathematics, Information, and Computer Science Office, and by NSF Grant CDA-9626370.

important in *scientific visualization*. For example, we might want to visualize the internal structure of a patient's brain from a dataset collected from a computed tomography (CT) scanner, or we might want to visualize the distribution of the density of the mass of an object, and so on. Therefore volume graphics is used in virtually all scientific visualization applications. Since the dataset consists of points sampling the entire volume rather than just vertices defining the surfaces, typical volume datasets are huge. This makes volume visualization an ideal application domain for I/O techniques.

Input/Output (I/O) communication between fast internal memory and slower external memory is the major bottleneck in many large-scale applications. Algorithms specifically designed to reduce the I/O bottleneck are called *external-memory* algorithms. In this paper, we survey our recent work that gives the first external memory techniques for one of the most important problems in volume graphics: *isosurface extraction* in scientific visualization.

1.1. Isosurface Extraction.

1.1. Isosurface Extraction. Isosurface extraction represents one of the most effective and powerful techniques for the investigation of volume datasets. It has been used extensively, particularly in visualization [**20, 22**], simplification [**14**], and implicit modeling [**23**]. Isosurfaces also play an important role in other areas of science such as biology, medicine, chemistry, computational fluid dynamics, and so on. Its widespread use makes efficient isosurface extraction a very important problem.

The problem of isosurface extraction can be stated as follows. The input dataset is a *scalar volume dataset* containing a list of tuples $(\mathbf{x}, \mathcal{F}(\mathbf{x}))$, where $\mathbf{x}$ is a 3D sample point and $\mathcal{F}$ is a scalar function defined over 3D points. The scalar function $\mathcal{F}$ is an unknown function; we only know the *sample value* $\mathcal{F}(\mathbf{x})$ at each sample point $\mathbf{x}$. The function $\mathcal{F}$ may denote temperature, density of the mass, or intensity of an electronic field, etc., depending on the applications. The input dataset also has a list of *cells* that are cubes or tetrahedra or of some other geometric type. Each cell is defined by its vertices, where each vertex is a 3D sample point $\mathbf{x}$ given in the list of tuples $(\mathbf{x}, \mathcal{F}(\mathbf{x}))$. Given an isovalue (a scalar value) q, to extract the isosurface of q is to compute and display the isosurface $C(q) = \{\mathbf{p} | \mathcal{F}(\mathbf{p}) = q\}$. Note that the isosurface point $\mathbf{p}$ may not be a sample point $\mathbf{x}$ in the input dataset: if there are two sample points with their scalar values smaller and larger than q, respectively, then the isosurface $C(q)$ will go between these two sample points via linear interpolation. Some examples of isosurfaces (generated from our experiments) are shown in Fig. 1, where the Blunt Fin dataset shows an airflow through a flat plate with a blunt fin, and the Combustion Chamber dataset comes from a combustion simulation. Typical use of isosurface is as follows. A user may ask: "display all areas with temperature equal to 25 degrees." After seeing that isosurface, the user may continue to ask: "display all areas with temperature equal to 10 degrees." By repeating this process interactively, the user can study and perform detailed measurements of the properties of the datasets. Obviously, to use isosurface extraction effectively, it is crucial to achieve fast interactivity, which requires efficient computation of isosurface extraction.

The computational process of isosurface extraction can be viewed as consisting of two phases (see Fig. 2). First, in the *search phase*, one finds all *active* cells of the dataset that are intersected by the isosurface. Next, in the *generation phase*, depending on the type of cells, one can apply an algorithm to actually generate the

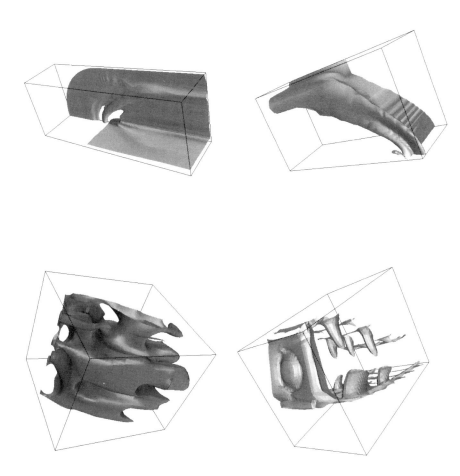

FIGURE 1. Typical isosurfaces are shown. The upper two are for
the Blunt Fin dataset. The ones in the bottom are for the Combustion Chamber dataset.

isosurface from those active cells (Marching Cubes [20] is one such algorithm for
hexahedral cells). Notice that the search phase is usually the bottleneck of the entire
process, since it searches the 3D dataset and produces 2D data. In fact, letting N
be the total number of cells in the dataset and K the number of active cells, it
is estimated that the typical value of K is $O(N^{2/3})$ [15]. Therefore an exhaustive
scanning of all cells in the search phase is inefficient, and a lot of research efforts
have thus focused on developing *output-sensitive* algorithms to speed up the search
phase.

In the rest of the paper we use N and K to denote the total number of cells in
the dataset and the number of active cells, respectively, and M and B to respectively
denote the numbers of cells fitting in main memory and in a disk block. Each I/O
operation reads or writes one disk block.

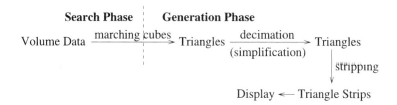

FIGURE 2. A pipeline of the isosurface extraction process.

1.2. Overview of Main Memory Isosurface Techniques. There is a very rich literature for isosurface extraction. Here we only briefly review the results that focus on speeding up the search phase. For an excellent and thorough review, see [**19**].

In Marching Cubes [**20**], all cells in the volume dataset are searched for isosurface intersection, and thus $O(N)$ time is needed. Concerning the main memory issue, this technique does not require the entire dataset to fit into main memory, but $\lceil N/B \rceil$ disk reads are necessary. Wilhems and Van Gelder [**30**] propose a method of using an octree to optimize isosurface extraction. This algorithm has worst-case time of $O(K + K \log(N/K))$ (this analysis is presented by Livnat et al. [**19**]) for isosurface queries, once the octree has been built.

Itoh and Kayamada [**15**] propose a method based on identifying a collection of *seed cells* from which isosurfaces can be propagated by performing local search. Basically, once the seed cells have been identified, they claim to have a nearly $O(N^{2/3})$ expected performance. (Livnat et al. [**19**] estimate the worst-case running time to be $O(N)$, with a high memory overhead.) More recently, Bajaj et al. [**3**] propose another contour propagation scheme, with expected performance of $O(K)$. Livnat et al. [**19**] propose NOISE, an $O(\sqrt{N} + K)$-time algorithm. Shen et al. [**27, 28**] also propose nearly optimal isosurface extraction methods.

The first *optimal* isosurface extraction algorithm was given by Cignoni et al. [**11**], based on the following two ideas. First, for each cell, they produce an interval $I = [\min, \max]$ where min and max are the minimum and maximum of the scalar values in the cell vertices. Then the active cells are exactly those cells whose intervals contain q. Searching active cells then amounts to performing the following *stabbing queries*: Given a set of 1D intervals, report all intervals (and the associated cells) containing the given query point q. Secondly, the stabbing queries are solved by using an internal-memory interval tree [**12**]. After an $O(N \log N)$-time preprocessing, active cells can be found in optimal $O(\log N + K)$ time.

All the isosurface techniques mentioned above are main-memory algorithms. Except for the inefficient exhaustive scanning method of Marching Cubes, all of them require the time and main memory space to read and keep the entire dataset in main memory, plus additional preprocessing time and main memory space to build and keep the search structure. Unfortunately, for (usually) very large volume datasets, these methods often suffer the problem of not having enough main memory, which can cause a major slow-down of the algorithms due to a large number of page faults. Another issue is that the methods need to load the dataset into main memory and build the search structure each time we start the running process. This start-up cost can be very expensive since loading a large volume dataset from disk is very time-consuming.

1.3. Summary of External Memory Isosurface Techniques. In [8] we give *I/O-filter*, the first I/O-optimal technique for isosurface extraction. We follow the ideas of Cignoni *et al.* [11], but use the I/O-optimal interval tree of Arge and Vitter [2] as an indexing structure to solve the stabbing queries. This enables us to find the active cells in optimal $O(\log_B N + K/B)$ I/O's. We give the first implementation of the I/O interval tree (where the *corner structure* is not implemented, which may result in non-optimal disk space and non-optimal query I/O cost in the worst case), and also implement our method as an *I/O filter* for the isosurface extraction routine of Vtk [24, 25] (which is one of the currently best visualization packages). The experiments show that the isosurface queries are faster than Vtk by a factor of two orders of magnitude for datasets larger than main memory. In fact, the search phase is no longer a bottleneck, and the performance is *independent* of the main memory available. Also, the preprocessing is performed only *once* to build an indexing structure in disk, and later on there is no start-up cost for running the query process. The major drawback is the overhead in disk scratch space and the preprocessing time necessary to build the search structure, and of the disk space needed to hold the data structure.

In [9], we give the second version of *I/O-filter*, by replacing the I/O interval tree [2] with the metablock tree of Kanellakis *et al.* [16]. We give the first implementation of the metablock tree (where the corner structure is not implemented to reduce the disk space; this may result in non-optimal query I/O cost in the worst case). While keeping the query time the same as in [8], the tree construction time, the disk space and the disk scratch space are all improved.

In [10], at the cost of slightly increasing the query time, we greatly improve all the other cost measures. In the previous methods [8, 9], the *direct vertex information* is duplicated many times; in [10], we avoid such duplications by employing a *two-level indexing* scheme. We use a new *meta-cell* technique and a new I/O-optimal indexing data structure (the *binary-blocked I/O interval tree*) that is simpler and more space-efficient in practice (where the corner structure is not implemented, which may result in non-optimal I/O cost for the stabbing queries in the worst case). Rather than fetching only the active cells into main memory as in *I/O-filter* [8, 9], this method fetches the set of *active meta-cells*, which is a *superset* of all active cells. While the query time is still at least one order of magnitude faster than Vtk, the disk space is reduced from 7.2–7.7 times the original dataset size to 1.1–1.5 times, and the disk scratch space is reduced from 10–16 times to less than 2 times. Also, instead of being a single-cost indexing approach, the method exhibits a smooth trade-off between disk space and query time.

1.4. Organization of the Paper. The rest of the paper is organized as follows. In Section 2, we review the I/O-optimal data structures for stabbing queries, namely the metablock tree [16] and the I/O interval tree [2] that are used in the two versions of our I/O-filter technique, and the binary-blocked I/O-interval tree [10] that is used in our two-level indexing scheme. The preprocessing algorithms and the implementation issues, together with the dynamization of the binary-blocked I/O interval tree (which is not given in [10] and may be of independent interest; see Section 2.3.3) are also discussed. We describe the I/O-filter technique [8, 9] and summarize the experimental results of both versions in Section 3. In Section 4 we survey the two-level indexing scheme [10] together with the experimental results. Finally we conclude the paper in Section 5.

2. I/O Optimal Data Structures for Stabbing Queries

In this section we review the metablock tree [16], the I/O interval tree [2], and the binary-blocked I/O interval tree [10]. The metablock tree is an external-memory version of the priority search tree [21]. The static version of the metablock tree is the first I/O-optimal data structure for *static* stabbing queries (where the set of intervals is fixed); its *dynamic* version only supports insertions of intervals and the update I/O cost is not optimal. The I/O interval tree is an external-memory version of the main-memory interval tree [12]. In addition to being I/O-optimal for the static version, the dynamic version of the I/O interval tree is the first I/O-optimal *fully dynamic* data structure that also supports insertions and deletions of intervals with optimal I/O cost. Both the metablock tree and the I/O interval tree have three kinds of secondary lists and each interval is stored up to three times in practice. Motivated by the practical concern on the disk space and the simplicity of coding, the binary-blocked I/O interval tree is an alternative external-memory version of the main-memory interval tree [12] with only two kinds of secondary lists. In practice, each interval is stored twice and hence the tree is more space-efficient and simpler to implement. We remark that the powerful dynamization techniques of [2] for dynamizing the I/O interval tree can also be applied to the binary-blocked I/O interval tree to support I/O-optimal updates (amortized rather than worst-case bounds as in the I/O interval tree — although we believe the techniques of [2] can further turn the bounds into worst-case, we did not verify the details. See Section 2.3.3). For our application of isosurface extraction, however, we only need the *static* version, and thus all three trees are I/O-optimal. We only describe the static version of the trees (but in addition we discuss the dynamization of the binary-blocked I/O interval tree in Section 2.3.3). We also describe the preprocessing algorithms that we used in [8, 9, 10] to build these static trees, and discuss their implementation issues.

Ignoring the cell information associated with the intervals, we now use M and B to respectively denote the numbers of intervals that fit in main memory and in a disk block. Recall that N is the total number of intervals, and K is the number of intervals reported from a query. We use Bf to denote the branching factor of a tree.

2.1. Metablock Tree.

2.1.1. *Data Structure.* We briefly review the metablock tree data structure [16], which is an external-memory version of the priority search tree [21]. The stabbing query problem is solved in the *dual space*, where each interval $[left, right]$ is mapped to a dual point (x, y) with $x = left$ and $y = right$. Then the query "find intervals $[x, y]$ with $x \leq q \leq y$" amounts to the following *two-sided orthogonal range query* in the dual space: report all dual points (x, y) lying in the intersection of the half planes $x \leq q$ and $y \geq q$. Observe that all intervals $[left, right]$ have $left \leq right$, and thus all dual points lie in the half plane $x \leq y$. Also, the "corner" induced by the two sides of the query is the dual point (q, q), so all query corners lie on the line $x = y$.

The metablock tree stores dual points in the same spirit as a priority search tree, but increases the branching factor Bf from 2 to $\Theta(B)$ (so that the tree height is reduced from $O(\log_2 N)$ to $O(\log_B N)$), and also stores $Bf \cdot B$ points in each tree node. The main structure of a metablock tree is defined recursively as follows (see Fig. 3(a)): if there are no more than $Bf \cdot B$ points, then all of them are assigned to

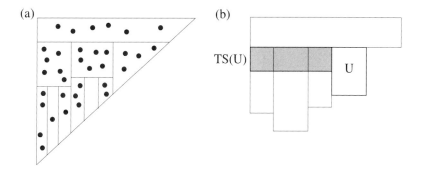

FIGURE 3. A schematic example of metablock tree: (a) the main structure; (b) the TS list. In (a), $Bf = 3$ and $B = 2$, so each node has up to 6 points assigned to it. We relax the requirement that each vertical slab have the same number of points.

the current node, which is a leaf; otherwise, the topmost $Bf \cdot B$ points are assigned to the current node, and the remaining points are distributed by their x-coordinates into Bf vertical slabs, each containing the same number of points. Now the Bf subtrees of the current node are just the metablock trees defined on the Bf vertical slabs. The $Bf - 1$ slab boundaries are stored in the current node as keys for deciding which child to go during a search. Notice that each internal node has no more than Bf children, and there are Bf blocks of points assigned to it. For each node, the points assigned to it are stored twice, respectively in two lists in disk of the same size: the *horizontal* list, where the points are horizontally blocked and stored sorted by decreasing y-coordinates, and the *vertical* list, where the points are vertically blocked and stored sorted by increasing x-coordinates. We use unique dual point ID's to break a tie. Each node has two pointers to its *horizontal* and *vertical* lists. Also, the "bottom" (*i.e.*, the y-value of the bottommost point) of the *horizontal* list is stored in the node.

The second piece of organization is the TS list maintained in disk for each node U (see Fig. 3(b)): the list $TS(U)$ has at most Bf blocks, storing the topmost Bf blocks of points from *all left siblings* of U (if there are fewer than $Bf \cdot B$ points then all of them are stored in $TS(U)$). The points in the TS list are horizontally blocked, stored sorted by decreasing y-coordinates. Again each node has a pointer to its TS list, and also stores the "bottom" of the TS list.

The final piece of organization is the *corner structure*. A corner structure can store $t = O(B^2)$ points in optimal $O(t/B)$ disk blocks, so that a two-sided orthogonal range query can be answered in optimal $O(k/B + 1)$ I/O's, where k is the number of points reported. Assuming all t points can fit in main memory during preprocessing, a corner structure can be built in optimal $O(t/B)$ I/O's. We refer to [16] for more details. In a metablock tree, for each node U where a query corner can possibly lie, a corner structure is built for the $(\le Bf \cdot B = O(B^2))$ points assigned to U (recall that $Bf = \Theta(B)$). Since any query corner must lie on the line $x = y$, each of the following nodes needs a corner structure: (1) the leaves, and (2) the nodes in the rightmost root-to-leaf path, including the root (see Fig. 3(a)). It is easy to see that the entire metablock tree has height $O(\log_{Bf}(N/B)) = O(\log_B N)$ and uses optimal $O(N/B)$ blocks of disk space [16]. Also, it can be seen that the corner structures are *additional* structures to the metablock tree; we can save some storage space by not implementing the corner structures (at the cost of increasing the worst-case query bound; see Section 2.1.2).

As we shall see in Section 2.1.3, we will slightly modify the definition of the metablock tree to ease the task of preprocessing, while keeping the bounds of tree height and tree storage space the same.

2.1.2. *Query Algorithm.* Now we review the query algorithm given in [**16**]. Given a query value q, we perform the following recursive procedure starting with *meta-query* (q, the root of the metablock tree). Recall that we want to report all dual points lying in $x \leq q$ and $y \geq q$. We maintain the invariant that the current node U being visited always has its x-range containing the vertical line $x = q$.

Procedure *meta-query* (query q, node U)
1. If U contains the corner of q, *i.e.*, the bottom of the *horizontal* list of U is lower than the horizontal line $y = q$, then use the corner structure of U to answer the query and stop.
2. Otherwise ($y(bottom(U)) \geq q$), all points of U are above or on the horizontal line $y = q$. Report all points of U that are on or to the left of the vertical line $x = q$, using the *vertical* list of U.
3. Find the child U_c (of U) whose x-range contains the vertical line $x = q$. The node U_c will be the next node to be recursively visited by *meta-query*.
4. Before recursively visiting U_c, take care of the left-sibling subtrees of U_c first (points in all these subtrees are on or to the left of the vertical line $x = q$, and thus it suffices to just check their heights):
   (a) If the bottom of $TS(U_c)$ is lower than the horizontal line $y = q$, then report the points in $TS(U_c)$ that lie inside the query range. Go to step 5.
   (b) Else, for each left sibling W of U_c, repeatedly call procedure *H-report* (query q, node W). (*H-report* is another recursive procedure given below.)
5. Recursively call *meta-query* (query q, node U_c).

H-report is another recursive procedure for which we maintain the invariant that the current node W being visited have all its points lying on or to the left of the vertical line $x = q$, and thus we only need to consider the condition $y \geq q$.

Procedure *H-report* (query q, node W)
1. Use the *horizontal* list of W to report all points of W lying on or above the horizontal line $y = q$.
2. If the bottom of W is lower than the line $y = q$ then stop.
   Otherwise, for each child V of W, repeatedly call *H-report* (query q, node V) recursively.

It can be shown that the queries are performed in optimal $O(\log_B N + \frac{K}{B})$ I/O's [**16**]. We remark that only one node in the search path would possibly use its corner structure to report its points lying in the query range since there is at most one node containing the query corner (q, q). If we do not implement the corner structure, then step 1 of Procedure *meta-query* can still be performed by checking the *vertical* list of U up to the point where the current point lies to the right of the vertical line $x = q$ and reporting all points thus checked with $y \geq q$. This might perform extra Bf I/O's to examine the entire *vertical* list without reporting any point, and hence is not optimal. However, if $K \geq \alpha \cdot (Bf \cdot B)$ for some constant

$\alpha < 1$ then this is still worst-case I/O-optimal since we need to perform $\Omega(Bf)$ I/O's to just report the answer.

2.1.3. *Preprocessing Algorithms.* Now we describe a preprocessing algorithm proposed in [9] to build the metablock tree. It is based on a paradigm we call *scan and distribute*, inspired by the *distribution sweep* I/O technique [6, 13]. The algorithm relies on a slight modification of the definition of the tree.

In the original definition of the metablock tree, the vertical slabs for the subtrees of the current node are defined by dividing the *remaining* points not assigned to the current node into Bf groups. This makes the distribution of the points into the slabs more difficult, since in order to assign the topmost Bf blocks to the current node we have to sort the points by y-values, and yet the slab boundaries (x-values) from the remaining points cannot be directly decided. There is a simple way around it: we first sort all N points by increasing x-values into a fixed set X. Now X is used to decide the slab boundaries: the root corresponds to the entire x-range of X, and each child of the root corresponds to an x-range spanned by consecutive $|X|/Bf$ points in X, and so on. In this way, the slab boundaries of the entire metablock tree is *pre-fixed*, and the tree height is still $O(\log_B N)$.

With this modification, it is easy to apply the *scan and distribute* paradigm. In the first phase, we sort all points into the set X as above and also sort all points by decreasing y-values into a set Y. Now the second phase is a recursive procedure. We assign the first Bf blocks in the set Y to the root (and build its *horizontal* and *vertical* lists), and scan the remaining points to distribute them to the vertical slabs of the root. For each vertical slab we maintain a temporary list, which keeps one block in main memory as a *buffer* and the remaining blocks in disk. Each time a point is distributed to a slab, we put that point into the corresponding buffer; when the buffer is full, it is written to the corresponding list in disk. When all points are scanned and distributed, each temporary list has all its points, automatically sorted by decreasing y. Now we build the TS lists for child nodes U_0, U_1, $\cdots$ numbered left to right. Starting from U_1, $TS(U_i)$ is computed by merging two sorted lists in decreasing y and taking the first Bf blocks, where the two lists are $TS(U_{i-1})$ and the temporary list for slab $i-1$, both sorted in decreasing y. Note that for the initial condition $TS(U_0) = \emptyset$. (It suffices to consider $TS(U_{i-1})$ to take care of all points in slabs $0, 1, \cdots, i-2$ that can possibly enter $TS(U_i)$, since each TS list contains up to Bf blocks of points.) After this, we apply the procedure recursively to each slab. When the current slab contains no more than Bf blocks of points, the current node is a leaf and we stop. The corner structures can be built for appropriate nodes as the recursive procedure goes. It is easy to see that the entire process uses $O(\frac{N}{B} \log_B N)$ I/O's. Using the same technique that turns the nearly-optimal $O(\frac{N}{B} \log_B N)$ bound to optimal in building the static I/O interval tree [1], we can turn this nearly-optimal bound to optimal $O(\frac{N}{B} \log_{\frac{M}{B}} \frac{N}{B})$. (For the metablock tree, this technique basically builds a $\Theta(M/B)$-fan-out tree and converts it into a $\Theta(B)$-fan-out tree during the tree construction; we omit the details here.) Another I/O-optimal preprocessing algorithm is described in [9].

2.2. I/O Interval Tree. In this section we describe the I/O interval tree [2]. Since the I/O interval tree and the binary-blocked I/O interval tree [10] (see Section 2.3) are both external-memory versions of the (main memory) binary interval tree [12], we first review the binary interval tree T.

Given a set of N intervals, such interval tree T is defined recursively as follows. If there is only one interval, then the current node r is a leaf containing that interval. Otherwise, r stores as a key the median value m that partitions the interval endpoints into two slabs, each having the same number of endpoints that are smaller (resp. larger) than m. The intervals that contain m are assigned to the node r. The intervals with both endpoints smaller than m are assigned to the left slab; similarly, the intervals with both endpoints larger than m are assigned to the right slab. The left and right subtrees of r are recursively defined as the interval trees on the intervals in the left and right slabs, respectively. In addition, each internal node u of T has two secondary lists: the *left list*, which stores the intervals assigned to u, sorted in *increasing left endpoint values*, and the *right list*, which stores the same set of intervals, sorted in *decreasing right endpoint values*. It is easy to see that the tree height is $O(\log_2 N)$. Also, each interval is assigned to exactly one node, and is stored either twice (when assigned to an internal node) or once (when assigned to a leaf), and thus the overall space is $O(N)$.

To perform a query for a query point q, we apply the following recursive process starting from the root of T. For the current node u, if q lies in the left slab of u, we check the left list of u, reporting the intervals sequentially from the list until the first interval is reached whose left endpoint value is larger than q. At this point we stop checking the left list since the remaining intervals are all to the right of q and cannot contain q. We then visit the left child of u and perform the same process recursively. If q lies in the right slab of u then we check the right list in a similar way and then visit the right child of u recursively. It is easy to see that the query time is optimal $O(\log_2 N + K)$.

2.2.1. *Data Structure.* Now we review the I/O interval tree data structure [2]. Each node of the tree is one block in disk, capable of holding $\Theta(B)$ items. The main goal is to increase the tree fan-out Bf so that the tree height is $O(\log_B N)$ rather than $O(\log_2 N)$. In addition to having left and right lists, a new kind of secondary lists, the *multi* lists, is introduced, to store the intervals assigned to an internal node u that *completely span* one or more vertical slabs associated with u. Notice that when $Bf = 2$ (*i.e.*, in the binary interval tree) there are only two vertical slabs associated with u and thus no slab is completely spanned by any interval. As we shall see below, there are $\Theta(Bf^2)$ *multi* lists associated with u, requiring $\Theta(Bf^2)$ pointers from u to the secondary lists, therefore Bf is taken to be $\Theta(\sqrt{B})$.

We describe the I/O interval tree in more details. Let E be the set of $2N$ endpoints of all N intervals, sorted from left to right in increasing values; E is *pre-fixed* and will be used to define the *slab boundaries* for each internal node of the tree. Let S be the set of all N intervals. The I/O interval tree on S and E is defined recursively as follows. The root u is associated with the entire range of E and with all intervals in S. If S has no more than B intervals, then u is a leaf storing all intervals of S. Otherwise u is an internal node. We evenly divide E into Bf slabs $E_0, E_1, \cdots, E_{Bf-1}$, each containing the same number of endpoints. The $Bf - 1$ *slab boundaries* are the first endpoints of slabs $E_1, \cdots, E_{Bf-1}$; we store these slab boundaries in u as keys. Now consider each interval I in S (see Fig. 4). If I crosses one or more slab boundaries of u, then I is assigned to u and is stored in the secondary lists of u. Otherwise I completely lies inside some slab E_i and is assigned to the subset S_i of S. We associate each child u_i of u with the slab E_i and with the set S_i of intervals. The subtree rooted at u_i is recursively defined as the I/O interval tree on E_i and S_i.

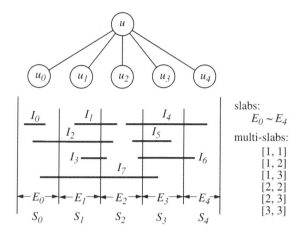

FIGURE 4. A schematic example of the I/O interval tree for the branching factor $Bf = 5$. Note that this is not a complete example and some intervals are not shown. Consider only the intervals shown here and node u. The interval sets for its children are: $S_0 = \{I_0\}$, and $S_1 = S_2 = S_3 = S_4 = \emptyset$. Its *left* lists are: $left(0) = (I_2, I_7)$, $left(1) = (I_1, I_3)$, $left(2) = (I_4, I_5, I_6)$, and $left(3) = left(4) = \emptyset$ (each list is an *ordered* list as shown). Its *right* lists are: $right(0) = right(1) = \emptyset$, $right(2) = (I_1, I_2, I_3)$, $right(3) = (I_5, I_7)$, and $right(4) = (I_4, I_6)$ (again each list is an *ordered* list as shown). Its *multi* lists are: $multi([1,1]) = \{I_2\}$, $multi([1,2]) = \{I_7\}$, $multi([1,3]) = multi([2,2]) = multi([2,3]) = \emptyset$, and $multi([3,3]) = \{I_4, I_6\}$.

For each internal node u, we use three kinds of secondary lists for storing the intervals assigned to u: the *left*, *right* and *multi* lists, described as follows.

For each of the Bf slabs associated with u, there is a *left* list and a *right* list; the *left* list stores all intervals belonging to u whose left endpoints lie in that slab, sorted in *increasing left endpoint values*. The *right* list is symmetric, storing all intervals belonging to u whose right endpoints lie in that slab, sorted in *decreasing right endpoint values* (see Fig. 4).

Now we describe the third kind of secondary lists, the *multi* lists. There are $(Bf - 1)(Bf - 2)/2$ *multi* lists for u, each corresponding to a *multi-slab* of u. A *multi-slab* $[i,j]$, $0 \le i \le j \le Bf - 1$, is defined to be the union of slabs $E_i, \cdots, E_j$. The *multi* list for the multi-slab $[i,j]$ stores all intervals of u that *completely span* $E_i \cup \cdots \cup E_j$, i.e., all intervals of u whose left endpoints lie in E_{i-1} and whose right endpoints lie in E_{j+1}. Since the *multi* lists $[0,k]$ for any k and the *multi* lists $[\ell, Bf-1]$ for any ℓ are always empty by the definition, we only care about multi-slabs $[1,1], \cdots, [1, Bf-2], [2,2], \cdots, [2, Bf-2], \cdots, [i,i], \cdots, [i, Bf-2], \cdots, [Bf-2, Bf-2]$. Thus there are $(Bf - 1)(Bf - 2)/2$ such multi-slabs and the associated *multi* lists (see Fig. 4).

For each *left*, *right*, or *multi* list, we store the entire list in consecutive blocks in disk, and in the node u (occupying one disk block) we store a pointer to the starting position of the list in disk. Since in u there are $O(Bf^2) = O(B)$ such pointers, they can all fit into one disk block, as desired.

It is easy to see that the tree height is $O(\log_{Bf}(N/B)) = O(\log_B N)$. Also, each interval I belongs to exactly one node, and is stored at most three times. If I belongs to a leaf node, then it is stored only once, if it belongs to an internal node, then it is stored once in some *left* list, once in some *right* list, and possibly one more time in some *multi* list. Therefore we need *roughly* $O(N/B)$ disk blocks to store the entire data structure.

Theoretically, however, we may need more disk blocks. The problem is because of the *multi* lists. In the worst case, a *multi* list may have only very few ($<< B$) intervals in it, but still requires one disk block for storage. The same situation may occur also for the *left* and *right* lists, but since each internal node has *Bf left/right* lists and the same number of children, these *underflow* blocks can be charged to the child nodes. But since there are $O(Bf^2)$ *multi* lists for an internal node, this charging argument does not work for the *multi* lists. In [2], the problem is solved by using the *corner structure* of [16] that we mentioned at the end of Section 2.1.1.

The usage of the corner structure in the I/O interval tree is as follows. For each of the $O(Bf^2)$ *multi* lists of an internal node, if there are at least $B/2$ intervals, we directly store the list in disk as before; otherwise, the ($< B/2$) intervals are maintained in a corner structure associated with the internal node. Observe that there are $O(B^2)$ intervals maintained in a corner structure. Assuming that all such $O(B^2)$ intervals can fit in main memory, the corner structure can be built with optimal I/O's (see Section 2.1.1). In summary, the height of the I/O interval tree is $O(\log_B N)$, and using the corner structure, the space needed is $O(N/B)$ blocks in disk, which is worst-case optimal.

2.2.2. *Query Algorithm.* We now review the query algorithm given in [2], which is very simple. Given a query value q, we perform the following recursive process starting from the root of the interval tree. For the current node u that we want to visit, we read it from disk. If u is a leaf, we just check the $O(B)$ intervals stored in u, report those intervals containing q and stop. If u is an internal node, we perform a binary search for q on the keys (slab boundaries) stored in u to identify the slab E_i containing q. Now we want to report all intervals belonging to u that contain q. We check the *left* list associated with E_i, report the intervals sequentially until we reach some interval whose *left* endpoint value is *larger* than q. Recall that each *left* list is sorted by increasing left endpoint values. Similarly, we check the *right* list associated with E_i. This takes care of all intervals belonging to u whose *endpoints* lie in E_i. Now we also need to report all intervals that *completely span* E_i. We carry out this task by reporting the intervals in the *multi* lists $multi([\ell, r])$, where $1 \le \ell \le i$ and $i \le r \le Bf - 2$. Finally, we visit the i-th child u_i of u, and recursively perform the same process on u_i.

Since the height of the tree is $O(\log_B N)$, we only visit $O(\log_B N)$ nodes of the tree. We also visit the *left, right,* and *multi* lists for reporting intervals. Let us discuss the theoretically worst-case situations about the underflow blocks in the lists. An underflow block in the *left* or *right* list is fine. Since we only visit one *left* list and one *right* list per internal node, we can charge this $O(1)$ I/O cost to that internal node. But this charging argument does not work for the *multi* lists, since we may visit $\Theta(B)$ *multi* lists for an internal node. This problem is again solved in [2] by using the corner structure of [16] as mentioned at the end of Section 2.2.1. The underflow *multi* lists of an internal node u are not accessed, but are collectively taken care of by performing a query on the corner structure of u. Thus the query algorithm achieves $O(\log_B N + K/B)$ I/O operations, which is worst-case optimal.

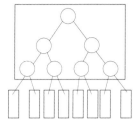

FIGURE 5. Intuition of a binary-blocked I/O interval tree $\mathcal{T}$: each circle is a node in the binary interval tree T, and each rectangle, which blocks a subtree of T, is a node of $\mathcal{T}$.

2.2.3. *Preprocessing Algorithms.* In [8] we used the *scan and distribute* paradigm to build the I/O interval tree. Since the algorithmic and implementation issues of the method are similar to those of the preprocessing method for the binary-blocked I/O interval tree [10], we omit the description here and refer to Section 2.3.4. Other I/O-optimal preprocessing algorithms for the I/O interval tree are described in [8].

2.3. Binary-Blocked I/O Interval Tree. Now we present our binary-blocked I/O interval tree [10]. It is an external-memory extension of the (main memory) binary interval tree [12]. Similar to the I/O interval tree [2] (see Section 2.2), the tree height is reduced from $O(\log_2 N)$ to $O(\log_B N)$, but the branching factor Bf is $\Theta(B)$ rather than $\Theta(\sqrt{B})$. Also, the tree does not introduce any *multi* lists, so it is simpler to implement and also is more space-efficient (by a factor of 2/3) in practice.

2.3.1. *Data Structure.* Recall the binary interval tree [12], denoted by T, described at the beginning of Section 2.2. We use $\mathcal{T}$ to denote our binary-blocked I/O interval tree. Each node in $\mathcal{T}$ is one disk block, capable of holding B items. We want to increase the branching factor Bf so that the tree height is $O(\log_B N)$. The intuition of our method is extremely simple — we *block* a subtree of the binary interval tree T into one node of $\mathcal{T}$ (see Fig. 5). In the following, we refer to the nodes of T as *small nodes.* We take Bf to be $\Theta(B)$. Then in an internal node of $\mathcal{T}$, there are $Bf - 1$ small nodes, each having a key, a pointer to its left list and a pointer to its right list, where all left and right lists are stored in disk.

Now we give a more formal definition of the tree $\mathcal{T}$. First, we sort all *left* endpoints of the N intervals in increasing order from left to right, into a set E. We use interval ID's to break ties. The set E is used to define the keys in small nodes. Then $\mathcal{T}$ is recursively defined as follows. If there are no more than B intervals, then the current node u is a leaf node storing all intervals. Otherwise, u is an internal node. We take $Bf - 1$ median values from E, which partition E into Bf slabs, each with the same number of endpoints. We store sorted, in non-decreasing order, these $Bf - 1$ median values in the node u, which serve as the keys of the $Bf - 1$ small nodes in u. We *implicitly* build a subtree of T on these $Bf - 1$ small nodes, by a *binary-search scheme* as follows. The root key is the median of the $Bf - 1$ sorted keys, the key of the left child of the root is the median of the lower half keys, and the right-child key is the median of the upper half keys, and so on. Now consider the intervals. The intervals that contain one or more keys of u are assigned to u. In fact, each such interval I is assigned to the *highest* small node (in the subtree in

u) whose key is contained in I; we store I in the corresponding left and right lists of that small node in u. For the remaining intervals, each has both endpoints in the same slab and is assigned to that slab. We recursively define the Bf subtrees of the node u as the binary-blocked I/O interval trees on the intervals in the Bf slabs.

Notice that with the above binary-search scheme for implicitly building a (sub)tree on the keys stored in an internal node u, Bf does not need to be a power of 2 — we can make Bf as large as possible, as long as the $Bf-1$ keys, the $2(Bf-1)$ pointers to the left and right lists, and the Bf pointers to the children, etc., can all fit into one disk block.

It is easy to see that $\mathcal{T}$ has height $O(\log_B N)$: $\mathcal{T}$ is defined on the set E with N left endpoints, and is perfectly balanced with $Bf = \Theta(B)$. To analyze the space complexity, observe that there are no more than N/B leaves and thus $O(N/B)$ disk blocks for the tree nodes of $\mathcal{T}$. For the secondary lists, as in the binary interval tree T, each interval is stored either once or twice. The only issue is that a left (right) list may have very few ($<< B$) intervals but still needs one disk block for storage. We observe that an internal node u has $2(Bf-1)$ left plus right lists, $i.e.$, at most $O(Bf)$ such $underfull$ blocks. But u also has Bf children, and thus we can charge the underfull blocks to the child blocks. Therefore the overall space complexity is optimal $O(N/B)$ disk blocks.

As we shall see in Section 2.3.2, the above data structure supports queries in non-optimal $O(\log_2 \frac{N}{B} + K/B)$ I/O's, and we can use the $corner\ structures$ to achieve optimal $O(\log_B N + K/B)$ I/O's while keeping the space complexity optimal.

2.3.2. $Query\ Algorithm$. The query algorithm for the binary-blocked I/O interval tree $\mathcal{T}$ is very simple and mimics the query algorithm for the binary interval tree T. Given a query point q, we perform the following recursive process starting from the root of $\mathcal{T}$. For the current node u, we read u from disk. Now consider the subtree T_u implicitly built on the small nodes in u by the binary-search scheme. Using the same binary-search scheme, we follow a root-to-leaf path in T_u. Let r be the current small node of T_u being visited, with key value m. If $q = m$, then we report all intervals in the left (or equivalently, right) list of r and stop. (We can stop here for the following reasons. (1) Even some descendent of r has the same key value m, such descendent must have empty left and right lists, since if there are intervals containing m, they must be assigned to r (or some small node higher than r) before being assigned to that descendent. (2) For any non-empty descendent of r, the stored intervals are either entirely to the left or entirely to the right of $m = q$, and thus cannot contain q.) If $q < m$, we scan and report the intervals in the left list of r, until the first interval with the left endpoint larger than q is encountered. Recall that the left lists are sorted by increasing left endpoint values. After that, we proceed to the left child of r in T_u. Similarly, if $q > m$, we scan and report the intervals in the right list of r, until the first interval with the right endpoint smaller than q is encountered. Then we proceed to the right child of r in T_u. At the end, if q is not equal to any key in T_u, the binary search on the $Bf-1$ keys locates q in one of the Bf slabs. We then visit the child node of u in $\mathcal{T}$ which corresponds to that slab, and apply the same process recursively. Finally, when we reach a leaf node of $\mathcal{T}$, we check the $O(B)$ intervals stored to report those that contain q, and stop.

Since the height of the tree $\mathcal{T}$ is $O(\log_B N)$, we only visit $O(\log_B N)$ nodes of $\mathcal{T}$. We also visit the left and right lists for reporting intervals. Since we always report the intervals in an $output\text{-}sensitive$ way, this reporting cost is roughly $O(K/B)$.

However, it is possible that we spend one I/O to read the first block of a left/right list but only very few ($<< B$) intervals are reported. In the worst case, all left/right lists visited result in such *underfull reported blocks* and this I/O cost is $O(\log_2 \frac{N}{B})$, because we visit one left or right list per small node and the total number of small nodes visited is $O(\log_2 \frac{N}{B})$ (this is the height of the balanced binary interval tree T obtained by "concatenating" the small-node subtrees T_u's in all internal nodes u's of $\mathcal{T}$). Therefore the overall worst-case I/O cost is $O(\log_2 \frac{N}{B} + K/B)$.

We can improve the worst-case I/O query bound by using the *corner structure* [16] mentioned at the end of Section 2.1.1. The idea is to check a left/right list from disk *only when* at least *one full block* is reported; the underfull reported blocks are collectively taken care of by the corner structure.

For each internal node u of $\mathcal{T}$, we remove the first block from each left and right lists of each small node in u, and collect all these removed intervals (with duplications eliminated) into a single corner structure associated with u (if a left/right list has no more than B intervals then the list becomes empty). We also store in u a "guarding value" for each left/right list of u. For a left list, this guarding value is the smallest left endpoint value among the *remaining* intervals still kept in the left list (*i.e.*, the $(B + 1)$-st smallest left endpoint value in the *original* left list); for a right list, this value is the largest right endpoint value among the remaining intervals kept (*i.e.*, the $(B + 1)$-st largest right endpoint value in the *original* right list). Recall that each left list is sorted by increasing left endpoint values and symmetrically for each right list. Observe that u has $2(Bf - 1)$ left and right lists and $Bf = \Theta(B)$, so the corner structure of u has $O(B^2)$ intervals, satisfying the restriction for the corner structure (see Section 2.1.1). Also, the overall space needed is still optimal $O(N/B)$ disk blocks.

The query algorithm is basically the same as before, with the following modification. If the current node u of $\mathcal{T}$ is an internal node, then we first query the corner structure of u. A left list of u is checked from disk only when the query value q is larger than or equal to the guarding value of that list; similarly for the right list. In this way, although a left/right list might be checked using one I/O to report very few ($<< B$) intervals, it is ensured that in this case the *original first block* of that list is also reported, from the corner structure of u. Therefore we can charge this one underflow I/O cost to the one I/O cost needed to report such first full block. This means that the overall underflow I/O cost can be charged to the K/B term of the reporting cost, so that the overall query I/O cost is optimal $O(\log_B N + K/B)$.

2.3.3. *Dynamization.* As a detour from our application of isosurface extraction which only needs the static version of the tree, we now discuss the ideas on how to apply the dynamization techniques of [2] to the binary-blocked I/O interval tree $\mathcal{T}$ so that it also supports insertions and deletions of intervals each in optimal $O(\log_B N)$ I/O's amortized (we believe that the bounds can further be turned into worst-case, as discussed in [2], but we did not verify the details).

The first step is to assume that all intervals have *left* endpoints in a fixed set E of N points. (This assumption will be removed later.) Each left/right list is now stored in a B-tree so that each insertion/deletion of an interval on a secondary list can be done in $O(\log_B N)$ I/O's. We slightly modify the way we use the guarding values and the corner structure of an internal node u of $\mathcal{T}$ mentioned at the end of Section 2.3.2: instead of putting the first B intervals of each left/right list into the corner structure, we put between $B/4$ and B intervals. For each left/right list L, we keep track of how many of its intervals are actually stored in the corner structure; we denote this number by $C(L)$. When an interval I is to be inserted to a node u of $\mathcal{T}$, we insert I to its destination left and right lists in u, by checking the corresponding guarding values to actually insert I to the corner structure or to the list(s). Some care must be taken to make sure that no interval is inserted twice to the corner structure. Notice that each insertion/deletion on the corner structure needs amortized $O(1)$ I/O's [16]. When a left/right list L has B intervals stored in the corner structure (i.e., $C(L) = B$), we update the corner structure, the list L and its guarding value so that only the first $B/2$ intervals of L are actually placed in the corner structure. The update to the corner structure is performed by rebuilding it, in $O(B)$ I/O's; the update to the list L is done by inserting the extra $B/2$ intervals (coming from the corner structure) to L, in $O(B \log_B N)$ I/O's. We perform deletions in a similar way, where an adjustment of the guarding value of L occurs when L has only $B/4$ intervals in the corner structure (i.e., $C(L) = B/4$), in which case we delete the first $B/4$ intervals from L and insert them to the corner structure to make $C(L) = B/2$ ($C(L) < B/2$ if L stores less than $B/4$ intervals before the adjustment), using $O(B \log_B N)$ I/O's. Observe that between two $O(B \log_B N)$-cost updates there must be already $\Omega(B)$ updates, so each update needs amortized $O(\log_B N)$ I/O's.

Now we show how to remove the assumption that all intervals have *left* endpoints in a fixed set E, by using the *weight-balanced B-tree* developed in [2]. This is basically the same as the dynamization step of the I/O interval tree [2]; only the details for the rebalancing operations differ.

A weight-balanced B-tree has a *branching parameter* a and a *leaf parameter* k, such that all leaves have the same depth and have weight $\Theta(k)$, and that each internal node on level l (leaves are on level 0) has weight $\Theta(a^l k)$, where the weight of a leaf is the number of items stored in the leaf and the weight of an internal node u is the sum of the weights of the leaves in the subtree rooted at u (items defining the weights are stored only in the leaves). For our application on the binary-blocked I/O interval tree, we choose the maximum number of children of an internal node, $4a$, to be $Bf (= \Theta(B))$, and the maximum number of items stored in a leaf, $2k$, to be B. The leaves collectively store all N *left* endpoints of the intervals to define the weight of each node, and each node is one disk block. With these parameter values, the tree uses $O(N/B)$ disk blocks, and supports searches and insertions each in $O(\log_B N)$ worst-case I/O's [2]. As used in [2], this weight-balanced B-tree serves as the base tree of our binary-blocked I/O interval tree. Rebalancing of the base tree during an insertion is carried out by *splitting* nodes, where each split takes $O(1)$ I/O's and there are $O(\log_B N)$ splits (on the nodes along a leaf-to-root path). A key property is that after a node of weight w splits, it will split again only after another $\Omega(w)$ insertions [2]. Therefore, our goal is to split a node of weight w, including updating the secondary lists involved, in $O(w)$ I/O's, so that each split uses amortized $O(1)$ I/O's and thus the overall insertion cost is $O(\log_B N)$ amortized I/O's.

Suppose we want to split a node u on level l. Its weight $w = \Theta(a^l k) = \Theta(B^{l+1})$. As considered in [2], we split u into two new nodes u' and u'' along a slab boundary b of u, such that all children of u to the left of b belong to u' and the remaining children of u, which are to the right of b, belong to u''. The node u is replaced by u' and u'', and b becomes a *new* slab boundary in $parent(u)$, separating the two new adjacent slabs corresponding to u' and u''. To update the corresponding secondary lists, recall that the intervals belonging to u are distributed to the left/right lists of the small nodes of u by a binary-search scheme, where the keys of the small nodes are the slab boundaries of u, and the small-node tree T_u of u is implicitly defined by the binary-search scheme. Here we slightly modify this, by just specifying the small-node tree T_u, which then guides the search scheme (so if T_u is not perfectly balanced, the search on the keys is not a usual binary search). Now, we have to re-distribute the intervals of u, so that those containing b are first put to $parent(u)$, and those to the left of b are put to u' and the rest put to u''. To do so, we first merge all left lists of u to get all intervals of u sorted by the left endpoints. Actually, we need one more list to participate in the merging, namely the list of those intervals stored in the corner structure of u, sorted by the left endpoints. This corner-structure list is easily produced by reading the intervals of the corner structure into main memory and performing an internal sorting. After the merging is done, we scan through the sorted list and re-distribute the intervals of u by using a *new* small-node tree T_u of u where b is the root key and the left and right subtrees of b are the small-node trees $T_{u'}$ and $T_{u''}$ inside the nodes u' and u''. This takes care of all left lists of u' and u'', each automatically sorted by the left endpoints, and also decides the set S_b of intervals that have to be moved to $parent(u)$. We construct the right lists in u' and u'' in a similar way. We build the corner structures of u' and u'' appropriately, by putting the first $B/2$ intervals of each left/right list (or the entire list if its size is less than $B/2$) to the related corner structure. Since each interval belonging to u has its left endpoint inside the slab associated with u, the total number of such intervals is $O(w)$, and thus we perform $O(w/B)$ I/O's in this step. Note that during the merging, we may have to spend one I/O to read an underfull left/right list (i.e., a list storing $<< B$ intervals), resulting in a total of $O(B)$ I/O's for the underfull lists. But for an internal node u, its weight $w = \Theta(B^{l+1})$ with $l \geq 1$, so the $O(B)$ term is dominated by the $O(w/B)$ term. (If u is a leaf then there are no left/right lists.) This completes the operations on the nodes u' and u''.

We also need to update $parent(u)$. First, the intervals in S_b have to be moved to $parent(u)$. Also, b becomes a new slab boundary in $parent(u)$, and we have to re-distribute the intervals of $parent(u)$, including those in S_b, against a *new* small-node tree $T_{parent(u)}$ in which b is the key of some small node. We again consider the left lists and the right lists separately as before. Notice that we get two lists for S_b from the previous step, sorted respectively by the left and the right endpoints. Since $parent(u)$ has weight $\Theta(a^{l+1}k) = \Theta(Bw)$, updating $parent(u)$ takes $O(w)$ I/O's. The overall update cost for splitting u is thus $O(w)$ I/O's, as desired.

We remark that in the case of the I/O interval tree [2], updating $parent(u)$ only needs $O(w/B)$ I/O's, which is better than needed. Alternatively, we can simplify our task of updating $parent(u)$ by always putting the new slab boundary b as a *leaf* of the small-node tree $T_{parent(u)}$. In this way, the original intervals of $parent(u)$ do not need to be re-distributed, and we only need to attach the left and right lists of S_b into $parent(u)$. Again, we need to insert the first $B/2$ intervals of the two lists (removing duplications) to the corner structure of $parent(u)$, using $O(B)$ I/O's by simply rebuilding the corner structure. For u being a leaf, the overall I/O cost for splitting u is $O(1+B) = O(w)$ (recall that $w = \Theta(B^{l+1})$), and for u being an internal node, such I/O cost is $O(w/B+B) = O(w/B)$. We remark that although the small-node trees (e.g., $T_{parent(u)}$) may become very unbalanced, our optimal query I/O bound and all other performance bounds are not affected, since they do not depend on the small-node trees being balanced. Finally, the deletions are performed by lazy deletions, with the rebalancing carried out by a global rebuilding, as described in [2]. The same amortized bound carries over. In summary, our binary-blocked I/O interval tree can support insertions and deletions of intervals each in optimal $O(\log_B N)$ I/O's amortized, with all the other performance bounds unchanged.

2.3.4. *Preprocessing Algorithm.* Now we return to the static version of the tree $\mathcal{T}$. In [10] we again use the *scan and distribute* preprocessing algorithm to build the tree. The algorithm follows the definition of $\mathcal{T}$ given in Section 2.3.1.

In the first phase, we sort (using external sorting) all N input intervals in increasing *left* endpoint values from left to right, into a set S. We use interval ID's to break a tie. We also copy the *left* endpoints, in the same sorted order, from S to a set E. The set E is used to define the median values to partition E into slabs throughout the process.

The second phase is a recursive process. If there are no more than B intervals, then we make the current node u a leaf, store all intervals in u and stop. Otherwise, u is an internal node. We first take the $Bf-1$ median values from E that partition E into Bf slabs, each containing the same number of endpoints. We store sorted in u, in non-decreasing order from left to right, these median values as the keys in the small nodes of u. We now scan all intervals (from S) to distribute them to the node u or to one of the Bf slabs. We maintain a temporary list for u, and also a temporary list for each of the Bf slabs. For each temporary list, we keep one block in main memory as a *buffer*, and keep the remaining blocks in disk. Each time an interval is distribute to the node u or to a slab, we put that interval to the corresponding buffer; when a buffer is full, it is written to the corresponding list in disk. The distribution of each interval I is carried out by the *binary-search scheme* described in Section 2.3.1, which implicitly defines a balanced binary tree T_u on the $Bf-1$ keys and the corresponding small nodes in u. We perform this binary search on these keys to find the highest small node r whose key is contained in I, in which case we assign I to small node r (and also to the current node u), by appending the small node ID of r to I and putting it to the temporary list for the node u, or to find that no such small node exists and both endpoints of I lie in the same slab, in which case we distribute I to that slab by putting I to the corresponding temporary list. When all intervals in S are scanned and distributed, each temporary list has all its intervals, automatically sorted in increasing left endpoint values. Now we sort the intervals belonging to the node u by the small node ID as the first key and the left endpoint value as the second key, in increasing order, so that intervals assigned to the same small node are put together, sorted in increasing left endpoint values. We read off these intervals to set up the left lists of all small nodes in u. Then we copy each such left list to its corresponding right list, and sort the right list by decreasing right endpoint values. The corner structure of u, if we want to construct, can be built at this point. This completes the construction of u. Finally, we perform the process recursively on each of the Bf slabs, using the intervals in the corresponding temporary list as input, to build each subtree of the node u.

We remark that in the above *scan and distribute* process, instead of keeping all intervals assigned to the current node u in *one* temporary list, we could maintain $Bf-1$ temporary lists for the $Bf-1$ small nodes of u. This would eliminate the subsequent sorting by the small node ID's, which is used to *re-distribute* the intervals of u into individual small nodes. But as we shall see in Section 2.4, our method is used to address the system issue that a process cannot open too many files simultaneously, while avoiding a blow-up in disk scratch space.

It is easy to see that the entire process uses $O(\frac{N}{B} \log_B N)$ I/O's, which is nearly optimal. To make a theoretical improvement, we can view the above algorithm as processing $\Theta(\log_2 B)$ levels of the binary interval tree T at a time. By processing $\Theta(\log_2 \frac{M}{B})$ levels at a time, we achieve the theoretically optimal I/O bound $O(\frac{N}{B} \log_{\frac{M}{B}} \frac{N}{B})$. This is similar to the tree-height conversion method of [1] that turns the nearly-optimal I/O bound to optimal for the *scan and distribute* algorithm that builds the I/O interval tree of [2].

2.4. Implementing the Scan and Distribute Preprocessing Algorithms. There is an interesting issue associated with the implementation of the *scan and distribute* preprocessing algorithms. We use the binary-blocked I/O interval tree as an illustrating example. Recall from Section 2.3.4 that during one pass of *scan and distribute*, there are Bf temporary lists for the Bf child slabs and one additional temporary list for the current node u; all these lists grow simultaneously while we distribute intervals to them. If we use one file for each temporary list, then all these $Bf + 1$ files (where $Bf = 170$ in our implementation) have to be open at the same time. Unfortunately, there is a hard limit, imposed by the OS, on the number of files a process can open at the same time. This number is given by the system parameter OPEN_MAX, which in older versions of UNIX was 20 and in many systems was increased to 64. Certainly we can not simultaneously open a file for each temporary list.

We solve this problem by using a single scratch file `dataset.intvl_temp` to collect all child-slab temporary lists, and a file `dataset.current` for the temporary list of the current node u. We use a file `dataset.intvl` for the set S of input intervals. Recall that we have a set E of the *left* endpoints of all intervals; this set is used to define the $Bf - 1$ median values that partition E into Bf slabs of $\lceil n/Bf \rceil$ blocks each, where n is the size of E in terms of integral blocks. We use an interval to represent its left endpoint, so that E is of the same size as S. Since an interval belongs to a slab if both endpoints lie in that slab, the size of each child-slab temporary list is no more than $\lceil n/Bf \rceil$ blocks. Therefore we let the i-th such list start from the block $i \cdot \lceil n/Bf \rceil$ in the file `dataset.intvl_temp`, for $i = 0, \cdots, Bf-1$. After the construction of all such lists is over, we copy them to the corresponding positions in the file `dataset.intvl`, and the scratch file `dataset.intvl_temp` is available for use again. Note that this scratch file is of size n blocks. Recall from Section 2.3.4 that the temporary list for the current node u is used to keep the intervals assigned to u. Thus the size of the scratch file `dataset.current` is also no more than n blocks. After constructing the left and right lists of u, this scratch file is again available for use. To recursively perform the process for each child slab i, we use the portion of the file `dataset.intvl` starting from the block $i \cdot \lceil n/Bf \rceil$ with no more than $\lceil n/Bf \rceil$ blocks as the new set S of input intervals.

As mentioned in Section 2.3.4, our algorithm of collecting all intervals assigned to the current node u in a *single* temporary list is crucial for keeping the disk scratch space small. If we were to maintain $Bf - 1$ temporary lists for the individual small nodes in u and collect all such lists into a scratch file as we do above for child slabs, then we would have a disk space blow-up. Observe that the *potential* size of each small-node list can be much larger than the actual size; in fact, *all* input intervals can belong to the topmost small node, or to the two small nodes one level below, or to any of the $\lceil \log_2 Bf \rceil$ levels in the subtree on the small nodes of u. Therefore to reserve enough space for each small-node list, the disk scratch space would blow up by a factor of $\lceil \log_2 Bf \rceil$. In addition, the varying sizes of the temporary lists inside the scratch file would complicate the coding. (Using a uniform list size instead in this method, then, would increase the blow-up factor to $Bf - 1$!) On the contrary, our method is both simple and efficient in solving the problem.

3. The I/O-Filter Technique

In this section we describe our I/O-filter technique [8, 9]. As mentioned in Section 1.3, we first produce an interval $I = [\min, \max]$ for each cell of the dataset so that searching active cells amounts to performing stabbing queries. We then use the I/O interval tree or the metablock tree as the indexing structure to solve the stabbing queries, together with the isosurface engine of Vtk [25].

It is well known that random accesses in disk following pointer references are very inefficient. If we keep the dataset and build a *separate* indexing structure where each interval has a pointer to its corresponding cell record in disk, then during queries we have to perform pointer references in disk to obtain the cell information, possibly reading one disk block per active cell, *i.e.*, the reporting I/O cost becomes $O(K)$ rather than $O(K/B)$, which is certainly undesirable. In the I/O-filter technique [8, 9], we store the cell information *together with* its interval in the indexing data structure, so that this kind of pointer references are avoided. Also, the cell information we store is the *direct cell information*, *i.e.*, the x-, y-, z- and the scalar values of the vertices of the cell, rather than pointers to the vertices in the vertex information list. (In addition to the direct cell information, we also store the cell ID and the left and right endpoint values for each interval.) In this way, the dataset is *combined* with the indexing structure, and the original dataset can be thrown away. Inefficient pointer references in disk are completely avoided, at the cost of increasing the disk space needed to hold the combined indexing structure. We will address this issue in Section 4 when we describe our two-level indexing scheme [10].

3.1. Normalization. If the input dataset is given in a format that provides direct cell information, then we can build the interval/metablock tree directly. Unfortunately, the datasets are often given in a format that contains indices to vertices[1]. In the I/O filter technique, we first de-reference the indices before actually building the interval tree or the metablock tree. We call this de-referencing process *normalization*.

Using the technique of [5, 7], we can efficiently perform normalization as follows. We make one file (the *vertex file*) containing the direct information of the vertices (the 3D coordinates and the scalar values), and another file (the *cell file*) of the cell records with the vertex indices. In the first pass, we externally sort the cell file by the indices (pointers) to the first vertex, so that the first group in the file contains the cells whose first vertices are vertex 1, the second group contains the cells whose first vertices are vertex 2, and so on. Then by scanning through the vertex file and the cell file simultaneously, we fill in the direct information of the first vertex of each cell in the cell file. In the next pass, we sort the cell file by the indices to the second vertices, and fill in the direct information of the second vertex of each cell in the same way. Actually, each pass is a *joint* operation

---

[1]The input is usually a Toff file, which is analogous to the Geomview "off" file. It has the number of vertices and tetrahedra, followed by a list of the vertices and a list of the tetrahedra, each of which is specified using the vertex location in the file as an index. See [29].

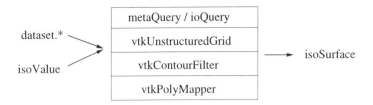

FIGURE 6. Isosurface extraction phase. Given the data structure files of the metablock/interval tree and an isovalue, `metaQuery/ioQuery` filters the dataset and passes to Vtk only those active cells of the isosurface. Several Vtk routines are used to generate the isosurface.

(commonly used in database), using the vertex ID's (the vertex indices) as the key on both the cell file and the vertex file. By repeating the joint process for each vertex of the cells, we obtain the direct information for each cell; this completes the normalization process.

3.2. Interfacing with Vtk. A full isosurface extraction pipeline should include several steps in addition to finding active cells (see Fig. 2). In particular, (1) the intersection points and triangles have to be computed; (2) the triangles need to be decimated [**26**]; and (3) the triangle strips have to be generated. Steps (1)–(3) can be carried out by the existing code in Vtk [**25**].

Our two pieces of isosurface querying code, `metaQuery` (for querying the metablock tree) and `ioQuery` (for querying the I/O interval tree), are implemented by linking the respective I/O querying code with Vtk's isosurface generation code, as shown in Fig. 6. Given an isovalue q, we use `metaQuery` or `ioQuery` to query the indexing structure in disk, and bring *only the active cells* to main memory; this much smaller set of active cells is treated as an input to Vtk, whose usual routines are then used to generate the isosurface. Thus we *filter out* those portions of the dataset that are not needed by Vtk. More specifically, given an isovalue, (1) all active cells are collected from disk; (2) a vtkUnstructuredGrid object is generated; (3) the isosurface is extracted with vtkContourFilter; and (4) the isosurface is saved in a file with vtkPolyMapper. At this point, memory is deallocated. If multiple isosurfaces are needed, this process is repeated. Note that this approach requires double buffering of the active cells during the creation of the vtkUnstructuredGrid data structure. A more sophisticated implementation would be to incorporate the functionality of `metaQuery` (resp. `ioQuery`) inside the Vtk data structures and make the methods I/O aware. This should be possible due to Vtk's pipeline evaluation scheme (see Chapter 4 of [**25**]).

3.3. Experimental Results. Now we present some experimental results of running the two implementations of *I/O-filter* and also Vtk's native isosurface implementation on real datasets. We have run our experiments on four different datasets. All of these are tetrahedralized versions of well-known datasets. The Blunt Fin, the Liquid Oxygen Post, and the Delta Wing datasets are courtesy of NASA. The Combustion Chamber dataset is from Vtk [**25**]. Some representative isosurfaces generated from our experiments are shown in Fig. 1.

Our benchmark machine was an off-the-shelf PC: a Pentium Pro, 200MHz with 128M of RAM, and two EIDE Western Digital 2.5Gb hard disk (5400 RPM, 128Kb

	Blunt	Chamber	Post	Delta
`metaQuery` – 128M	9s	17s	19s	26s
`ioQuery` – 128M	7s	16s	18s	31s
`vtkiso` – 128M	15s	22s	44s	182s
`vtkiso` I/O – 128M	3s	2s	12s	40s
`metaQuery` – 32M	9s	19s	21s	31s
`ioQuery` – 32M	10s	19s	22s	32s
`vtkiso` – 32M	21s	54s	1563s	3188s
`vtkiso` I/O – 32M	8s	28s	123s	249s

TABLE 1. Overall running times for the extraction of the 10 isosurfaces using `metaQuery`, `ioQuery`, and `vtkiso` with different amounts of main memory. These include the time to read the datasets and write the isosurfaces to files. `vtkiso` I/O is the fractional amount of time of `vtkiso` for reading the dataset and generating a vtkUnstructuredGrid object.

cache, 12ms seek time). Each disk block size is 4,096 bytes. We ran Linux (kernels 2.0.27, and 2.0.30) on this machine. One interesting property of Linux is that it allows during booting the specification of the exact amount of main memory to use. This allows us to *fake* for the isosurface code a given amount of main memory to use (after this memory is completely used, the system will start to use disk swap space and have page faults). This has the exact same effect as removing physical main memory from the machine.

In the following we use `metaQuery` and `ioQuery` to denote the entire isosurface extraction codes shown in Fig. 6, and `vtkiso` to denote the Vtk-only isosurface code. There are two batteries of tests, each with different amount of main memory (128M and 32M). Each test consists of calculating 10 isosurfaces with isovalues in the range of the scalar values of each dataset, by using `metaQuery`, `ioQuery`, and `vtkiso`. We did not run X-windows during the isosurface extraction time, and the output of vtkPolyMapper was saved in a file instead.

We summarize in Table 1 the *total* running times for the extraction of the 10 isosurfaces using `metaQuery`, `ioQuery`, and `vtkiso` with different amounts of main memory. Observe that both `metaQuery` and `ioQuery` have significant advantages over `vtkiso`, especially for large datasets and/or small main memory. In particular, from the Delta entries in 32M, we see that both `metaQuery` and `ioQuery` are about 100 times faster than `vtkiso`!

In Fig. 7, we show representative benchmarks of calculating 10 isosurfaces from the Delta dataset with 32M of main memory, using `ioQuery` (left column) and `vtkiso` (right column). For each isosurface calculated using `ioQuery`, we break the running time into four categories. In particular, the bottommost part is the time to perform I/O search in the I/O interval tree to find the active cells (the search phase), and the third part from the bottom is the time for Vtk to actually generate the isosurface from the active cells (the generation phase). It can be seen that the search phase always takes *less* time than the generation phase, *i.e.*, the search phase is no longer a bottleneck. Moreover, the cost of the search phase can be hidden by

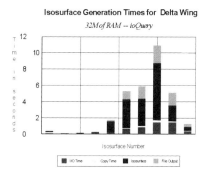
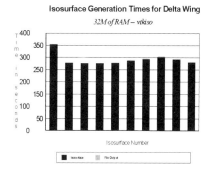

FIGURE 7. Running times for extracting isosurfaces from the Delta dataset, using `ioQuery` (left column) and `vtkiso` (right column) with 32M of main memory. Note that for `vtkiso` the cost of reading the entire dataset into main memory is not shown.

overlapping this I/O time with the CPU (generation phase) time. The isosurface query performance of `metaQuery` is similar to that of `ioQuery`.

In Table 2, we show the performance of querying the metablock tree and the I/O interval tree on the Delta dataset. In general, the query times of the two trees are comparable. We are surprised to see that sometimes the metablock tree performs much more disk reads but the running time is faster. This can be explained by a better locality of disk accesses of the metablock tree. In the metablock tree, the *horizontal, vertical,* and *TS* lists are always read sequentially during queries, but in the I/O interval tree, although the *left* and *right* lists are always read sequentially, the *multi* lists reported may cause scattered disk accesses. Recall from Section 2.2.2 that for a query value q lying in the slab i, all *multi* lists of the multi-slabs spanning the slab i are reported; these include the multi-slabs $[1, i], [1, i+1], \cdots, [1, Bf-2]$, $[2, i], [2, i+1], \cdots, [2, Bf-2], \cdots, [i, i], [i, i+1], \cdots, [i, Bf-2]$. While $[\ell, \cdot]$'s are in consecutive places of a file and can be sequentially accessed, changing from $[\ell, Bf-2]$ to $[\ell+1, i]$ causes non-sequential disk reads (since $[\ell+1, \ell+1], \cdots, [\ell+1, i-1]$ are skipped) — we store the *multi* lists in a "row-wise" manner in our implementation, but a "column-wise" implementation would also encounter a similar situation. This

Delta (1,005K cells)											
Isosurface ID		1	2	3	4	5	6	7	8	9	10
Active Cells		32	296	1150	1932	5238	24788	36738	55205	32677	8902
`metaQuery`	Page Ins	3	8	506	503	471	617	705	1270	1088	440
	Time (sec)	0.05	0	0.31	0.02	0.02	0.89	0.59	1.24	1.88	0.29
`ioQuery`	Page Ins	6	31	35	46	158	578	888	1171	765	271
	Time (sec)	0.1	0.05	0.05	0.09	0.2	0.67	0.85	1.44	1.43	0.35

TABLE 2. Searching active cells on the metablock tree (using `metaQuery`) and on the I/O interval tree (using `ioQuery`) in a machine with 32M of main memory. This shows the performance of the query operations of the two trees. (A "0" entry means "less than 0.01 before rounding".)

leads us to believe that in order to correctly model I/O algorithms, some cost should be associated with the *disk-head movements*, since this is one of the major costs involved in disk accesses.

Finally, without showing the detailed tables, we remark that our metablock-tree implementation of I/O-filter improves the practical performance of our I/O-interval-tree implementation of I/O-filter as follows: the tree construction time is twice as fast, the average disk space is reduced from 7.7 times the original dataset size to 7.2 times, and the disk scratch space needed during preprocessing is reduced from 16 times the original dataset size to 10 times. In each implementation of I/O-filter, the running time of preprocessing is the same as running external sorting a few times, and is linear in the dataset size even when it exceeds the main memory size, showing the *scalability* of the preprocessing algorithms.

We should point out that our I/O interval tree implementation did not implement the corner structure for the sake of simplicity of the coding. This may result in non-optimal disk space and non-optimal I/O query cost in the worst case. Our metablock tree implementation did not implement its corner structure either; this is to reduce the disk space overhead (unlike the I/O interval tree, the corner structure is not needed to achieve the worst-case optimal space bound; implementing the corner structure in the metablock tree would only increase the space overhead by some constant factor), but may also result in non-optimal query I/O cost in the worst case. Since the I/O query time already took less time than the CPU generation time, our main interest for implementing the metablock tree was in reducing the disk space overhead and hence there was no need to implement the corner structure. However, from the view point of data-structure experimentation, it is an interesting open question to investigate the effects of the corner structure into the practical performance measures of the two trees.

4. The Two-Level Indexing Scheme

In this section we survey our two-level indexing scheme proposed in [10]. The goal is to reduce the large disk space overhead of I/O-filter, at the cost of possibly increasing the query time. In particular, we would like to have a flexible scheme that can exhibit a smooth trade-off between disk space and query time.

We observe that in the I/O-filter technique [8, 9], to avoid inefficient *pointer references* in disk, the *direct cell information* is stored with its interval, in the indexing data structure (see Section 3). This is very inefficient in disk space, since the vertex information (*i.e.*, the x-, y-, $z-$ and the scalar values of the vertex) is duplicated many times, once for each cell sharing the vertex. Moreover, in the indexing structures [2, 16] used, each interval is stored three times in practice, increasing the duplications of vertex information by another factor of three. To eliminate this inefficiency, the new indexing scheme uses a two-level structure. First, we partition the original dataset into clusters of cells, called *meta-cells*. Secondly, we produce *meta-intervals* associated with the meta-cells, and build our indexing data structure on the meta-intervals. We *separate* the cell information, kept only in meta-cells in disk, from the indexing structure, which is also in disk and only contains pointers to meta-cells. Isosurface queries are performed by first querying the indexing structure, then using the reported meta-cell pointers to read from disk the *active* meta-cells intersected by the isosurface, which can then be generated from active meta-cells.

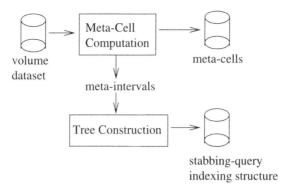

FIGURE 8. The preprocessing pipeline of the two-level indexing scheme isosurface technique.

While we need to perform *pointer references* in disk from the indexing structure to meta-cells, the *spatial coherences* of isosurfaces and of our meta-cells ensure that each meta-cell being read contains *many* active cells, so such pointer references are efficient. Also, a meta-cell is always read as a whole, hence we can use pointers *within* a meta-cell to store each meta-cell compactly. In this way, we obtain efficiencies in *both* query time and disk space.

4.1. Main Techniques. We show the preprocessing pipeline of the overall algorithm in Fig. 8. The main tasks are as follows.

1. Group spatially neighboring cells into *meta-cells*. The total number of vertices in each meta-cell is roughly the same, so that during queries each meta-cell can be retrieved from disk with approximately the same I/O cost. Each cell is assigned to exactly one meta-cell.
2. Compute and store in disk the meta-cell information for each meta-cell.
3. Compute *meta-intervals* associated with each meta-cell. Each meta-interval is an interval [min, max], to be defined later.
4. Build in disk a stabbing-query indexing structure on meta-intervals. For each meta-interval, only its min and max values and the meta-cell ID are stored in the indexing structure, where the meta-cell ID is a pointer to the corresponding meta-cell record in disk.

We describe the representation of meta-cells. Each meta-cell has a list of vertices, where each vertex entry contains its x-, y-, z- and scalar values, and a list of cells, where each cell entry contains pointers to its vertices in the vertex list. In this way, a vertex shared by many cells in the same meta-cell is stored just *once* in that meta-cell. The only duplications of vertex information occur when a vertex belongs to two cells in *different* meta-cells; in this case we let both meta-cells include that vertex in their vertex lists, so that each meta-cell has *self-contained* vertex and cell lists. We store the meta-cells, one after another, in disk.

The purpose of the meta-intervals for a meta-cell is analogous to that of an interval for a cell, *i.e.*, a meta-cell is *active* (intersected by the isosurface of q) if and only if one of its meta-intervals contains q. Intuitively, we could just take the minimum and maximum scalar values among the vertices to define the meta-interval (as cell intervals), but such big range would contain "gaps" in which no

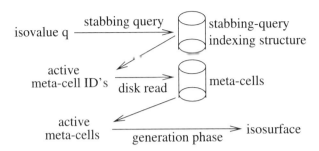

FIGURE 9. The query pipeline of the two-level indexing scheme isosurface technique.

cell interval lies. Therefore, we break such big range into pieces, each a meta-interval, by the gaps. Formally, we define the *meta-intervals* of a meta-cell as the *connected components* among the intervals of the cells in that meta-cell. With this definition, searching active meta-cells amounts to performing stabbing queries on the meta-intervals.

The query pipeline of our overall algorithm is shown in Fig. 9. We have the following steps.

1. Find all meta-intervals (and the corresponding meta-cell ID's) containing q, by querying the stabbing-query indexing structure in disk.
2. (Internally) sort the reported meta-cell ID's. This makes the subsequent disk reads for active meta-cells *sequential* (except for skipping inactive meta-cells), and minimizes the disk-head movements.
3. For each active meta-cell, read it from disk to main memory, identify active cells and compute isosurface triangles, throw away the current meta-cell from main memory and repeat the process for the next active meta-cell. At the end, patch the generated triangles and perform the remaining operations in the generation phase to generate and display the isosurface.

Now we argue that in step 3 the pointer references in disk to read meta-cells are efficient. Intuitively, by the way we construct the meta-cells, we can think of each meta-cell as a cube, with roughly the same number of cells in each dimension. Also, by the *spatial coherence* of an isosurface, the isosurface usually cannot cut too many meta-cells *through corners only*. Thus by a dimension argument, if an active meta-cell has C cells, for most times the isosurface cuts through $C^{2/3}$ cells. This is similar to the argument that usually there are $\Theta(N^{2/3})$ active cells in an N-cell volume dataset. This means that we read C cells (a whole meta-cell) for every $C^{2/3}$ active cells, *i.e.*, we traverse a "thickness" of $C^{1/3}$ layers of cells, for one layer of isosurface. Therefore we read $C^{1/3} \cdot (K/B)$ disk blocks for K active cells, which is a factor of $C^{1/3}$ from optimal. Notice that when the size of the meta-cells is increased, the number of duplicated vertices is decreased (less vertices in meta-cell boundaries), and the number of meta-intervals is also decreased (less meta-cells), while the number C is increased. Hence we have a *trade-off* between space and query time, by varying the meta-cell size. Since the major cost in disk reads is in *disk-head movements* (*e.g.*, reading two disk blocks takes approximately the same time as reading one block, after moving the disk head), we can increase meta-cell sizes while keeping the effect of the factor $C^{1/3}$ negligible.

4.1.1. *Stabbing Query Indexing Structure.* In the two-level indexing scheme, any stabbing-query indexing structure can be used. For the sake of simplicity of coding and being more space-efficient, we implemented the binary-blocked I/O interval tree [**10**] (see Section 2.3) as the indexing structure. Since our goal is to reduce the disk space overhead, we do not implement the corner structure (whose only purpose is to improve the query I/O cost to optimal; see Section 2.3.2).

4.1.2. *Meta-Cell Computation.* The efficient subdivision of the dataset into meta-cells lies at the heart of the overall isosurface algorithm. The computation is similar to the partition induced by a k-d-tree [**4**], but we do not need to compute the multiple levels. The computation is essentially carried out by performing external sorting a few times.

We assume that the input dataset is in a general "index cell set" (ICS) format, *i.e.*, there is a list of vertices, each containing its x-, y-, z- and scalar values, and a list of cells, each containing pointers to its vertices in the vertex list. We want to partition the dataset into H^3 meta-cells, where H is a parameter that we can adjust to vary the meta-cell sizes, which are usually several disk blocks. The final output of the meta-cell computation is a single file that contains all meta-cells, one after another, each an *independent* ICS file (*i.e.*, the pointer references from the cells of a meta-cell are *within* the meta-cell). We also produce meta-intervals for each meta-cell. The meta-cell computation consists of the following steps.

1. Partition vertices into clusters of equal size. This is the *key* step in constructing meta-cells. We use each resulting cluster to define a meta-cell, whose vertices are those in the cluster, plus some *duplicated* vertices to be constructed later. Observe that meta-cells may differ dramatically in their volumes, but their numbers of vertices are roughly the same. The partitioning method is very simple. We first externally sort all vertices by the x-values, and partition them into H consecutive chunks. Then, for each such chunk, we externally sort its vertices by the y-values, and partition them into H chunks. Finally, we repeat the process for each refined chunk, except that we externally sort the vertices by the z-values. We take the final chunks as clusters. Clearly, each cluster has spatially neighboring vertices. The computing cost is bounded by three passes of external sorting. This step actually *assigns* vertices to meta-cells. We produce a *vertex-assignment* list with entries (v_{id}, m_{id}), indicating that the vertex v_{id} is assigned to the meta-cell m_{id}.

2. Assign cells to meta-cells and duplicate vertices. Our assignment of cells to meta-cells attempts to minimize the wasted space. The basic coverage criterion is to see how a cell's vertices have been mapped to meta-cells. A cell whose vertices all belong to the same meta-cell is assigned to that meta-cell. Otherwise, the cell is in the boundary, and a simple voting scheme is used: the meta-cell that contains the "most" vertices owns that cell, and the "missing" vertices of the cell have to be duplicated and inserted to this meta-cell. We break ties arbitrarily. In order to determine this assignment, we need to obtain for each cell, the destination meta-cells of its vertices. This pointer de-referencing in disk is carried out by performing the *joint* operation a few times as described in the *normalization* process in Section 3.1. At the end, we have a list for assigning cells to meta-cells, and also a *vertex-duplication* list with entries (v_{id}, m_{id}), indicating that the vertex v_{id} has to be duplicated and inserted to the meta-cell m_{id}.

3. Compute the vertex and cell lists for each meta-cell. To actually duplicate vertices and insert them to appropriate meta-cells, we first need to de-reference the

vertex ID's (to obtain the *complete* vertex information) from the vertex-duplication list. We can do this by using one join operation, using vertex ID as the key, on the original input vertex list and the vertex duplication list. Now the vertex-duplication list contains for each entry the complete vertex information, together with the ID of the meta-cell to which the vertex must be inserted. We also have a list for assigning cells to meta-cells. To finish the generation of meta-cells, we use a main join operation on these lists, using the meta-cell ID's as the main key. To avoid possible replications of the same vertex inside a meta-cell, we use vertex ID's as the secondary key during the sorting for the join operation. Finally, we update the vertex pointers for the cells *within* each meta-cell. This can be easily done since each meta-cell can be kept in main memory.

4. Compute meta-intervals for each meta-cell. Since each meta-cell can fit in main memory, this step only involves in-core computation. First, we compute the interval for each cell in the meta-cell. Then we sort all interval endpoints. We scan through the endpoints, with a counter initialized to 0. A left endpoint encountered increases the counter by 1, and a right endpoint decreases the counter by 1. A "$0 \to 1$" transition gives the beginning of a new meta-interval, and a "$1 \to 0$" transition gives the end of the current meta-interval. We can easily see that the computation is correct, and the computing time is bounded by that of internal sorting.

4.2. Experimental Results. The experimental set up is similar to the one described in Section 3.3. We considered five datasets in the experimental study; four of them were used in [8, 9] as shown in Section 3.3. A new, larger dataset, Cyl3 with about 5.8 million cells has been added to our test set. This new dataset was originally a smaller dataset (the Cylinder dataset) from Vtk [25] but was subdivided to higher resolution by breaking the tetrahedra into smaller ones.

Tables 3 and 4 summarize some important statistics about the performance of the preprocessing. In Table 3, a global view of the performance of the technique can be seen on four different datasets. Recall that the indexing structure used is the binary-blocked I/O interval tree [10] (see Section 2.3), abbreviated the BBIO tree here. It is interesting to note that the size of the BBIO tree is quite small. This is because we separate the indexing structure from the dataset. Also, previously we have one interval stored for each cell, but now we have only a few (usually no more than 3) meta-intervals for a meta-cell, which typically contains more than 200 cells. We remark that if we replace the I/O interval tree or the metablock tree with the BBIO tree in the I/O-filter technique (described in Section 3), then the average tree size (combining the direct cell information together) is about 5 times the original dataset size, as opposed to 7.2–7.7 times, showing that the BBIO tree *along* can improve the disk space by a factor of 2/3 (and much more improvements are obtained by employing the two-level indexing scheme and the meta-cell technique).

In Table 4 we vary the number of meta-cells used for the Delta dataset. This table shows that our algorithm scales well with increasing meta-cell sizes. The most important feature is the linear dependency of the querying accuracy versus the disk overhead. That is, using 146 meta-cells (at 7% disk space overhead), for a given isosurface, we needed 3.34s to find the active cells. When using 30,628 meta-cells (at 63% disk space overhead), we only need 1.18s to find the active cells. Basically, the more meta-cells, the more accurate our active cell searchers, and less data we need to fetch from disk. An interesting point is that the more data fetched, the

	Blunt	Chamber	Post	Cyl3
# of meta-cells	737	1009	1870	27896
Const. Time	50s	60s	154.8s	3652s
Original Size	3.65M	4.19M	10M	152M
Meta-Cell Size	4.39M	5M	12.2M	271M
Size Increase	20%	21%	22%	78%
Avg # of Cells	254.2	213.1	274.5	208
BBIO_Tree (size)	29K	28K	84K	1.7M
BBIO_Tree (time)	0.35s	0.67s	1.23s	43s

TABLE 3. Statistics for the preprocessing on different datasets. First, we show the number of meta-cells used for partitioning the dataset, followed by the total time for the meta-cell computation. Secondly, the original dataset size and the size of the meta-cell file are shown. We also show the overall increase in storage, and the average number of cells in a meta-cell. Next, we show the size (in bytes) of the indexing structure, the binary-blocked I/O interval tree (the BBIO tree) and its construction time.

# of meta-cells	146	361	1100	2364	3600	8400	30628
Size Increase	7%	10%	16%	22%	26.9%	37.9%	63%
Act. Meta-Cells	59%	52%	37%	31%	30%	23%	16%
Query Time	3.34s	2.76s	2.09s	1.82s	1.73s	1.5s	1.18s

TABLE 4. Statistics for preprocessing and querying isosurfaces on the Delta dataset (original size 19.4M). We show the size increase in the disk space overhead. Also, we show the performance of a representative isosurface query: percentage of the meta-cells that are active (intersected by the isosurface), and the time for finding the active cells (the time for actual isosurface generation is not included here).

more work (and more main memory usage due to larger individual meta-cell sizes) that the isosurface generation engine has to do. By paying 63% disk overhead, we only need to fetch 16% of the dataset into main memory, which is clearly a substantial saving.

In summary, we see that the disk space is reduced from 7.2–7.7 times the original dataset size in the I/O-filter technique to 1.1–1.5 times, and the disk scratch space is reduced from 10–16 times to less than 2 times. The query time is still at least one order of magnitude faster than Vtk. Also, rather than being a single-cost indexing approach, the method exhibits a smooth trade-off between disk space and query time.

5. Conclusions and Future Directions

In this paper we survey our recent work on external memory techniques for isosurface extraction in scientific visualization. Our techniques provide cost-effective methods to speed up isosurface extraction from volume data. The actual code can be made much faster by fine tuning the disk I/O. This is an interesting but hard and time-consuming task, and might often be non-portable across platforms, since the interplay among the operating system, the algorithms, and the disk is non-trivial to optimize. We believe that a substantial speed-up can be achieved by optimizing the external sorting and the file copying primitives.

The two-level indexing scheme is also suitable for dealing with *time-varying* datasets, in which there are several scalar values at each sample point, one value for each time step. By separating the indexing data structure from the meta-cells, one can keep a single copy of the geometric data (in the meta-cells), and have multiple indexing structures for indexing different time steps.

The technique we use to compute the meta-cells has a wider applicability in the preprocessing of general cell structures larger than main memory. For example, one could use our technique to break polyhedral surfaces larger than main memory into spatially coherent sections for simplification, or to break large volumetric grids into smaller ones for rendering purposes.

We believe the two-level indexing scheme [10] brings efficient external-memory isosurface techniques closer to practicality. One remaining challenge is to improve the preprocessing times for large datasets, which, even though is much lower than the ones presented in [8, 9], is still fairly costly.

Acknowledgements

We thank William Schroeder, who is our co-author in the two-level indexing scheme paper [10] and also one of the authors of Vtk [24, 25]. We thank James Abello, Jeff Vitter and Lars Arge for useful comments.

References

[1] L. Arge. Personal communication, April, 1997.

[2] L. Arge and J. S. Vitter. Optimal interval management in external memory. In *Proc. IEEE Foundations of Comp. Sci.*, pages 560–569, 1996.

[3] C. L. Bajaj, V. Pascucci, and D. R. Schikore. Fast isocontouring for improved interactivity. In *1996 Volume Visualization Symposium*, pages 39–46, October 1996.

[4] J. L. Bentley. Multidimensional binary search trees used for associative searching. *Commun. ACM*, 18(9):509–517, 1975.

[5] Y.-J. Chiang. Dynamic and I/O-efficient algorithms for computational geometry and graph problems: theoretical and experimental results. Ph.D. Thesis, Technical Report CS-95-27, Dept. Computer Science, Brown University, 1995.

[6] Y.-J. Chiang. Experiments on the practical I/O efficiency of geometric algorithms: Distribution sweep vs. plane sweep. *Computational Geometry: Theory and Applications*, 9(4):211–236, 1998.

[7] Y.-J. Chiang, M. T. Goodrich, E. F. Grove, R. Tamassia, D. E. Vengroff, and J. S. Vitter. External-memory graph algorithms. In *Proc. ACM-SIAM Symp. on Discrete Algorithms*, pages 139–149, 1995.

[8] Y.-J. Chiang and C. T. Silva. I/O optimal isosurface extraction. In *Proc. IEEE Visualization*, pages 293–300, 1997.

[9] Y.-J. Chiang and C. T. Silva. Isosurface extraction in large scientific visualization applications using the I/O-filter technique. Submitted for a journal publication. Preprint: Technical Report, University at Stony Brook, 1997.

[10] Y.-J. Chiang, C. T. Silva, and W. J. Schroeder. Interactive out-of-core isosurface extraction. In *Proc. IEEE Visualization*, 1998 (to appear).

[11] P. Cignoni, C. Montani, E. Puppo, and R. Scopigno. Optimal isosurface extraction from irregular volume data. In *1996 Volume Visualization Symposium*, pages 31–38, October 1996.

[12] H. Edelsbrunner. A new approach to rectangle intersections, Part I. *Internat. J. Comput. Math.*, 13:209–219, 1983.

[13] M. T. Goodrich, J.-J. Tsay, D. E. Vengroff, and J. S. Vitter. External-memory computational geometry. In *IEEE Foundations of Comp. Sci.*, pages 714–723, 1993.

[14] T. He, L. Hong, A. Varshney, and S. Wang. Controlled topology simplification. *IEEE Transactions on Visualization and Computer Graphics*, 2(2):171–184, June 1996.

[15] T. Itoh and K. Koyamada. Automatic isosurface propagation using an extrema graph and sorted boundary cell lists. *IEEE Transactions on Visualization and Computer Graphics*, 1(4):319–327, December 1995.

[16] P. C. Kanellakis, S. Ramaswamy, D. E. Vengroff, and J. S. Vitter. Indexing for data models with constraints and classes. In *Proc. ACM Symp. on Principles of Database Sys.*, pages 233–243, 1993.

[17] A. Kaufman. Volume visualization. In J. G. Webster, editor, *Encyclopedia of Electrical and Electronics Engineering*, pages 163–169. Wiley Publishing, 1997.

[18] A. Kaufman, D. Cohen, and R. Yagel. Volume graphics. *IEEE Computer*, 26:51–64, July 1993.

[19] Y. Livnat, H.-W. Shen, and C.R. Johnson. A near optimal isosurface extraction algorithm using span space. *IEEE Transactions on Visualization and Computer Graphics*, 2(1):73–84, March 1996.

[20] W. E. Lorensen and H. E. Cline. Marching cubes: A high resolution 3D surface construction algorithm. In Maureen C. Stone, editor, *Computer Graphics (SIGGRAPH '87 Proceedings)*, volume 21, pages 163–169, July 1987.

[21] E. M. McCreight. Priority search trees. *SIAM J. Comput.*, 14:257–276, 1985.

[22] G. M. Nielson and B. Hamann. The asymptotic decider: Removing the ambiguity in marching cubes. In *Visualization '91*, pages 83–91, 1991.

[23] W. Schroeder, W. Lorensen, and C. Linthicum. Implicit modeling of swept surfaces and volumes. In *IEEE Visualization '94*, pages 40–45, October 1994.

[24] W. Schroeder, K. Martin, and W. Lorensen. The design and implementation of an object-oriented toolkit for 3D graphics and visualization. In *IEEE Visualization '96*, October 1996.

[25] W. Schroeder, K. Martin, and W. Lorensen. *The Visualization Toolkit*. Prentice-Hall, 1996.

[26] W. Schroeder, J. Zarge, and W. Lorensen. Decimation of triangle meshes. In Edwin E. Catmull, editor, *Computer Graphics (SIGGRAPH '92 Proceedings)*, volume 26, pages 65–70, July 1992.

[27] H.-W. Shen, C. D. Hansen, Y. Livnat, and C. R. Johnson. Isosurfacing in span space with utmost efficiency (ISSUE). In *IEEE Visualization '96*, October 1996.

[28] H.-W. Shen and C.R. Johnson. Sweeping simplices: A fast iso-surface extraction algorithm for unstructured grids. In *IEEE Visualization '95*, pages 143–150, October 1995.

[29] C. T. Silva, J. S. B. Mitchell, and A. E. Kaufman. Fast rendering of irregular grids. In *1996 Volume Visualization Symposium*, pages 15–22, October 1996.

[30] J. Wilhelms and A. Van Gelder. Octrees for faster isosurface generation. In *Computer Graphics (San Diego Workshop on Volume Visualization)*, volume 24, pages 57–62, November 1990.

DEPARTMENT OF COMPUTER AND INFORMATION SCIENCE, POLYTECHNIC UNIVERSITY, 6 METROTECH CENTER, BROOKLYN, NY 11201
E-mail address: yjc@poly.edu

P.O. BOX 704, IBM T. J. WATSON RESEARCH CENTER, YORKTOWN HEIGHTS, NY 10598
E-mail address: csilva@watson.ibm.com

DIMACS Series in Discrete Mathematics
and Theoretical Computer Science
Volume **50**, 1999

R-tree Retrieval of Unstructured Volume Data For Visualization

Scott T. Leutenegger and Kwan-Liu Ma

ABSTRACT. This paper addresses techniques for supporting volume visualization of disk resident data sets. We assume the data sets to be visualized are too large to fit in memory and hence too expensive to visualize in its entirety. Our solution is to retrieve and visualize data subsets of interest. We investigate how to efficiently retrieve, in particular, unstructured volume data subsets from disk. We experimentally compare both a brute force approach and a packed R-tree multidimensional indexing algorithm. For the data considered, our test results demonstrate that retrieval using R-trees requires significantly fewer disk accesses than using brute force. Finally, we describe how to integrate the indexing into a visualization system architecture utilizing multi-resolution data representations and an approximate rendering method.

1. Introduction

The use of high performance multiprocessor machines and networks of workstations for scientific computation has become common. The data sets produced by these simulations are often large in size, exceeding the memory capacity of workstations by many times, yet scientists wish to interactively explore the produced data. Thus, tools for visualization of disk resident data are needed. In this paper we consider disk resident unstructured-grid data sets from aerodynamics calculations. The disk resident problem is compounded by the fact that the associated meshes are irregular in both shape and resolution.

Since the entire data set can not fit in memory, subregions of the data set are retrieved from disk and then visualized. We show through analysis and experimental results that R-trees constructed using a special packing algorithm support significantly faster subset retrieval than brute force approaches. We then show how this subset retrieval processes can be used within a general visualization system architecture utilizing multi-resolution data representation of the base data to enable efficient exploration of the disk resident data. This general architecture allows for

1991 *Mathematics Subject Classification.* Primary 68P05.

The work of Leutenegger has been supported under the National Aeronautics and Space Administration under NASA Contract No. NAS1-19480, NSF under grant IRI-9610240, and by the Colorado Advanced Software Institute under Grant CASI-97-05.

The work of Ma is supported under the National Aeronautics and Space Administration under NASA Contract No. NAS1-19480.

the user to find which subregions of the base data set to visualize. Hence, the architecture allows the user to specify the boundaries of the subregion to be retrieved from disk.

To find a subset in a memory resident data set one would normally employ k-d or quad/oct trees to reduce the search space. These memory based indexing techniques are not appropriate for our disk based data sets since they have poor paging behavior. When dealing with disk based data, the primary objective is to minimize the number of pages read from disk since a disk access is several orders of magnitude slower than memory accesses. Of the main memory data structures, the most likely candidates for disk based data would be quad/octrees, but they suffer from the following deficiencies:

- Octrees will be imbalanced for unstructured data. Schemes to preprocess the data to determine where to put partitioning lines could be constructed to provide balance to some degree.
- Packing the octree nodes into disk pages to get good paging behavior is difficult.
- Octrees have a low fan-out of 8. *R-trees* have a much larger fan-out, typically 50 to 200 for 4096 byte pages depending the size of the data stored in the pages. The internal nodes contain no data, and hence typically have a fan out of 200. If the number of items indexed is N, the number of nodes needing to be accessed for a small search is $O(log_8 N)$ for octree versus $O(log_{200} N)$ for the R-tree. How the number of nodes accessed turns into the number of disk pages accessed is dependent on the mapping from nodes to pages for octrees, and is 1-1 for *R-trees*.
- Octrees are designed for point data, not region data like tetrahedra.

Thus, initially, we do not consider octrees.

The data sets used in this paper are 3D point data (vertices) and region data (tetrahedra). How to support efficient retrieval of multidimensional disk resident point and region data is an active area of research in the database community [1, 2, 3, 4, 6, 11, 18, 19, 21]. Of these techniques, the R-tree is one of the most promising approaches for indexing both point and region data. In addition the algorithms community has recently begun to consider disk resident multidimensional data [7, 20, 25]. These algorithms offer the advantage of provable near optimal complexity whereas many of the database algorithms are heuristics that have poor worst case performance.

We choose to focus on R-tree variants since in practice they provide excellent performance, are easy to implement, are easily extendible to higher dimensions, and also readily support nearest neighbor queries and spatial joins. Furthermore, when the data is static, i.e. all known a priori as in our scenario, R-trees can be loaded to 100% disk utilization. We propose a specific R-tree based structure and packing scheme to efficiently retrieve disk based computation fluid dynamics (CFD) data.

The rest of this paper is organized as follows. First we describe R-trees and R-tree packing in sections 2 and 3. In section 4 we provide an experimental comparison of the brute force and R-tree methods. In section 5 we outline an overall visualization architecture utilizing multi-resolution date representation and R-tree retrieval. We conclude the paper in section 6.

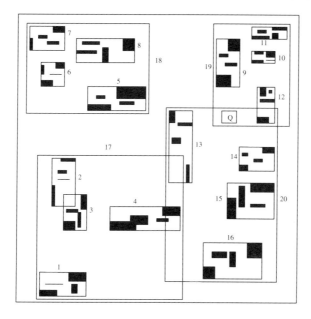

FIGURE 1. A sample R-tree. Input rectangles are shown solid.

2. R-trees

An *R-tree* is a hierarchical data structure derived from the *B-tree* and designed for efficient execution of intersection queries. *R-trees* can be used for any number of dimensions. For clarity and brevity we limit our discussion to the two-dimensional case but our results are for the three dimension case. *R-trees* store a collection of rectangles which can change over time through insertions and deletions. Arbitrary geometric objects are handled by representing each object by the smallest upright rectangle which encloses the object.

Each node of the *R-tree* stores a maximum of n entries. Each entry consists of a rectangle R and a pointer P. For nodes at the leaf level, R is the bounding box of an actual object pointed to by P. Instead of adding this level of indirection the objects themselves may be stored in the leaf nodes instead of a pointer to the object. This is the approach we use in this work. At internal nodes, R is the minimum bounding rectangle of all rectangles stored in the subtree pointed to by P. Every path down through the tree corresponds to a sequence of nested rectangles, the last of which contains an actual data object. Note also that rectangles at any level may overlap and that an *R-tree* created from a particular set of objects is by no means unique.

Figure 1 illustrates a 3-level R-tree where a maximum of 4 rectangles fit per node. The rectangles are the minimum bounding rectangle (MBR) of the data items, triangles in 2D or tetrahedra in 3D for our data. We assume that the levels are numbered 0 (root), 1, and 2 (leaf level). There are 64 rectangles represented by the small dark boxes. The 64 rectangles are grouped into 16 leaf level pages, numbered 1 to 16. The MBR enclosing each leaf node is the smallest box that fully contains the rectangles within the node. The MBRs of the leaf nodes are the rectangles stored in the nodes at the next higher level of the tree.

For example, leaf nodes 1 through 4 are placed in node 17 which is at level 1. The MBR of node 17 (and nodes 18,19,20) is purposely drawn slightly larger than needed for clarity. The root node contains the four level 1 nodes: 17, 18, 19, and 20.

To perform a query Q, all rectangles that intersect the query region must be retrieved and examined (regardless of whether they are stored in an internal node or a leaf node). This retrieval is accomplished by using a simple recursive procedure that starts at the root node and which may follow several paths down through the tree. A node is processed by first retrieving all rectangles stored at that node which intersect Q. If the node is an internal node, the subtrees corresponding to the retrieved rectangles are searched recursively. Otherwise, the node is a leaf node and the retrieved rectangles (or the data objects themselves) are simply returned.

For illustration, consider the query Q in the example of Figure 1. After examining the root node, we determine that nodes 19 and 20 of level 1 must be searched. The search then proceeds to each of these nodes. It is then determined that the query region does not intersect any rectangles stored in node 19 or node 20 and each of these two sub-queries are terminated. See [3] for a more detailed description of R-tree structures and searching.

3. R-tree Packing Algorithms

R-trees and variants allow for dynamic insertion and deletion at the expense of efficient search times. When the data is all present at load time and non-changing, as in our CFD data sets, preprocessing can be done to create more efficient *R-trees*. Such preprocessing is known as *R-tree packing* [23], and result in well structured trees for efficient queries and 100% disk utilization. Several algorithms exist [5, 9, 23]. We choose to focus on the Sort and Tile Recursive (STR) algorithm since it is the easiest to implement and has been shown to provide more efficient point query support and at least as efficient region query support as the others [9].

In the following text we assume a data file of r rectangles and that each *R-tree* node can hold n rectangles. The general process is similar to building a B-tree from a collection of keys by creating the leaf level first and then creating each successively higher level until the root node is created [22].

General Algorithm:

1. Preprocess the data file so that the r rectangles are ordered in $\lceil r/n \rceil$ consecutive groups of n rectangles, where each group of n is intended to be placed in the same leaf level node. The last group may contain less than n rectangles.
2. Load the $\lceil r/n \rceil$ groups of rectangles into pages and output the pair (MBR, page-number) for each leaf level page into a temporary file. The page-numbers are used as the child pointers in the nodes of the next higher level.
3. Recursively pack these MBRs into nodes at the next level and continue proceeding upward until the root node is created.

Packing algorithms differ only in how the rectangles at each level are ordered. The STR algorithm orders rectangles as follows:

"Tile" the data using $\sqrt{r/n}$ rectangular buckets of various sizes so that each bucket contains roughly $\sqrt{r/n}$ input rectangles. Once again we assume coordinates are for the center points of the rectangles. First sort the rectangles based on x-coordinate. Determine the number of leaf level pages $P = \lceil r/n \rceil$

and let $S = \lceil\sqrt{P}\rceil$. Now divide the rectangles into S vertical slices so that each vertical slice contains $S*n$ rectangles. Sort the rectangles from each slice based on y-coordinate and pack them into nodes (the first n rectangles into the first node, the next n into the second node, and so on).

The sorting mentioned above is disk based sorting using merge sort if the files are too large to be sorted in main memory. We call an *R-tree* created by STR packing an STR-tree.

4. Experimental Comparison

To compare subset retrieval time using our STR-tree based method versus brute force, we have performed tests on an SGI Indigo2 with an R4400 250MHZ processor, 128 MB of main memory, and a dedicated disk.

4.1. Methodology. The data set used as a test case is an unstructured 3-d CFD grid consisting of 804,056 nodes and 4,607,888 tetrahedra. The data is stored in two binary files. The node file stores vertex coordinates, 5 CFD solution values and a node id $(x, y, z, q_0, q_1, q_2, q_3, q_4, id)$. The tetrahedra file stores the values that delimit the lower and upper points of the smallest upright 3D region enclosing the tetrahedra and the indices of the nodes that make up the vertices of the tetrahedra $(xmin, ymin, zmin, xmax, ymax, zmax, v_1, v_2, v_3, v_4)$.

A more concise format, used in main memory implementations, would be to have the node file of the form: $(x, y, z, q_0, q_1, q_2, q_3, q_4)$ and the tetrahedra file be of the form $(v1, v2, v3, v4)$. The node number is inferred in the node file from the offset of the node from the beginning. The coordinates of the tetrahedra can then be looked up to determine if the tetrahedra is in the subregion. Such a file structure would be inefficient for disk based data since not only must all data be scanned to find the subset, but each tetrahedra may require up to four disk accesses to find the coordinates of the vertices. Thus, for disk resident data this format is not acceptable.

Relative to the main memory format, our file format requires 12.5% additional disk space for the node file, and 2.5 times as much disk space for the tetrahedra file. We reiterate though, that the more compact format would perform abysmally. Thus, improved access time (orders of magnitude faster for large files) comes at the expense of additional disk space.

Another choice for the R-tree structure would be to store only the vertices for the tetrahedra file in the leaf level nodes of the R-tree. The data is first sorted using the location information, but once packed into a leaf level node the only boundary information needed would be in the parent node and all the data in the node would be returned. This would result in no additional disk space being needed but more tetrahedra would be returned. We intend to investigate this tradeoff in the future.

The brute force method is as follows: a) read through the node file and return all nodes that are contained in the specified subset region; b) read through the tetrahedra file and return all tetrahedra that are contained in the specified subset. The code has been optimized to read in 16K chunks at a time. We ran some experiments with different blocking factors and found that a buffer of 16K to minimize input time.

The STR-tree based method creates two STR-trees, one for the node file and one for the tetrahedra file. The trees are then used for the searches, hence only the relevant data plus a small portion of overhead nodes are read in from memory.

The STR-trees are stored using the normal UNIX file system, no attempt to tune performance by clustering the nodes on raw disk was made

The absolute and relative performance of the two policies is highly dependent on many factors, the most important being main memory size, whether the disk is local or across a network (such as when using NFS), and disk speed versus processor speed. Main memory size is especially relevant since the file system will cache pages of files in main memory between runs. Thus if main memory is sufficiently large, subsequent runs will incur no actual disk I/O, only *soft* page faults.

Our primary comparison metric was the wall clock time (obtained from the *rusage* system call) to run a subset retrieval. CPU time is not sufficient since most of the time is spent in disk retrieval. We also present the number of disk reads for the experiments. There was some variance in measured response times. All data points presented are the average of 40 runs for the same subset region. No attempt at generating confidence intervals was made.

4.2. Experimental Results. In the first set of experiments we alternate running the brute force method with the *R-tree* method. This has the effect of the brute force method flushing some of the remaining pages from the buffer pool from the previous STR-tree retrieval. Thus, performance of the STR-tree method is worse than it would be in a setting of repeated queries. Figure 2 presents results for retrievals of subsets where the regions are cubes centered on the origin and increasing in size.

For a region boundary containing 164,251 tetrahedra (about 3.6% of the overall data) or less, response time is less than 5 seconds for the STR-tree method but 44 seconds or greater for the brute force method. The STR-tree method results in access times that are 9.7 to 16.4 times faster than the brute force method.

As the query region increases, the response time for the brute force method is constant whereas that of the STR-tree method increases. The brute force method reads in the entire data set each time, regardless of the region size. The STR-method only brings in data for the desired region plus some extra data from around the surface of the subset region. According to our test results, the number of actual page reads for the STR tree method is significantly smaller than the brute force method until retrieving more than 40% of the data. We surmise that typical data exploration would retrieve significantly less that 40% of the data thus justifying the use of multi-dimensional indexing.

So far we assume the entire memory contents from the previous STR-tree query are flushed before re-running. In a real interactive setting the user would likely select sub-regions repetitively. Since the file cache or virtual memory system will buffer some of the data between runs, we would expect better performance due to memory hits. How much better depends on how much of the data is re-used in subsequent queries. We would certainly expect the top levels of the STR-tree to be in memory [**10**].

Devising a string of query regions as a benchmark is not easy to do in an unbiased way. Instead, we re-run the same query 41 times and remove the response time of the first run. This provides a lower bound on the expected response time since we would expect only some of the data to be reused, not all of it. Figure 2 also includes the results for these repetitive optimistic bounds. The brute force method does not benefit since it must still read in the same number of disk pages, but the STR-tree based method experiences a substantial reduction in response time. The

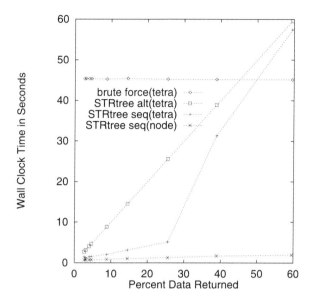

FIGURE 2. Experimental Results, Local Disk

actual response times would fall somewhere between these two extremes and be dependent on the query string. Finally, if splatting rendering is used, we only need to bring back the node data. In this case, the retrieval time is very small and could support interactive visualization.

4.3. Anticipated Effect of Larger Data Sets. The data set used in these experiments is much smaller than anticipated future data sets. When a data set is 10-100 times larger, searches using the brute force method will require 10-100 times more time, whereas we expect the STR-tree method to only require a modest amount (0-1%) more time. The reason is that with a branching factor of 200, one can increase the data size by a factor of 200 and only add one level to the STR-tree, and hence one more page access, Thus, we expect for near future data sets the STR-tree method will be yet another 10-100 times faster relative to brute force.

5. A Visualization System Architecture for Data Exploration

One way subset retrieval can be integrated into a general visualization system is as follows. The proposed visualization process includes two steps as shown in Figure 3. The first step attempts to derive desirable viewing and rendering parameters and to locate regions of interest, a sub-volume. This may be performed using a fast but less accurate rendering algorithm on a coarse representation, thus a much smaller version, of the original data stored in the main memory of the workstation. Once a region of interest is identified at a particular resolution, the user may switch to viewing at a higher-resolution for further exploration. This exploration process continues until the region of interest and viewing as well as rendering parameters are completely determined.

5.1. Multiresolution Representation. We need a multi-resolution representation of the original data (*mrrd*) [14] to make possible interactive exploration of the data on an average workstation. For general three-dimensional unstructured

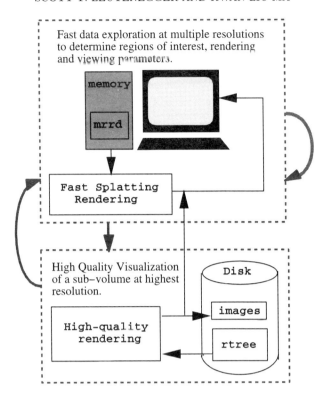

FIGURE 3. The proposed visualization process and system architecture.

grids, the construction of a sequence of nested coarser grid levels is still an open research problem, though refining a coarse grid into finer ones is straightforward. On the other hand, techniques such as decimation which remove a subset of the original grid points and re-triangulate the remaining points have the problem of conforming to the original domain boundaries during coarsening.

Since we use the splatting rendering method to be described in the following section, we can avoid the re-triangulation step. We use an agglomeration technique to construct coarse levels [16]. Essentially, we produce coarse grid point sets that are maximal independent sets of the original fine grid points. The deleted points can be thought of as fused or agglomerated into their respective seed point. The inferred graph for the coarse level is then obtained by deleting all edges within an agglomerated group of points, and replacing all edges between neighboring agglomerated groups with a single edge. In general, for each level of coarsening the data size is reduced by a factor of five.

5.2. Fast Splatting Rendering. We use a fast, approximated splatting algorithm for rendering data at resolutions according to visualization requirements. Splatting was first introduced by Westover [27] and has been mostly used as a fast approximation technique for rendering on uniformly-spaced rectilinear grids. More recently, splatting rendering for data on curvilinear grids has been investigated by other researchers [15]. In addition, Mueller, et al. [17] have studied splatting errors and the accompany anti-aliasing problems. With splatting rendering, an image is formed by determining the screen space contribution of each grid point (a foot-print) and compositing the footprints on top of each other in the visibility order. For parallel projection, a single footprint table can be pre-calculated and shared by all the voxels.

Splatting the node data allows us to ignore the type of computational cells we are dealing with. However, because of the unstructured nature of the grid, a separate footprint must be constructed for each node. Using parallel projection, further approximation has been taken by always representing a footprint with a circle. So each footprint is now defined by the scalar value (e.g. density or pressure) and coordinates of the corresponding grid point, and a radius value which is the average distance from the point to all other immediately neighboring nodes. To take advantage of graphics hardware, we can further approximate each footprint, for example, as an octagon, with a set of hardware Gouraud-shaded triangles as described in [8]. Compositing is then done with the hardware blending support. In this way, we can achieve interactive rendering rates. Presently, only software splatting has been implemented for this work and we can thus use any general-purpose workstations.

The multiple levels of approximation taken certainly degrade the quality and accuracy of visualization results. Frequently, the rough pictures generated with this approach provide enough information for the user to identify the size, shape, and location of particular features in the data in order to refine visualization parameters. For example, Figure 4 shows show multi-resolution visualization results generated by using a software splatting approximation technique. Both images show flow phenomena near the tip of an aircraft wing. The data visualized contains about 1.8 million tetrahedra. It takes about 5 seconds to generate the top image on an SGI Indigo2 with an R4400 250MHZ processor. It only takes about 1 second to generate the bottom image which displays a lower resolution result. The lower resolution image is usually sufficient to allow the user to identify areas of interest for more detailed exploration. By taking advantage of the graphics hardware support, sub-second rendering rates can be achieved, making this setup even more attractive.

5.3. Enhancing Clarity and Interactivity with Feature Lines. Some-times the user needs to locate regions of interest based on geometric features and thus visualize the computational mesh for selecting an ideal view position. For a large data set, real-time display of the corresponding mesh may not be possible even with graphics hardware support. A technique for extracting geometric features from computational meshes can then be used to display the feature lines instead in an highly interactive manner [12]. In Figure 5, the top image shows the surface mesh of the overall domain explored; the dark, dense area corresponds to the re-gion of interest which is near the tip of the wing as already shown in Figure 4. The bottom image shows the corresponding feature lines representation. Displaying the feature lines in conjunction with ordinary visualization results allow the clarity of the visualization to be enhanced since the viewer can then relate solution features with geometric features.

5.4. High Quality Rendering. The second step of the proposed visualiza-tion process takes the parameters derived, extracts the selected sub-volume out of the original data using the R-tree indexing, and invokes a more accurate rendering program to produce high quality visualization results [28]. The sub-volume repre-sents a spatial region of interest usually much smaller than the overall grid domain

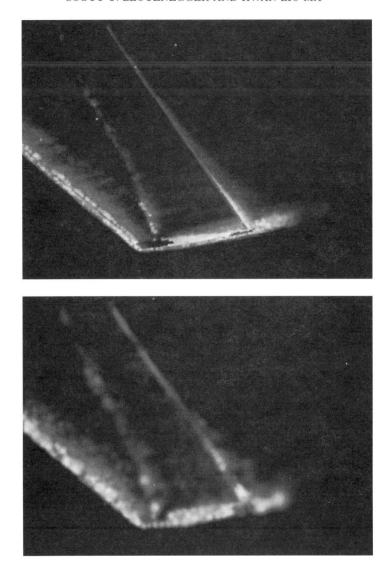

FIGURE 4. Top: software splatting approximation of higher resolution data. Bottom: software splatting approximation of lower resolution data.

and thus may be rendered more efficiently on a workstation. If the sub-volume chosen is too large acceptable rendering times can only achieved by parallel rendering techniques [**12**].

Data exploration is inherently iterative. The visualization process and system architecture developed in this research allow computational scientists to study their data at the highest possible resolution in a more efficient manner, rather than reducing the data or operating in an inefficient batch-mode.

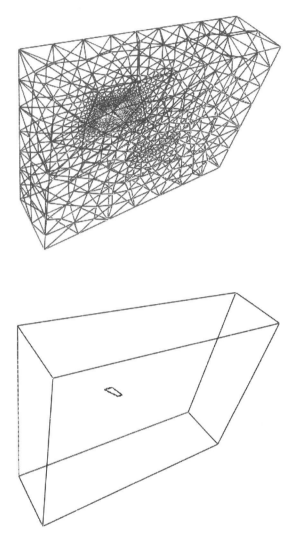

FIGURE 5. Top: overall domain as a surface mesh. Bottom: feature lines visualization.

6. Conclusions

We have shown that the *R-tree* is an efficient data structure for retrieving subsets of disk-resident data. Our experimental comparison with a brute force method clearly demonstrates that in order to achieve reasonable interactive visualization a multi-dimensional indexing algorithm such as R-trees is essential.

Furthermore, we have proposed an architecture to provide interactive data exploration by using multi-resolution data representations, subset retrieval, and rendering methods.

We intend to conduct more extensive comparison tests of the data retrieval methods involved. First we are implementing well balanced octrees for comparison. Next we intend to test the techniques on much larger data sets. We also intend to conduct tests using a remote disk and using multiple disks to distribute extremely large data sets.

Acknowledgments

The authors would like to thank Dimitri Mavripilis for the test datasets and useful discussions about this work.

References

[1] Beckmann, N. , Kriegel, H.P. , Schneider, R. and Seeger, B., *The R*-tree: an Efficient and robust Access Method for Points and Rectangles*, Proc. ACM SIGMOD, May, 1990, pp. 323-331

[2] Faloutsos, C., and Roseman, S., *Fractals for Secondary Key Retrieval*, Eighth Symposium on Principles of Database Systems (PODS-89), March, 1989, pp. 247-252

[3] Guttman, A., *R-trees: a Dynamic Index Structure for Spatial Searching*, Proc. of the ACM SIGMOD, 1984, pp. 47-57

[4] Jagadish, H.V., *Linear Clustering of Object with Multiple Attributes*, Proc. ACM SIGMOD, 1990, pp. 332-342

[5] Kamel, I. and Faloutsos, C., *On Packing R-trees*, Proc. of the 2nd International Conference on Information and Knowledge Management (CIKM-93), Nov, 1993, pp. 490-499

[6] Kamel, I. and Faloutsos, C., *Hilbert R-tree: An Improved R-tree Using Fractals*, Proc. International Conference on Very Large Databases, 1994, pp. 500-509

[7] Kanellakis, P.C. , Ramaswamy, S. , Vengroff, D.E. and Vitter, J.S., *Indexing for Data Models with Constraints and Classes*, Proc. 12th Symposium on Principles of Database Systems (PODS), 1994, pp. 233-243

[8] Laur, D. and Hanrahan, P., *Hierarchical Splatting: A Progressive Refinement Algorithm for Volume Rendering*, Proc. of SIGGRAPH, 1991, pp. 285-288

[9] Leutenegger, S.T., Lopez, M.A., and Edgington, J.M., *STR: A Simple and Efficient Algorithm for R-Tree Packing*, Proc. of 13th International Conference on Data Engineering, Apr, 1997, pp. 497-506

[10] Leutenegger, S.T., and Lopez, M.A., *The Effect of Buffering on the Performance of R-Trees*, Proc. of 14th International Conference on Data Engineering, Feb, 1998, pp. 164-171

[11] Lomet, D.B. and Salzberg, B., *The HB-tree: A Multiatrribute Indexing Method With Good Guaranteed Performance*, ACM Transactions on Database Systems, volume 15, number 4, Mar, 1990, pp. 625-658

[12] Ma, K.-L. and Crockett, T., *A Scalable, Cell-Projection Volume Rendering Algorithm for 3D Unstructured Data*, Proc. of Parallel Rendering '97 Symposium", October, 1997, pp. 95-104

[13] Ma, K.-L. and Interrante, V., *Extracting Features from 3D Unstructured Meshes for Interactive Visualization*, Proc. of the Visualization '97 Conference, 1997, pp. 285-292

[14] Ma, K.-L., Leutenegger, S. and Mavriplis, D., *Interactive Exploration of Large 3D Unstructured-Grid Data*, Institute for Computer Applications in Science and Engineering Technical Report, ICASE Report No. 96-63, 1996 .

[15] Mao, X., Hong, L. and Kaufman, A., *Splatting of Curvilinear Volumes*, Proc. IEEE Visualization '95, Oct, 1995 pp. 61-68

[16] Mavriplis, D. and Venkatakrishnan, V, *Agglomeration Multigrid for Viscous Turbulent Flows*, Institute for Computer Applications in Science and Engineering Technical Report, ICASE Report No. 96-2, 1994.

[17] Mueller, K., Moller, T., Swan, J. E., Crawfis, R., Shareef, N. and Yagel, R., *Splatting Errors and Antialiasing*, IEEE Transactions on Visualization and Computer Graphics, volume 4, number 2, 1998, pp. 178-191

[18] Nievergelt, J. , Hinterberger, H. and Sevcik, K.C. , *The Grid File: An Adaptable, Symmetric Multikey File Structure*, ACM Transactions on Database Systems, volume 9, number 1, Mar, 1984, pp. 38-71

[19] Orenstein, J., *Spatial Query Processing in an Object-oriented Database System*, Proc. ACM SIGMOD, 1986, pp. 326-335

[20] Ramaswamy, S. and Subramanian, S., *Path Caching: A Technique for Optimal External Searching*, Proc. 13th Symposium on Principles of Database Systems (PODS), 1994, pp. 25-35

[21] Robinson, J.T. , *The k-d-b-tree: A Search Structure for Larger Multidimensional Dynamic Indexes*, Proc. ACM SIGMOD, 1981, pp. 10-18

[22] Rosenberg, A.L., and Snyder, L., *Time and Space Optimality in B-Trees*, ACM Transactions on Database Systems, volume 6, number 1, Mar, 1981

[23] Roussopoulos, N. and Leifker, D., *Direct Spatial Search on Pictorial Databases Using Packed R-trees*, Proc. of the ACM SIGMOD, May, 1985, pp. 17-31

[24] Samet, H., *The Design and Analysis of Spatial Data Structures*, 1989, Addison Wesley

[25] Vengroff, D.E., and Vitter, J.S., *Efficient 3-D Range Searching in External Memory*, Proc. 28th ACM Symposium on Theory of Computing (STOC), 1996, pp. 192-201

[26] Westermann, R. and Ertl, T. *Efficiently Using Graphics Hardware in Volume Rendering Applications*, Computer Graphics Proceedings, Annual Conference Series, SIGGRAPH '98, Jul, 1998 pp. 169-177

[27] Westover, Lee, *Footprint Evaluation for Volume Rendering*, Proc. SIGGRAPH '90, Aug, 1990, pp. 267-276

[28] Williams, P. L., Max, N. L. and Stein, C. M., *A High Accuracy Volume Renderer for Unstructured Data*, IEEE Transactions on Visualization and Computer Graphics, volume 4, number 1, 1998, pp. 21-36

SCOTT T. LEUTENEGGER, MATHEMATICS AND COMPUTER SCIENCE DEPARTMENT, UNIVERSITY OF DENVER, DENVER, COLORADO 80208-0189
E-mail address: leut@cs.du.edu

KWAN-LIU MA, INSTITUTE FOR COMPUTER APPLICATIONS IN SCIENCE AND ENGINEERING, NASA LANGLEY RESEARCH CENTER, HAMPTON, VIRGINIA 23681-2199
E-mail address: kma@icase.edu

Author Index

Index

upper triangular matrix, 114
useful blocks, 229, 230
user-level paging, 170

variance, 49
vector probing, 115
vector reversal permutations, 13
vertex cover, 119
vertex degree, 108, 121
vertical stripes, 94
vicinity initialization, 213
virtual file system, 217
virtual memory, 169, 170
virtual memory invariant, 151
virtual reality systems, 16
Vmalloc library, 182, 187–189, 199, 201, 204
volume data, 276, 279
volume dataset, 272
volume graphics, 247
volume visualization, 248
volumetric grids, 276
Voronoi diagrams, 17, 137
voxels, 287
Vtk, 267, 275
Vtk isosurface engine, 266

Walmarts's warehouse, 41
wavelet decompositions, 18
wavelets, 66
web hyperlink, 108
web usage, 181
WEBSQL, 108
weight balanced B-tree, 23
weight balanced B-trees, 20, 21, 24
weight-balanced B-tree, 103, 262
weight-balanced tree, 96
well balanced octree, 289
Western Digital hard disk, 267
wilderness heuristic, 193
windowing problems, 90
WINDOWS, 189, 208, 215
work efficient parallel algorithms, 32
work tape, 143, 144, 151
workspace, 108
write nodes, 157
write-only output memory, 43

X-windows, 268

Zipf distribution, 57

Selected Titles in This Series